W9-BTI-526

# Wildlife and Landscape Ecology

# Springer

*New York*
*Berlin*
*Heidelberg*
*Barcelona*
*Budapest*
*Hong Kong*
*London*
*Milan*
*Paris*
*Santa Clara*
*Singapore*
*Tokyo*

John A. Bissonette

Editor

# Wildlife and Landscape Ecology

## Effects of Pattern and Scale

With 115 Illustrations

 Springer

John A. Bissonette
Utah Cooperative Fish and Wildlife Research Unit
(U.S. Geological Survey—Biological Resources Devision)
Department of Fisheries and Wildlife
College of Natural Resources
Utah State University
Logan, UT 84322-5290, USA
E-mail: fajohn@cc.usu.edu.

Library of Congress Cataloging-in-Publication Data
Wildlife and landscape ecology : effects of pattern and scale / John
   A. Bissonette, editor.
      p.   cm.
   Some of these papers were originally presented at the Second
Annual Meeting of the Wildlife Society in Portland, Oregon in
September of 1995.
   Includes index.
   ISBN 0-387-94789-2
   1. Wildlife management—Congresses.   2. Landscape ecology—
Congresses.   I. Bissonette, John A.   II. Wildlife Society.
Meeting (2nd : 1995 : Portland, Or.)
SK352.W54   1997
639.9—dc21                                             97-7739

Printed on acid-free paper.

Production coordinated by Chernow Editorial Services, Inc. and managed by Terry Kornak;
manufacturing supervised by Jacqui Ashri.
Typeset by Best-set Typesetter Ltd., Hong Kong.
Printed and bound by Maple-Vail Book Manufacturing Group, York, PA.
Printed in the United States of America.

9 8 7 6 5 4 3 2 1

ISBN 0-387-94789-2 Springer-Verlag New York Berlin Heidelberg SPIN 10538673

# Preface

The idea for this book began in 1992 when I began teaching a class in landscape ecology for Fisheries and Wildlife graduate students at Utah State University. During the summer as I was scanning the literature and gathering material for the class, I was impressed that although there was a growing body of excellent research in the discipline, there was no single source where a graduate student or a wildlife manager could go for enlightenment on the underlying concepts, precepts, or theory that undergird ecology writ large. The voluminous body of ecosystem-level literature seemed to concentrate on but a few observation sets, i.e., nutrient cycling and energy flow, setting the stage for much discordant discourse with organism-based ecologists, much of which provided more heat than light and contributed little towards a more pluralistic view of system dynamics. It seemed to me that a significant contribution might be made by assembling a series of papers that addressed scale issues in particular and landscape ecology in general and that was directed towards the needs of wildlife biologists and resource managers.

With this in mind I organized a symposium at the second Annual Meeting of The Wildlife Society in Portland, Oregon in September of 1995. Attendance was excellent, and I received numerous comments requesting a written record of the papers presented. Although the papers presented at the symposium concentrated on theory (Bissonette: Emergent Properties; Milne: Fractal Landscapes; Peak: Taming Chaos; Pickett and Rogers: Patch Dynamics; White and Harrod: Disturbance Regimes) and metrics (Hargis et al.: Behavior of Landscape Metrics; Pearson and Gardner: Neutral Models), upon reflection, I decided to invite additional papers on theory (Ritchie: Metapopulation Dynamics; King: Hierarchy Theory) and to add a section on application, i.e., how scale concepts have been used in real-life wildlife studies (Bowyer, Van Ballenberghe and Kie: Foraging Behavior of Moose; Cooke: Spatial View of Population Dynamics; Storch: Importance of Scale Issues in the Conservation of Capercaillie; Turner, Pearson, Romme, and Wallace: Important Spatial Scales for Ungulate Dynamics; Krausman: Scale and Desert Bighorn Sheep Management; Bissonette, Harrison,

Hargis and Chapin: Scale-sensitive Properties Influence Marten Demographics). Carl Walters consented to contribute a paper on Adaptive Policy Thinking to bridge the gap between ecology and management at larger scales.

I realize that in this book, we have not covered all of the theoretical bases for ecology conducted at larger scales. I doubt that any single volume can do so. Rather, I have included papers on those issues that wildlife biologists and managers will find important because they are confronted with the management implications of them every day. The concepts, when digested and integrated into the manager's ecological view of the world, should provide a deeper understanding of how scale considerations impact how resource management is done.

I am sensitive to the perception of many field biologists and managers who are confronted with "putting out brush fires every day," i.e., crisis management, that "theory is for academics; it provides little enlightenment for my day-to-day activities." At the same time, it is evident that managers come to the table with whatever academic learning they have gleaned in college as well as a good deal of on-the-job knowledge. The school of hard knocks, most often in a highly politicized atmosphere, makes realists out of field biologists who may have been at one time, budding philosophers. I sincerely respect and admire my colleagues in both federal and state agencies who are the front line biologists and managers, and it is because I realize that their job is not an easy one that this book came into being.

In a recent National Public Radio broadcast, aired in Logan, Utah in October 1996, the commentator suggested that the useable knowledge base that people need to make enlightened everyday decisions changes every 18 months. Even by the most conservative estimates, scientific knowledge is doubling at a prodigious rate, perhaps every 3 to 5 years. Biologists who left school only five years ago will be challenged by the current advancement in ecology. The emergence of chaos theory and fractal geometry as way to look at ecological complexity (chapters by Peak, Milne in this book) will be new to most field biologists and managers, as will the ideas encompassed by patch dynamics and disturbance regimes (chapters by Pickett and Rogers, White and Harrod). As one example of new insights that mangers can gain from understanding these new developments in ecology, Peak demonstrates that a signature pattern, borrowed from chaos theory and recognized in long-term population data, can provide a window of opportunity for wildlife population control that can promote stability.

The new ideas encompassed by metapopulation dynamics (Ritchie chapter) include explicit implications of scale and will change the way that the field biologist and manager think about population dynamics. An underlying premise of ecology conducted at larger scales is that landscape pattern and processes mean something: they have impact on how individual organisms interact with their environment. However, landscape ecology is not strictly synonymous with ecology at larger scales. There is much confusion regarding scale and hierarchy level as the chapter by King points out. The implications relate directly to interpolation across scales and across hierarchical levels and the emergence of scale-sensitive properties. King's chapter on hierarchy theory provides a most important conceptual foundation for anyone interested in understanding ecology and management. Coupled with my chapter on scale-sensitive properties, the message is about how we can put confidence in the data we collect. Wildlife managers make decisions regarding management based on studies and data bases relevant to the problem. We try to provide some insight into what scale-sensitive measurements and properties mean. It is perhaps most evident to wildlife managers that adequate experimentation is difficult if not impossible to attain. The chapters by Hargis et al. and by Pearson and Gardner demonstrate an approach on how to understand the effects of landscape pattern.

I hope that my agency colleagues, in particular, will appreciate the applications chapters. They represent examples by well-known and respected wildlife biologists and ecologists to integrate scale concepts into field studies of wildlife species—the very kinds of studies that managers need to incorporate into resource management decisions. The chapters by Bowyer et al., Turner et al., Cooke, and Krausman address scale issues of ungulates; the chapter by Storch provides a very nice assessment of the scale issues involved in Capercaillie conservation in Central Europe; and I and my coauthors Harrison, Hargis, and Chapin have written about the multiscale influences on marten demographics. Because all natural resource managers are involved in policy decisions at some level, I have included in this book the chapter by Walters that addresses a rigorous approach to making management decisions.

Although this book is intended as an introduction to the underlying concepts of landscape ecology for graduate students as well as managers who work for both federal, state, and private resource agencies, it is written at a level that will require serious reading. It will not be like reading a newspaper or magazine, where what is read is instantly understood. Rather, the intent of this book is to provide introductory but sophisticated chapters that, in the words of Adler and Van Doren in their book *How to Read a Book*, are intended to overcome the "initial inequality in understanding" between writer and reader. As such, for some these chapters will take concentrated and active reading. Hitting the intended target is always difficult with a diverse readership. The judgement of whether we have been successful will rest with each individual reader.

JOHN A. BISSONETTE

# Contents

## Section 2    Landscape Metrics

## Section 3    Applications and Large-Scale Management

# Contributors

John A. Bissonette
Utah Cooperative Fish and Wildlife Research Unit (U.S. Geological Survey–Biological Resources Division), Department of Fisheries and Wildlife, College of Natural Resources, Utah State University, Logan, Utah 84322-5290, USA, E-mail: fajohn@cc.usu.edu

R.T. Bowyer
Institute of Arctic Biology, and Department of Biology and Wildlife, University of Alaska Fairbanks, Fairbanks, Alaska 99775-7000 USA, E-mail: tbowyer@redback.lter.alaska.edu

Jerry L. Cooke
Texas Parks and Wildlife Department, 4200 Smith School Road, Austin, Texas 78744-3292 USA, E-mail: jerry.cooke@tpwd.state.tx.us

T.G. Chapin
Ecology and Environment, Inc., 368 Pleasant View Drive, Lancaster, New York, 14086 USA, E-mail: chapin@buffnet.net

J.L. David
FERMA, 3213 Montreal N.E., Albuquerque, New Mexico, 87110 USA, E-mail: ebo@arc.unm.edu, or ebo@sandien.com

R.H. Gardner
Appalachian Environmental Laboratory, University of Maryland, Gunter Hall, Frostburg, Maryland 21532 USA, E-mail: gardner@alnfs.al.umces.edu

C.D. Hargis
U.S. Forest Service, Rocky Mountain Research Station, Southwest Forest Science Complex, 2500 South Pine Knoll Drive, Flagstaff, Arizona, 86001, USA, E-mail: chargis@juno.com

D.J. Harrison

Department of Wildlife Ecology, University of Maine, 5755 Nutting Hall, Orono, Maine 04469-5755 USA, E-mail: Harrison@apollo. umenfa.maine.edu

Jonathan Harrod

Department of Biology, University of North Carolina at Chapel Hill, Chapel Hill, North Carolina 27599-3280 USA, E-mail: jon_ harrod@unc.edu

J.G. Kie

U.S. Forest Service, Pacific Northwest Research Station, Forestry and Range Sciences Laboratory, 1401 Gekeler Lane, La Grande, Oregon 97850 USA, E-mail: kien@eosc.osshe. edu

Anthony W. King

Environmental Sciences Division, Oak Ridge National Laboratory, P. O. Box 2008, Oak Ridge, Tennessee 37831-6335 USA, E-mail: awk@ornl.gov

Paul R. Krausman

325 Biological Sciences East, School of Renewable Natural Resources, Wildlife Resources Program, The University of Arizona, Tucson, Arizona 85721 USA, E-mail: krausman@ag. arizona.edu

Bruce T. Milne

Department of Biology, University of New Mexico, Albuquerque, New Mexico 87131 USA, E-mail: bmilne@sevilleta.unm.edu; http:/ /algodones.unm.edu/~bmilne

David Peak

Physics Department, Utah State University, Logan, Utah 84322-4415 USA, E-mail: PeakD@cc.usu.edu

S.M. Pearson

Biology Department, Mars Hill College, Mars Hill, North Carolina 28754 USA, E-mail: spearson@craggy.mhc.edu

S.T.A. Pickett

Institute of Ecosystem Studies, Mary Flager Cary Arboretum, Box AB (Route 44A) Millbrook, New York 12545-0129 USA, E-mail: stapickett@aol.com

Mark E. Ritchie    Department of Fisheries and Wildlife, College of Natural Resources, Utah State University, Logan, Utah 84322-5200 USA, E-mail: ritchie@cc.usu.edu

Kevin H. Rogers    Centre for Water in the Environment, Department of Botany, University of the Witwatersrand, P. Bag 3 Wits 2050, Johannesbury, South Africa, E-mail: kevinr@gecko.biol.wits.ac.za

W.H. Romme    Biology Department, Fort Lewis College, Durango, Colorado 81301 USA, E-mail: romme_w@fortlewis.edu

Ilse Storch    Munich Wildlife Society, Linderhof 2, D-82488 Ettal, Germany, E-mail: WGM.ev@t-online.de

Monica G. Turner    Department of Zoology, University of Wisconsin, Madison, Wisconsin 53706 USA, E-mail: MGT@macc.wisc.edu

V. Van Ballenberghe    USDA Forest Service, Pacific Northwest Research Station, 3301 C St., Suite 200, Anchorage, AK 99503-3954, DG Mail: /S=v.vanballenberg/ OU1=R10F04A@mhs.fswa.attmail.com

L.L. Wallace    Department of Botany and Microbiology, University of Oklahoma, Norman, Oklahoma 73019 USA, E-mail: lwallace@ou.edu

Carl J. Walters    University of British Columbia Fisheries Centre, 2204 Main Mall, Vancourver, British Columbia V6T 1Z4 Canada. E-mail: walters@ fisheries.com, walters@zoology.ubc.ca

Peter S. White    Department of Biology, University of North Carolina at Chapel Hill, Chapel Hill, North Carolina 27599-3280 USA, E-mail: pswhite@ unc.edu

# Section 1
## Underlying Concepts

# 1
# Scale-Sensitive Ecological Properties: Historical Context, Current Meaning

John A. BISSONETTE

## 1.1 Introduction

In landscape ecology, attention to scale is most important (Wiens 1989). Indeed, Levin (1992) has argued that the "problem of pattern and scale is the central problem in ecology". A Special Features section in the journal *Ecology* with the title "Space: the final frontier for ecological theory" (Kareiva 1994) echoed the same theme. Clearly, the choice of appropriate time and space scales is important if understanding in ecology is to advance (Bissonette 1996). This issue is central to current dialog in landscape ecology (Golley 1989, Kareiva 1994) because different patterns and properties tend to emerge at different scales (May 1994).

Until recently, the importance of scale has not been widely recognized by wildlife biologists or explicitly by population, community, or ecosystem ecologists, and studies have been conducted as if they were scale free. For example, in 1988, Kareiva and Anderson surveyed the community ecology literature and showed that about half of the surveyed studies were conducted by using plots no larger than 1 m, despite considerable differences in size and distribution of the organisms studied. And yet, the effects of sample unit size (quadrat size) have been known for some time. Kershaw (1964) showed clearly the effects of quadrat size, the relationship between quadrat size and size of the organism being sampled, and the effect of pattern on sampling results, given a standard quadrat size (Figure 1.1). For

John A. Bissonette is Leader of the Utah Cooperative Fish and Wildlife Research Unit, and a professor in the Department of Fisheries and Wildlife at Utah State University. He is interested in scale effects in ecological research as they relate to natural resource management. His research has centered on the effects of habitat disturbance and fragmentation on core-sensitive species, especially marten. He and his students have worked in the province of Newfoundland, Canada studying the effects of forest harvesting on marten populations since 1981. He is senior editor of the 1995 book titled *Integrating People and Wildlife for a Sustainable Future*. He teaches a graduate class in landscape ecology at Utah State University.

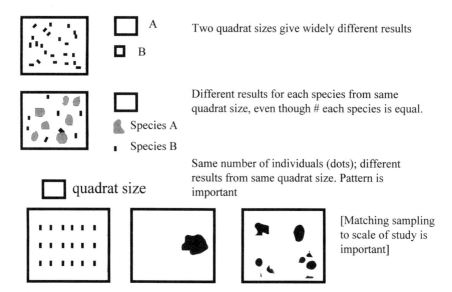

FIGURE 1.1. Quadrat size and scale. Quadrat size will highly influence the outcome of the sampling procedure. Likewise, scale of organism will influence sampling results, given a single quadrat sample frame. Attention must be given to the distribution patterns of the organisms being sampled if results are to approximate the true state of nature. Adapted from Kershaw (1964)

example, quadrat sampling units of different size can give widely different results, depending upon the relationship between quadrat size and size of the organisms being measured. Similarly, a specified quadrat size will give very different results if the organisms being measured are of very different size. Organism size varies by more than 20 orders of magnitude, from $10^{-13}$ to $10^8$ g (Brown 1995) and there is similar variation in life span and use of space. Finally, the distribution pattern of organisms being sampled will influence results obtained by using a specified quadrat size (Figure 1.1). This latter idea is most important because it has larger scale, landscape implications. Clearly, spatial and temporal scale (King et al. 1990) must be a consideration when designing methods to study this range of life forms and their spatial arrangement across the landscape (Bissonette 1996, in press). May (1994, p. 2) provides a nice summary of the history and reasons for the seeming avoidance by biologists of research efforts at large spatial scales.

Furthermore, we have not fully appreciated the implications of the new philosophy of science (Pickett et al. 1994) for ecology in general and landscape ecology in particular; and yet, how we conduct science—the domain and observation set we employ—always constrains the outcome (O'Neill et al. 1986). In this paper, I provide definitions for important landscape terms and discuss some of the underpinnings for landscape ecology. I first briefly describe the changing philosophy of science that provides an underlayment for understanding ecological complexity. I discuss why the concept of observation set is important and how choice of observation set constrains the answers one can expect from ecological studies. I then center the rest of the discussion on scale-sensitive (emergent and collective) properties and suggest that much additional dialog, including discussion on what emergence implies, is necessary if ecologists are to appreciate its implications fully. I conclude by arguing that the challenge for landscape ecology is to build conceptual frameworks that explicitly and simultaneously incorporate multiple scales.

## 1.2  A Scale-Sensitive Philosophy of Science

The intent of research at any scale is to provide a parsimonious explanation of the observed phenomena. At smaller scales, experimental mechanistic studies that are often but not always reductionist in nature (see Box 1.1 for a discussion of the distinction between the terms *reductionist* and *mechanistic*) and characterized by good study design, adequate replication, and controls are considered the sine qua non of rigor (Platt 1964). Usually, dichotomous hypotheses are established and tested, with the intent of disproving the null. This approach has led to the development of logical inductive trees (Platt 1964, Price 1986). When coupled with what has been termed *strong inference* (Platt 1964) (Box 1.2) that is systematic and uses multiple working hypotheses (Chamberlin 1890), the intent has been to produce reliable results. This approach is appealing, but it is not easily applied to ecological studies. Brown (1995) has suggested that no science has succeeded in understanding the structure and dynamics of a complex system by using a reductionist viewpoint alone. Further, the strictly mechanistic methods for achieving explanation usually are more difficult at larger spatial scales (Table 1.1). Pickett et al. (1994) argue that the traditionally accepted philosophical underpinnings of science, i.e., the classical philosophy of science as espoused by the logical positivists (Boyd 1991) and modified by Popper (1968) with his emphasis on falsifiability as the demarcation between science and something less, is far too restrictive if employed as *the* method of conducting ecological studies or indeed most any science (Hacking 1983, Boyd et al. 1991). Falsifiability refers to the ability of a scientific test to determine noncongruence between a scientific statement and the true state of nature, viz., the ecological pattern or process under examination. Pickett et al. (1994) have explained elegantly that even though falsifiability is a seemingly easy criterion to satisfy and, in some scientific endeavors, has been very successful in advancing knowledge, the prerequisites for its use, i.e., universality and simple causality, are much more restrictive than ecologists have previously imagined. They and others, i.e., Carnap (1966) and Thompson (1989), argue that the classical philosophy of

**Box 1.1. The Distinction Between Mechanistic vs. Reductionist Approaches**

Nagel (1961) argued for a classical, restrictive definition of reductionism, implying that living systems could be explained in terms of the articulation of components at the *lowest* hierarchical level, i.e., genes and molecules, and by the disciplines of physics and chemistry. He wrote (Nagel 1961, p. 398) that "outstanding biologists as well as physical scientists have therefore concluded that the methods of the physical sciences are fully adequate to the materials of biology, and many of these scientists have been confident that eventually the whole of biology would become simply a chapter of physics and chemistry." Schoener (1986) stated that most philosophers now advocate a broader view of reductionism, and Wimsatt (1976) suggested that the restricted view does not reflect modern usage among scientists.

Distinguishing between the *methods* used for achieving explanation and the *kinds* of explanation that are used to solve problems can help in understanding the differences between mechanistic and reductionist approaches. The following distinctions should prove helpful.

## Methods for Achieving Explanation

mechanisms, descriptions, definitions, generalizations, laws, syllogisms, statistical inferences, pattern recognition.

## Kinds of Explanations

reductionist, correlation, historical, spiritual, folklore, conventional wisdom.

A definition of the two terms, *reductionism* and *mechanism*, will help clarify the distinction.

*Reductionism*: explanation of level L phenomena by examination of the parts or components at level L–1. Explanations involving reductionism attempt to explain the whole by its components.

*Mechanism*: articulation of the parts. A mechanistic explanation shows how a phenomenon arises by explaining the interactions of the parts.

*Example*: If one asks the question: How does a piston rod move?, a mechanistic but nonreductionist explanation would involve explanations of how the rod is attached to the piston, the crankshaft, and other parts in the engine, i.e., how the parts articulate. The explanation is not reductionist because all the parts are at the same hierarchical level. If we asked the question: How does the engine run?, then the explanation of how the parts or components at level L–1 articulate to explain how the engine runs at level L is both mechanistic and reductionist. It is important to note that mechanistic approaches can be applied at any level of organization. It is only the constraints inherent at larger spatial scales (i.e., difficulty in replication, experimenting, controls) that pose problems and force ecologists to be inventive in their study approaches.

## Box 1.2. Elements of a Rigorous Approach

Strong inference is characterized by:

1. Multiple working hypotheses
2. Logical induction trees
3. Exclusionary experiments
4. Good study design
5. Replication and controls
6. Systematic approach
7. Iterative process

TABLE 1.1. Attribute comparison of small and large scale studies (adapted from Wiens et al. 1989)

| | Scale | |
|---|---|---|
| Attribute | Small | Large |
| 1. Detail resolution | High | Low |
| 2. Sampling adequacy | Good | Poor |
| 3. Effects of sampling error | Large | Small |
| 4. Experimental manipulation | Possible | Difficult |
| 5. Replication | Possible | Difficult |
| 6. Rigor | High | Low |
| 7. Generalizable | Low | High |
| 8. Model Form | Mechanistic | Correlative |
| 9. Survey type | Quantitative | Qualitative |
| 10. Study length | Short | Long |
| 11. Testability of $H_0$ | High | Low |

---

**Box 1.3. Components of Theory of the New Philosophy of Science**


1. domain
2. assumptions
3. concepts
4. definitions
5. facts
6. confirmed generalizations
7. laws
8. models
9. translation modes
10. hypotheses
11. frameworks

From Pickett et al. 1994

---

science took as its model the relatively well-developed science of physics where the "statement" view of theory, e.g., theory viewed as a series of mathematical statements, prevailed. The laws of physics were considered to be universal; if any exception was found, then the law did not apply. The notion of falsifiability seemed especially appropriate where universality and simple causality were the norm. However, ecological systems are characterized by organisms with a history and most often by multicausal effects. Seldom do ecologists deal with single causal events, except perhaps at very small scales and within limited domains. The classical view of science is often an inappropriate foundation for the conduct of ecological studies, especially at larger spatial scales.

The new philosophy does not rely solely on the statement view of theory (Pickett et al. 1994) but suggests that theory can take many forms, including multiple causality. As Pickett et al. (1994, p. 19) state, the new philosophy "admits tactics other than direct causality" for achieving ecological understanding. They detail nicely the increasingly recognized components of theory (i.e., domain, assumptions, concepts, definitions, facts, confirmed generalizations, laws, models, translation modes, hypotheses, and frameworks) that richly expand the conceptual and philosophical basis for the conduct of ecological investigations (Box 1.3). Inherent in the new philosophy of science is that ecological complexity often will not be understood by reductionist (Box 1.1) approaches alone. Studies conducted at larger spatial and temporal scales have a role to play by providing context that bridges understanding at smaller scales with understanding at broader scales.

Larger scale studies also can be used in a deductive fashion (Mauer 1994, Brown 1995). Patterns in large-scale data can be cast as hypotheses that make testable predictions. Consider the assumption that the general relationships between processes that influence the abundance, distribution, and diversity of organisms are reflected in the statistical distributional patterns of organisms (Brown 1995, p. 10). Then the classic pattern of changes in organism density from the center of its range to the edges (Whittaker 1956, Whittaker and Niering 1965) may be cast as a hierarchical problem (see King, Chapter 7) with several levels of hypothetical explanation (Brown 1995), all relating processes to distribution. For example, one might ask, Why does organism density change from the center of its range to the edge? The triadic structure with the associated variables of the pertinent hierarchy might look like: level L–1 = forage base, competition, predation; level L = organism density; level L+1 = landscape pattern. A level L+1 explanation might invoke the constraints of landscape pattern, while a level L–1 explanation might look at component interactions (among forage base, competition, predation) that influence organism density. Additionally, hypotheses can be generated by the elucidation of landscape models. For example, Gardner et al. (1987), Dale et al. (1989), and Pearson and Gardner (Chapter 8) discussed examples of cross-scale interpolations by using neutral models, which are, in essence, spatially cognizant models that are ecologically process free. Extrapolations to real landscapes from results obtained from simulation runs of neutral models (Hargis et al., Chapter 9, Pearson and Gardner, Chapter 8) are wonderfully heuristic. At the same time, a caveat is necessary. Extrapolations from neutral models are especially susceptible to transmutation effects (see section 1.4.1 for a definition and Figures 3 and 4); however, Naveh and Lieberman (1994, S2–8) suggest they are justified if the "scale of observations does not change the basic landscape patterns." This may be the case in landscapes undisturbed by land use practices and where the area is characterized by a single scale domain (Wiens 1989, Naveh and Lieberman 1994).

Studies conducted at larger scales seldom have the luxury of adequate replication and controls, and as a result, often are correlative rather than experimental, and pattern, rather than process oriented (Table 1.1). Hence, elucidation of how higher level components articulate is often more difficult. Wiens et al. (1993) have argued that individual level mechanisms operating in heterogeneous environments can produce patterns that are spatially dependent and thus require a larger scale approach and that under certain circumstances, experimental model systems (EMS) can be used to address larger scale phenomena. Essentially, EMS are small-scale systems that occupy "microlandscapes" and are amenable to manipulation and replication, but also serve as analogs to the larger system. The approach is realistic only if small scale processes can be scaled upwards to the larger system without significant transmutation of data. Figure 1.2 represents a

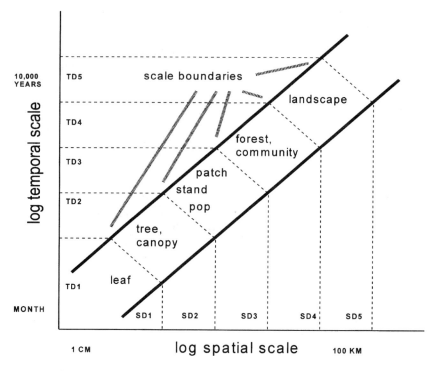

FIGURE 1.2. Approximate matching of spatial and temporal scales in ecological studies. Domains of scale are represented by the dotted lines. SD1 refers to spatial domain 1, TD1 refers to temporal domain 1. In general, longer time horizons are matched with larger spatial scales in ecological studies to insure reliable data collection. Adapted from Delcourt et al. (1983), Wiens (1989) and Holling (1992)

typical progression from small to large spatial and temporal scales using a forest community as the example and is meant to show the approximate correlation of spatial extent with temporal resolution required when conducting ecological studies. Although not necessarily hierarchical in nature (see King, Chapter 7), processes at the scale of the leaf are much more rapid, e.g., rate of stomatal opening and closing, than are process rates associated with disturbance regimes at the scale of the landscape, e.g., return interval of disturbances of a specified extent and intensity (White and Pickett 1985). Matching spatial and temporal scale domains perhaps by space-for-time substitutions at larger temporal scales, would seem to be appropriate to maintain data reliability (Delcourt et al. 1983, Wiens 1989, Bissonette 1996). Whether larger scale properties are truly emergent (in the sense I propose in section 1.4.5) would seem to bear directly on the utility of EMS, as does the fractal nature of the environment. Johnson et al. (1992) and Milne (Chapter 2) have suggested a fractal approach, and Milne et al. (1996) discussed a new approach involving multiscale association assessments and percolation, to help one recognize where domains of scale exist. I discuss scale-sensitive "emerge.˄ce" in section 1.4.

## 1.3 Domain and Observation Set

Larger spatial and temporal scales, in particular, require attention to domain and observation set. Rarely does one see overt recognition or mention of the domain, or bounded universe (Pickett et al. 1994), of a particular study (Box 1.4). One aspect of domain involves areal extent. For example, investigations into density dependence usually assume that increasing population density negatively affects reproductive or recruitment rate. The measurements taken include population density within a specific area, usually with little overt consideration of areal extent (Kawata 1995). In this case, the bounded universe or domain is left unspecified so that the areal extent over which the organisms in the population actually influence one another is unclear. However, not every individual in a specified site or population influences reproductive rate of the whole group. Sessile organisms, i.e., plants, trees, and even sessile barnacles and clams (Connell 1961) often are influenced only by closely located individuals, although Gaines and Roughgarden (1985), Roughgarden et al. (1985), Roughgarden et al.

---

**Box 1.4. Domain**


Wiens (1989, p. 392) defined domain of scale for a particular biological phenomenon as "a combination of elements of a natural system, the questions we ask of it, and the way we gather observations." This usage is reminiscent of the concept of *observation set* (O'Neill et al. 1986), but explicit to scale domain. Pickett et al. (1994, p. 32) defined *domain* as the "bounded universe in which the dialog between conceptual constructs and reality is conducted." For example, the domain for community succession includes: a species assembly;  specific characteristics of the species, e.g., life form, competitive ability; the sites where species interactions occurs and the resource levels associated with those sites; as well as a suite of general processes, including effects of disturbance and the resulting species responses and interactions (Pickett et al. 1994). All of these considerations, as well as any additional concerns that may be relevant to the problem, appropriately define the dimensions for ecological discourse on community succession, i.e., its domain. The reader is advised to refer to Pickett et al. (1994) for further explanation and examples.

---

(1987) showed that larval settlement rate of barnacles (*Balanus glandula*) was influenced by distant offshore processes, thereby influencing population dynamics in the intertidal zone. Here the scale of ecological neighborhoods (Antonovics and Levin 1980, Addicott et al. 1987) is the relevant scale for investigation and may be small or large.

Similarly, little mention is made of the particular observation set used (O'Neill et al. 1986), and yet, failure to recognize and reconcile observation set has led what Pickett et al. (1994) have referred to as dichotomous debate, which I characterize as argument without agreement either on assumptions used for viewing the natural world or even on the language used to explain the phenomenon under debate. Observation set includes "the phenomena of interest, the specific measurements taken, and the techniques used to analyze the data" (O'Neill et al. 1986, p. 7) as well as the terminology or lexicon used (Kuhn 1970). For example, if one views the world through the glass of element cycling, then the units of measurement, the actual measurements taken, and the analysis used are fundamentally different from those employed by a scientist whose unit of measurement is the individual organism, species, population, or community, e.g., an evolutionary biologist. Because measurements often are unique, e.g., Kcal or Joules vs. litter size or number of reproductives, the lexicon used is often distinct. In a sense, each view is limited by the observation set selected (O'Neill et al. 1986). Differences in observation set or the domain of the study, or both, predispose study results to different interpretation and dichotomous debate. As a result, there is great potential for ecological theory to be confused (Jackson 1991) by disagreement regarding the assumptions underlying the problem at hand. May (1994) has provided some useful examples from the literature where the notion of scale, i.e., the domain assumed for the particular study, determines the conclusions that are drawn.

Jackson (1991) and May (1994) suggested that confusion in ecological theories is the result of lack of consideration of important scale issues. I suggest that confusion also results from not realizing that different languages are, by necessity, spoken at each scale. It is as if a different dictionary is required with each different observation set, with the not so surprising result that each lexicon is largely unintelligible to the other. For example, ecological genetics and evolutionary biology define *fitness* in different ways (Brown et al. 1993), e.g., ecological genetics keeps track of allele frequency and evolutionary biology keeps track of phenotypes.

We can think of language as scale sensitive. Words make little sense unless they are appropriate to the scale of the questions being asked. It may be that knowledge gained at larger scales may be expressed clearly only with language appropriate to that scale. Likewise, answers for smaller scale questions often do not make sense at larger scales, because a different lexicon is being used. Different lexicons involve the use of level- or scale-specific terminology for the different properties that arise at different hierarchical levels and different scales. For example, the downward causation (Campbell 1974) or constraint imposed by landscape pattern is unrecognizable from a purely L–1 assembly of components. The concepts of patch size, shape, and isolation are not used in population or community dynamics investigations and make no sense unless space is considered explicitly.

These difficulties in defining domain, observation sets, and language at different scales set the central questions in landscape ecology. How is understanding at each scale achieved? How do we translate between hierarchical levels (see King, Chapter 7). Part of the answer may lie in an understanding of the nature of patterns that emerge at each level. What then is the nature of scale-sensitive patterns? Is the pattern emergent in the sense that a synergism is operating, so that knowledge of the components does not convey complete understanding of the whole (Allen and Starr 1982), or are they collective (Salt 1979) and thereby merely the sum of the component parts? What is the nature of transmutation and what relationship does it bear to emergence? How is complexity composed?

Understanding the nature of the pattern that emerges at a specified scale would seem to provide the "translation" necessary to understand the ecological phenomena in question. The issue of emergence will be discussed at length in the next section. For now, it is sufficient to say that choice of observation set and domain bracket the range of possible questions and answers for a research effort, and arguments between protagonists approaching the issue at hand from different observation sets are unlikely to produce even tacit agreement on a potential explanation. To illustrate, Wilson (1994) discussed the communication barrier that existed in biology in the 1960s, because organismal biologists were largely dismissed by cellular and molecular biologists who believed that all answers in biology could be answered by solely reductionist approaches.

## 1.4 Scale-Sensitive Properties

### 1.4.1 What Happens to "Knowledge" and "Explanation" When Scale Is Changed?

The quick answer to this question is that extrapolation across scale domains and hierarchical levels often involve the appearance of qualitatively different patterns, which in turn lead us to question the reliability of our explanation. To put it into perspective, O'Neill et al. (1986), O'Neill et al. (1989), Urban et al. (1987), and King (1991, 1993, Chapter 7) have argued persuasively that hierarchy theory provides a valuable framework for considering ecological problems, especially questions of scale. Predicting across scales (Turner, O'Neill et al. 1989) and hierarchical levels (O'Neill et al. 1989, King 1991, 1993, Chapter 7) is key to understanding complexity. From a strict reductionist viewpoint, biological phenomena at level L are explained

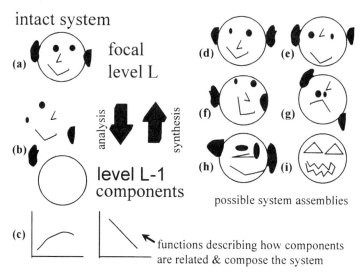

FIGURE 1.3. Transmutation: an explanation using caricatures. Transmutation occurs as a process or function changes qualitatively as one moves from one hierarchical level to another or across scale boundaries. In this figure, the intact system (a) is the focal level L, and its components (b) are level L–1. In much the same way that ecologists analyze a system (population, community, ecosystem, etc.) to understand the functional relationships and processes that compose the system, the face (or intact system) is analyzed and decomposed, i.e., functional relationships or assembly rules (c) are discovered that govern the way its component parts (eyes, ears, mouth, nose, head) are composed. Depending upon how successful the ecologist is in deciphering the assembly rules, composing the system is more or less accurate (d). However, transmutations may be more common than we realize as we try to interpolate simple rules across hierarchical levels and across scale. The caricatures represented by (e), (f), (g), and (h) are increasing different from the intact system (a) they are meant to represent and begin to become unrecognizable as representing the intact system. What this means ecologically is that the functional relationships or assembly rules used to compose the system, although they may be adequate representations at the L–1 level, do not adequately describe the intact system at level L. The caricature (i) is qualitatively so different (a pumpkin jack-o-lantern), that few would recognize it as a legitimate representation of (a). Box 4 gives a realistic ecological example.

when they are shown to be the results of interactions of the components at the L–1 level (i.e., a bottom-up explanation). However, O'Neill (1979), King et al. (1991), Wiens et al. (1993), and many others have suggested that extrapolation across scales often involves transmutation (Figure 1.3). Transmutation is the phenomenon that occurs "when a process (or the mathematical function describing the process) is changed as one moves from one hierarchical level to the next" (O'Neill 1979, p. 60). Transmutation involves qualitatively different explanations at different hierarchical levels and across scales (Wiens et al. 1993). In other words, scale-sensitive patterns or properties *emerge* that are difficult or impossible to explain solely on the basis of component interaction at level L–1.

In a fundamental sense, some level of imperfect explanation often occurs whenever hierarchical levels are crossed. For example, a common assumption with some modeling exercises is that the population response is equivalent to the response of the mean individual in the population. O'Neill (1979) has shown that a threshold function involving the influence of a critical temperature on an arbitrary process rate involves loss of information and transmutation as the relationship is scaled from the individual to population level (Figure 1.4). The discourse by Welsh et al. (1988) on the fallacy of averages is especially appropriate here. Similar difficulties in extrapolations occur between the population, community, and ecosystem levels (O'Neill 1979), and when "one moves from a local small scale model to a model of the aggregate expression of that process for a larger spatial extent," i.e., spatial transmutation (King et al. 1991). If some information is lost when hierarchical levels are crossed, at what point is "some information" too much information lost?; i.e., when is transmutation error large enough to be important? The answer to this question depends on the resolution and precision of explanation required by the investigator or manager. For example, a wildlife manager may have only limited and rather course-grained management options available, hence, exact representation of the system by the particular functional relationships or assembly rules may not be vitally important. A rough approximation of the relevant relationships may suffice.

It may be instructive to think about transmutation as variance or error to be explained. I suggest that because of our tradition, history, and success in using mechanistic and reductionist approaches that we consider L–1 components as the first level of explanation required to achieve ecological understanding. To the extent that some "variance" is left unexplained, we then need to look to the focal level components (mechanistically) and to the L+1 level for additional explanation and for context. Viewed in this light, the basis of the explanation is provided by the L–1 components, but a more complete understanding is achieved by understanding the effects of scale, at least some of which may not be reducible in the traditional sense we have employed in the past. Both Weinberg (1987) and Mayr (1988) argued that as one goes to successively higher levels of organization, new concepts are required to understand system behavior at that level.

For example, with marten (*Martes americana*), a pattern of response to fragmented landscapes emerges at larger spatial scales (Bissonette et al. 1989, Chapter 15, Hargis 1996, Chapin and Harrison in press, Harrison et al. 1997) that is only partially explained mechanistically by reference to natural history characteristics and response to selection pressures of homeothermy (Buskirk 1984, Buskirk et al. 1988, 1989), prey availability (Bissonette and Sherburne 1993, Sherburne and Bissonette 1993, 1994), and predator pres-

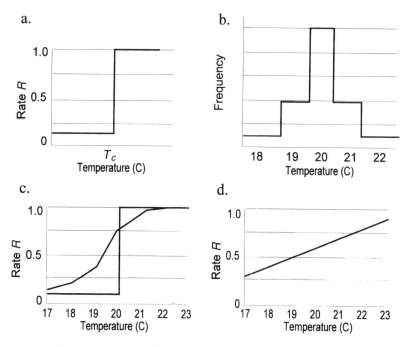

FIGURE 1.4. Transmutation: qualitative differences across hierarchical levels and across scales. Imagine a population whose individual members respond to a critical temperature $T_c$ with a change in a certain process rate $R$ by the step function shown in (a). Below $T_c$, the rate is 0.1, above $T_c$, the rate is 1.0. Further, individual response to the critical temperature is genetically controlled and is normally distributed in the population as shown in (b). The distribution in (b) shows that 50% of the population reaches the threshold at 20°C. The population rate is the weighted sum of the rates of the five subgroups and is shown in (c). Note that the population relationship is transmuted from the individual rate, which is superimposed over the population response (c). This example was take from O'Neill (1979), who provides a second population response (d) which is the transmuted response if the genetic variability in $T_c$ is distributed uniformly over the population, i.e., 10% of the population reaching a threshold at each temperature between 16 and 25°C. King et al. (1991) use a variation of this same example to show how temperature differences in individual landscape patches is transmuted as data are extrapolated from the patch to the landscape.

sure (Drew 1995, Drew and Bissonette, unpublished data). However, the spatial arrangement of elements in the landscape mosaic, i.e., patch isolation—the distance between contiguous patches of remnant forest (Snyder and Bissonette 1987)—as well as patch size and shape and the size and extent of clearcuts or open habitats, results in a negative marten population response at fragmentation levels of 20–35% (Hargis 1996, Bissonette et al., Chapter 15, Chapin and Harrison 1997, Harrison et al. in press) that cannot be explained fully by percolation theory or by response of marten to smaller scale selection pressures. Marten response is sensitive to both spatial scale and landscape pattern and, hence, more completely understood from a multiscale perspective.

## 1.4.2 Historical Context

The concept of *emergence* has a long history dating to the 19th century. It is not my intention in this chapter to describe the complete history of the concept. However, a brief journey into the recent literature will help to put the concept into context to understand its importance in landscape ecology. The reader will recognize from the following that there is little agreement regarding the meaning of *emergence* or its importance. Lewes (1874–1875, cited in Mayr 1982) is credited with developing the philosophical basis of emergence (Mandelbaum 1971, p. 330; Mayr 1982, p. 863). Mayr (1982, p. 63) stated that "systems almost always have the peculiarity that the characteristics of the whole cannot . . . . be deduced from the most complete knowledge of the components, taken separately or in other partial combinations." He distinguished between *constitutive reductionism* (reduction of the studied object into its most basic constituents), *theory reductionism* (a theory reduced to some other more inclusive theory) and *explanatory reductionism* (the view that knowledge of the ultimate components is sufficient to explain a complex system) (Mayr 1988). This paper deals with explanatory reductionism (Box 1.1). Mayr (1982) suggested that Morgan (1923, 1933) was perhaps preeminent in recognizing the importance of emergence. Two important characteristics of assemblages or patterns is that they can become parts of still higher level hierarchical entities (King, Chapter 7) and that the "wholes" can affect properties of their components, i.e., in a "downward causation" (Mayr 1982, p. 64, Kawata 1995).

The use of the term *emergent* and the recognition of the existence of higher level ecological patterns reappeared in the scientific literature in the 1970s and 1980s, and discussion has ensued relating the ideas of parsimony, emergence, transmutation, and aggregation error to landscape ecology, e.g., O'Neill (1979), Dunbar (1980), Gardner et al. (1982), Cale et al. (1983), Wikander (1985), Iwasa et al. (1987), Dale et al. (1989), O'Neill et al. (1989), Turner, Dale et al. (1989), Wiens (1989), King et al. (1990), King (1991, 1993), Johnson et al. (1992), Wiens et al. (1993), May (1994), Kawata (1995), Keitt and Johnson (1995), Milne et al. (1996) to mention only a sampling of the discourse available. One problem with the concept of *emergence* is that there has been less than full agreement on its definition. Indeed, Salt (1979, p. 145) referred to the term as a pseudocognate, i.e., each individual who uses the term "feels that all readers share his own intuitive definition." Harré (1972) suggested that emergence occurs when the property of the whole is a product of the properties of the parts, but is not qualitatively similar. Beckner (1974, p. 166) stated that "a common philosophical strategy is to define *emergence* in terms of reducibility. In the special case of hierarchically organized systems, an orthodox definition, neglecting refinement, would be something like this: *i*-level phenomena are "emergent" with respect to lower level theories when, and only when the *i*-level theories are not reducible to the theories of the lower level." Salt (1979, p. 145) argued that for ecologists, an emergent property is one that is "wholly unpredictable from observation of the components" of the entity in question. This view may assume a crippled hierarchy, i.e., a view from level L−1 to level L only, and ignore a fully engaged and certainly more appropriate hierarchical approach with refinement gained from both upward and downward (i.e., from level L+1 to level L) approaches (S.T.A. Pickett, personal communication). Edson et al. (1981) countered that predictability is always relative and to say that something is unpredictable may mean that we simply do not have enough knowledge at that point in time to understand the phenomenon, even though with additional knowledge, the components may explain the whole. Kareiva (1994) echoed this same point. Edson et al. (1981) argued that the distinction that is needed is between aggregate properties with simple, additive composition functions and those with more complicated unknown ones. The implication is that all aggregate functions, whether additive or complicated, are knowable from reductionist approaches, a position about which there is strong disagreement (Allen and Starr 1982, May 1994, Brown 1995). Kawata (1995) has suggested a somewhat different view of the distinction between types of higher level properties. He recognized the distinction between observed higher level patterns unrelated to unique properties and patterns characterized by properties that affect and cause change in ecological and evolutionary processes. Kawata (1995) referred to the latter as effective properties. Higher level patterns with no influence on processes were termed noneffective properties. Medawar (1974) argued for a geometric view of emergence.

## 1.4.3 Scale-Sensitive Properties or Measurements?

Some biologists have used the term *emergent property* almost synonymously with measurements that are appropriate only to a certain hierarchical level, e.g., composition, connectedness, interpatch fluxes (Lidicker 1995, p. 6); others have written of "emergent patterns" (Keitt and Johnson 1995). Lidicker (1995) listed the following as landscape emergent properties: the composition and diversity of community types and their spatial configuration; patch characteristics and edge effects; connectedness; interpatch fluxes of energy, nutrients, and organisms; dominance relationships among community types; stability, including resilience, constancy, predictability, succession, and degradation; and an anthropogenic index measuring human disturbance. Wiens et al. (1993) listed size distribution, boundary form, perimeter-area ratios, patch orientation, context, contrast, connectivity, richness, evenness, dispersion, and predictability as measurable features of the landscape. Few would disagree that emergent *measurements* exist that can be measured only at the specific level in the biological hierarchy. However, patterns, properties, and measurements are not the same. *Pattern* refers to a "combination of qualities, . . . forming a consistent or characteristic arrangement," a *property* is "an essential or distinctive attribute or quality of a thing", and a measurement is the extent, size, quantity, or dimension as determined by measurement (Dictionary of the English Language 1971).

In population ecology, mean generation time, recruitment rate, birth and death rates, and rate of intrinsic increase (Cole 1954, Ricklefs 1973, Emmel 1976) are but a few of the measurements that emerge at the level of populations, while species number, composition, relative abundance, and various trophic measurements emerge at the community level and cannot be measured legitimately at a lower level, i.e. at the level of the individual. Brown (1995) suggests that ecosystem ecology approaches that study whole-system patterns and processes of energy and material exchange are examples that focus on the emergent properties of complex ecological systems. Food chains, food webs, and trophic levels (Odum 1971, Ricklefs 1973) are some of the conceptual constructs (Golley 1993) that emerge at the level of the ecosystem, as are the measurements we take to determine quantitative and qualitative aspects of nutrient cycling (Bormann and Likens 1967), primary and secondary productivity, and community respiration (Odum 1971). Lidicker (1988) suggested that landscapes can be viewed as the next level of organization after communities (but see King, Chapter 7). As such, they may be characterized by a host of scale-dependent patterns (Wiens et al. 1993, Lidicker 1995), generally lumped into the following classes of landscape metrics (McGarigal and Marks 1995): area; patch density, size and variability; edge; shape; core area; nearest-neighbor; diversity; and contagion and interspersion metrics. Whether and under what circumstances emergent measurements measure emergent properties is not immediately clear.

## 1.4.4 Emergent or Collective?

Physicists and mathematicians have worked with what they commonly refer to as emerging patterns for some time (von Neumann 1966, Gardner 1970, Wuensche and Lesser 1992, Laplante 1994, Peak 1994). A favorite vehicle for exploring these patterns involved automata theory (von Neumann 1966) and the use of cellular automata. A cellular automaton is a dynamical system composed of a collection of identical cells occupying a specified space and whose states are changed over successive time steps by deterministic rules. When left to propagate over multiple generations according to simple rules that are iterated at each time step, unusual and complex patterns emerge from very simple starting conditions (Peak 1994). Wolfram (1984) observed that four classes of behavior resulted from different rule sets (Peak 1994), including three patterns uninteresting to ecologists. A fourth class of behavior was characterized by a pattern of cells that "grew" and "contracted" in a complicated way (Peak 1994), imitating life processes. The Game of Life (Gardner 1970) is a cellular automaton where complex patterns emerge with lifelike characteristics; but however lifelike the behavior, the patterns and dynamics that emerge do so according to known, deterministic rules. In this sense, by my definition, these are collective properties (Figure 1.5), and not emergent, and lead to the following proposal (section 1.4.5) for differentiating properties that scale in a different way.

An appreciation of the distinction between hierarchical level and scale (King, this volume) may provide insight to understand why some properties are emergent and others are collective. As one increases the scale of the observation, one does not necessarily change hierarchical level. For example, as King (Chapter 7) suggests, a decaying log may be the relevant ecological neighborhood (Addicott et al. 1987) to describe *small*-scale ecosystem dynamics. In this case, a higher hierarchical level is not coincident with a larger scale. This is the idea of natural scales, which differ among systems. Emergent properties may involve explanation across hierarchical levels, rather than necessarily across a range of scales. Whether the hierarchy is nested and "perfect" (Box 1.5) or not may bear on the question of emergence.

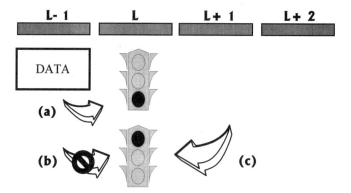

FIGURE 1.5. Emergent or collective? Given the definition of emergence proposed in this paper, at least two possibilities arise. If the property is collective (arrow *a*), data at a lower level in the hierarchy may be scaled upwards (a bottom-up explanation) without loss of information; i.e., level L–1 components explain biological phenomena completely at level L. If the property is emergent (arrow *b*), then level L–1 components cannot *fully* explain patterns and properties that appear at level L, and additional scale-sensitive and scale-specific information is needed for ecological understanding, i.e., downward causation (Campbell 1974, Kawata 1995) is operating, and level L+1 must be invoked (arrow *c*) to more completely explain the phenomena at level L.

---

**Box 1.5. Conditions for a "Perfect Hierarchy"**

The idea of a *perfect hierarchy* implies, by the definitions I propose for scale-sensitive properties, that components at the L–1 level can perfectly explain level L phenomena. Beckner's (1974) conditions for a perfect hierarchy are:

1. every component of the hierarchy is assigned to only one level
2. every component is a part of only one component at each level above it (excluding components at the highest level)
3. every component is composed only of components at the next lower level (excluding components at the lowest level)

---

## *1.4.5 A Proposal*

In the previous section 1.4.2, the message is that the term *emergence* has been used so loosely in biological and ecological literature so as to convey little meaning. Further, it is often associated with the ideas of *holism* and *emergent evolution* (Smuts 1926), ideas that in their original conception, "required a kind of 'soul', an *élan vital*, an entelechy" (O'Neill et al. 1986), i.e., ideas that are unacceptable to scientists. This is why I have used the more general term *scale-sensitive properties* to refer to the general class of properties that may be either collective or emergent, and that appear at different hierarchical levels and at different scales. I do not propose that special explanations are needed to understand scale-sensitive properties. Rather, I suggest that more complete explanations for scale-sensitive properties can be obtained by considering not only L–1 level explanations, but also the insight gained by considering the effects of L+1 constraints, such as landscape pattern. Further, I suggest that ecologists must agree and adhere to a rigorous definition in order for a scale-sensitive property to have significant biological meaning. I propose that for a property to be emergent means that the property cannot be *fully* explained by level L–1 components. If a property can be explained fully by mechanisms at the next lower level, complexity is explained by a reductionist approach. In these cases, I suggest we call the property collective (Figure 1.5). Beckner's (1974) formal conditions for a perfect hierarchy (Box 1.5) would appear to be the sine qua non for collective properties. To the extent that any significant components at level $L_i$ are not: (1) assigned to exactly one level, (2) a part of exactly one part at the next higher level ($L_{i+1}$), and (3) exhaustively composed of parts at the next lower level ($L_{i-1}$) (Beckner 1974, Schoener 1986), then the property would appear to be emergent. Scale-sensitive properties may be emergent or collective. I make a distinction between collective properties and emergent properties at any level of organization and at any scale. Collective properties may be simple or complex functions. Context conveys sensibility.

Brown (1995, p. 16) has suggested that diffusion is an emergent property. The movement dynamics of single molecules of a gas are characterized as stochastic and random, i.e., Brownian movement. At this small scale each molecule acts independently. At the next larger scale, one predictable movement of a "population" of gas molecules is referred to as diffusion. Diffusion rate is a measurement taken to characterize the property of diffusion. But diffusion can be explained by random motion, i.e., as the probability distribution of individual gas molecules. Rosen (1969) referred to this level of explanation as a "micro-description" of the system. If a gas is viewed as a fluid, the state variables are different—i.e., pressure, volume, temperature—and correspond to a "macro-description" of the system (Rosen 1969), where the lexicon used is different. In a sense, our view of the nature of scale-sensitive properties depends upon which aspects of the system interest us. Brown (1995, p. 20) recognized the statistical average aspect of diffusion. By way of example, a scale-sensitive and emergent property is the "aquosity" (Mayr 1982, p. 63) or liquidity of water. This property cannot be deduced from an understanding of the properties of oxygen and hydrogen atoms. I have found great difficulty in identifying a large number of emergent (by my definition) properties in ecology (but see Pearson 1993 and Andrén 1994), yet the persuasive argument has been made that complexity cannot be understood by strictly reductionist approaches alone (Lidicker 1988, May 1994, Brown 1995, Bissonette et al. Chapter 15), suggesting that emergent properties are present in nature. There is no apriori need to believe that a property that is emergent in one context cannot be collective in another. The question under consideration and the system being investigated will determine the context. To further complicate the issue, both collective and emergent properties have characteristics at multiple scales that together determine their behavior. Landscape pattern, a presumably "emergent" property is determined by both the grain and extent at which it is perceived (Milne 1991, Chapter 2). These examples suggest that the concept of emergence may indeed be a pseudocognate (Salt 1979), and that general agreement on a rigorous definition is required if the term is to have real meaning and currency in ecology.

The importance of the concept of scale-sensitive properties to landscape ecology would appear to be related to: a) whether smaller scale measurements, i.e., mechanistic level data, can be scaled upwards meaningfully, and b) whether bottom-up (L−1 → L) and top-down (L+1 → L) explanation is needed to explain the phenomena. Anderson (1972) suggested that the ability to reduce a problem to its components, i.e., to fundamental relationships does not necessarily imply "the ability to start from those laws and

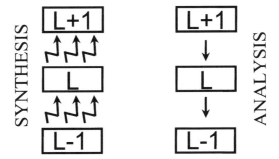

FIGURE 1.6. A "broken symmetry." Analysis of the biological phenomena of interest at focal level L yield components at level L–1 that are deemed to be *the* important variables that, when assembled by whatever functional relationships have been discovered, will explain the phenomena. However, synthesis, i.e., a proper articulation of the parts (components) is more onerous than we might expect, and is influenced by the appearance of scale-sensitive properties, hence the idea of a broken symmetry.

reconstruct the universe" (p. 393). He referred to this anomaly as "broken symmetry" (Figure 1.6). The implication is that the components at level L, derived by analysis and mechanistic methods, should not necessarily be expected to fully explain complexity at level L, especially given the complexity of ecology and attendant spatial complications. Pickett (personal communication) suggests that complexity would not even be known under strictly bottom-up reductionism. Rosen (1969) argued that observation and theory at higher organizational levels necessarily precede effective synthesis at lower organization levels. The message is that biological views need not be "crippled," i.e., either strictly bottom up or top down, but rather, a more effective approach involves a multiscaled view of the problem at hand. I think that this is the approach ecologists generally use, whether we recognize it overtly and admit it or not.

The arguments presented above suggest that the discourse regarding the importance of collective vs. emergent properties has been carried out from the perspective of restricted observation sets. Perhaps more to the point is that ecological questions have multiscale properties and characteristics that ultimately influence how a system appears and behaves. The challenge for landscape ecology is to build conceptual frameworks that explicitly and simultaneously incorporate multiple scales, e.g., Johnson et al. 1992, Bissonette et al. Chapter 15, Milne 1996, this volume, Peak, Chapter 3, Pearson and Gardner, Chapter 8, Ritchie 1997, Chapter 6. To the extent that this is done, the distinction between how collective and emergent properties behave will assume less importance.

## 1.5  Landscape and Scale-Related Patterns

Studies involving changing landscape patterns may be an appropriate place to look for emergent properties. For example, changes in habitat fragmentation result in changing characteristics of isolation and patch size; i.e., patch size frequency distribution and relative position in the landscape (isolation) change abruptly as a critical level of fragmentation is reached, demonstrating a nonlinearity of the relationship (Gardner et al. 1987, Turner 1989, Gardner and O'Neill 1991, Gustafson and Parker 1992, Andrén 1994, With and Crist 1995, Milne et al. 1996), although nonlinearity in itself is not sufficient to infer the presence of an emergent property. To the extent that birds and mammals respond to patch size, shape, and isolation, the relationship cannot be completely understood by examining a single fragmentation level. Here, the composite property associated with landscape pattern is in some sense an explanatory variable predicting animal response and appears to fit the definition of an emergent property, i.e., level L+1 causation.

Our conception of looking at the spatial configurations of the landscape as explanatory variables leads one to consider Holling's (1992) linking of pattern with process and with community structure. If indeed, a "small set of plant, animal, and abiotic processes structure ecosystems across scales in time and space" (Holling 1992, p. 447, but see Brown 1995, p. 80), and if the resulting discontinuous hierarchical structures and textures of the landscape do entrain and strongly influence the clumping of body size of organisms in these ecosystems, then we have the elements of a predictive theory to assess community, landscape, and ecosystem changes across scales (Holling, 1992). A hierarchical structure (O'Neill et al. 1986, King, Chapter 7) underlies current ideas in ecology and is appropriate for the analysis of scale issues (O'Neill et al. 1989). It is not too surprising to note that emergent properties are hierarchically ordered. Hierarchies need not conform to the traditional hierarchies (Pickett et al. 1994) used in ecology, and emergent properties may not either. The developing message is: (1) not all properties at a particular scale are truly emergent; (2) for those properties that are collective, scale domain boundaries appear flexible and not fixed; and (3) in these cases, one would expect a disjunction between scale domains and fractal (ecotone) boundaries. Few data address this issue, but if the logic holds, finding these disjuncts will become a central focus in ecology (Johnson et al. 1992).

## 1.6  Relevance to Wildlife Biology and Management

Wildlife management activities in North America traditionally have had a species focus with state and provincial programs focused on high-profile species, e.g., deer, elk, and bighorn sheep, as well as groups of species of related taxa, e.g., waterfowl, upland game birds, and furbearers. However, since the early 1990s, federal agencies in the United States (Forest Service, Bureau of Land Management, Fish and Wildlife Service) recognized that resource problems transcend political and agency boundaries and increasingly have emphasized ecosystem management. Grumbine (1994, p. 27) suggests that ecosystem management "offers a fundamental reframing of how humans . . . work with nature." Implied is that anthropogenic disturbances and impacts are important, large scale, and widespread. If wildlife biologists and managers are to address larger scale issues effectively, an appreciation of how larger scale patterns and properties are generated is important. In an increasingly contentious atmosphere regarding natural resource management, there is added emphasis and reliance on data to direct agency management activities. As the problems grow in scope and complexity, faith in the traditional methods of science to answer the pertinent questions has dwindled, and single scale approaches often do not suffice (May 1994, Brown 1995). Understanding the scalar aspects of the resource problem at hand is a prerequisite to mounting significant efforts to provide alternative solutions. Whether the patterns that emerge in large-scale resource issues are collective or emergent will assume more importance than they perhaps should, unless we realize that properties and characteristics at every scale influence our perception of how the system behaves. Efforts need be directed to understand ecological phenomena from a multiscale perspective.

*1.7 Acknowledgments.* I thank G. Belovsky, J.W. Haefner, C.D. Hargis, A.W. King, M. McClure, B. Milne, D. Peak, S.T.A. Pickett, J. Powell, M.E. Ritchie, and I. Storch for helpful criticism on the manuscript. Conversations with J. Haefner, C.D. Hargis, D. Keppie, M.E. Ritchie, and I. Storch were especially illuminating for me, and I appreciate and thank them for their interest.

## 1.8 References

Addicott, J.F., J.M. Abo, M.F. Antolin, D.F. Padilla, J.S. Richardson, and D.A. Soluk. 1987. Ecological neighborhoods: scaling environmental patterns. Oikos 49:340–346.

Allen, T.F.H., and T.B. Starr. 1982. Hierarchy: perspectives for ecological complexity. University of Chicago Press, Chicago, Illinois, USA.

Anderson, P.W. 1972. More is different. Science 177:393–396.

Andrén, H. 1994. Effects of habitat fragmentation on birds and mammals in landscapes with different proportions of suitable habitat: a review. Oikos 71:355–366.

Antonovics, J., and D.A. Levin. 1980. The ecological and genetic consequences of density-dependent regulation in plants. Annual Review of Ecology and Systematics 11:411–452.

Beckner, M. 1974. Reductionism, hierarchies and organicism. p. 163–177 in F.J. Ayala and T.G. Dobzhansky, editors. Studies in the philosophy of biology: reductionism and related problems. University of California Press, Berkeley, California, USA.

Bissonette, J.A. 1996. Linking temporal with spatial scales in wildlife research. Transactions of the 61st North American Wildlife and Natural Resources Conference 61:161–168.

Bissonette, J.A. In press. American marten habitat choice: scaled multiple hypotheses provide some answers. Third biennial scientific conference on the Greater Yellowstone Ecosystem: ecology and conservation in a changing landscape. National Park Service.

Bissonette, J.A., R.J. Fredrickson, and B.J. Tucker. 1989. American marten: a case for landscape-level management. Transactions of the 54th North American Wildlife and Natural Resources Conference 54:89–101.

Bissonette, J.A., and S.S. Sherburne. 1993. Subnivean access: the prey connection. Pp. 225–228 in I. Thompson, editor. Proceedings of the International Union of Game Biologists XXI Congress; August 15–20; Halifax, Nova Scotia, Canada.

Bormann, F.H., and G.E. Likens. 1967. Nutrient cycling. Science 155:424–429.

Boyd, R. 1991. Confirmation, semantics, and the interpretation of scientific theories. In R. Boyd, P. Gasper, and J.D. Trout, editors. The philosophy of science. MIT Press, Cambridge, Massachusetts, USA.

Boyd, R., P. Gasper, and J.D. Trout. Editors. 1991. The philosophy of science. MIT Press, Cambridge, Massachusetts, USA.

Brown, J.H. 1995. Macroecology. University of Chicago Press, Chicago, Illinois, USA.

Brown, J.H., P.A. Marquet, and M.L. Taper. 1993. Evolution of body size: consequences of an energetic definition of fitness. American Naturalist 142:573–584.

Buskirk, S.W. 1984. Seasonal use of resting sites by marten in south-central Alaska. Journal of Wildlife Management 48:950–953.

Buskirk, S.W., H.J. Harlow, and S.C. Forrest. 1988. Temperature regulation in American marten, (*Martes americana*) in winter. National Geographic Research 4:208–218.

Buskirk, S.W., H.J. Harlow, and S.C. Forest. 1989. Winter resting site ecology of marten in the central Rocky Mountains. Journal of Wildlife Management 53:191–196.

Cale, W.G., R.V. O'Neill, and R.H. Gardner. 1983. Aggregation error in nonlinear ecological models. Journal of Theoretical Biology 100:539–550.

Campbell, D.T. 1974. 'Downward causation' in hierarchically organized biological systems. Pp. 179–186 in F.J. Ayala and T.G. Dobzhansky, editors. Studies in the philosophy of biology: reductionism and related problems. University of California Press, Berkeley, California, USA.

Carnap, R. 1966. An introduction to the philosophy of science. Basic Books, New York, New York, USA.

Chamberlin, T.C. 1890. The method of multiple working hypotheses. Journal of Geology 5:837–848. [Reprinted in Science (1965)148:754–759.]

Chapin, T.G., and D.J. Harrison. In press. Marten use of residual stands in an industrial forest landscape in Maine. In G. Proulx, H. Bryant, and P. Woodard, editors. Proceedings of the Second International Martes Symposium, Provincial Museum of Alberta, Edmonton, Alberta, Canada.

Cole, L.C. 1954. The population consequences of life history phenomena. Quarterly Review of Biology 29:103–137.

Connell, J.H. 1961. The influence of interspecific competition and other factors on the distribution of the barnacle *Chthamalus stellatus*. Ecology 42:710–723.

Dale, V.H., R.H. Gardner, and M.G. Turner. 1989. Predicting across scales—comments of the guest editors of Landscape Ecology. Landscape Ecology 3:147–152.

Delcourt, H.R., P.A. Delcourt, and T. Webb III. 1983. Dynamic plant ecology: the spectrum of vegetation change in space and time. Quaternary Science Reviews 1:153–175.

Dictionary of the English Language. 1971 (unabridged). Random House, Inc., New York, New York, USA.

Drew, G.S. 1995. Winter habitat selection by American marten (*Martes americana*) in Newfoundland: why old growth? Dissertation, Utah State University, Logan Utah, USA.

Dunbar, M.J. 1980. The blunting of Occam's razor, or to hell with parsimony. Canadian Journal of Zoology 58:123–128.

Edson, M.M., T.C. Foin, and C.M. Knapp. 1981. "Emergent properties" and ecological research. American Naturalist 118:593–596.

Emmel, T.C. 1976. Population biology. Harper and Row Publishers, New York, New York, USA.

Gaines, S., and J. Roughgarden. 1985. Larval settlement rate: a leading determinant of structure in an ecological community of the marine intertidal zone. Proceedings of the National Academy of Sciences (USA) 82(11):3707–3711.

Gardner, M. 1970. Mathematical games. Scientific American 223(4):120–123.

Gardner, R.V., W.G. Cale, and R.V. O'Neill. 1982. Robust analysis of aggregation error. Ecology 63:1771–1779.

Gardner, R.H., B.T. Milne, M.G. Turner, and R.V. O'Neill. 1987. Neutral models for the analysis of broad-scale landscape pattern. Landscape Ecology 1:19–28.

Gardner, R.H., and R.V. O'Neill. 1991. Pattern, process, and predictability: the use of neutral models for landscape analysis. Pp. 289–307 in M.G. Turner and R.H. Gardner, editors. Quantitative methods in landscape ecology, Springer-Verlag, New York, New York, USA.

Golley, F.B. 1989. A proper scale. Editor's comment. Landscape Ecology 2:71–72.

Golley, F.B. 1993. A history of the ecosystem concept in ecology: more than the sum of the parts. Yale University Press, New Haven, Connecticut, USA.

Grumbine, R.E. 1994. What is ecosystem management? Conservation Biology 8:27–38.

Gustafson, E.J., and G.R. Parker. 1992. Relationships between landscape proportions and indices of landscape spatial pattern. Landscape Ecology 7:101–110.

Hacking, I. 1983. Representing and intervening: introductory topics in the philosophy of natural science. Cambridge University Press, Cambridge, UK.

Hargis, C.D. 1996. The influence of habitat fragmentation and landscape pattern on American marten and their prey. Dissertation, Utah State University, Logan, Utah, USA.

Harré, R. 1972. The philosophies of science. Oxford University Press, Oxford, UK.

Harrison, D.J., D.M. Phillips, T.G. Chapin, D.P. Katnik, and T.P. Hodgman. In press. Population performance and habitat selection by American marten: a need to reassess accepted paradigms and conservation practices. In G. Proulx, H.

Bryant, and P. Woodard, editors. Proceedings of the Second International Martes Symposium, Provincial Museum of Alberta, Edmondton, Alberta, Canada.

Holling, C.S. 1992. Cross-scale morphology, geometry, and dynamics of ecosystems. Ecological Monographs 62:447–502.

Iwasa, Y., V. Andreasen, and S. Levin. 1987. Aggregation in model ecosystems. I. Perfect aggregation. Ecological Modeling 37:287–302.

Jackson, J.B.C. 1991. Adaptation and diversity of reef corals. Bioscience 41:475–482.

Johnson, A.R., B.T. Milne, and J.A. Wiens. 1992. Diffusion in fractal landscapes: simulations and experimental studies of tenebrionid beetle movements. Ecology 73:1968–1983.

Kareiva, P. 1994. Space: the final frontier for ecological theory. Ecology 75:1.

Kareiva, P., and M. Anderson. 1988. Spatial aspects of species interactions: the wedding of models and experiments. Pp. 38–54 in A. Hastings, editor. Community ecology. Springer-Verlag, New York, New York, USA.

Kawata, M. 1995. Emergent and effective properties in ecology and evolution. Researches in Population Ecology 37:93–96.

Keitt, T.H., and A.R. Johnson. 1995. Spatial heterogeneity and anomalous kinetics: emergent patterns in diffusion-limited predator-prey interaction. Journal of Theoretical Biology 172:127–139.

Kershaw, K. 1964. Quantitative and dynamic ecology. Edward Arnold Limited, London, UK.

King, A.W. 1991. Translating models across scales in the landscape. Pp. 479–517 in M.G. Turner and R.H. Gardner, editors. Quantitative methods in landscape ecology. Springer-Verlag, New York, New York, USA.

King, A.W. 1993. Considerations of scale and hierarchy. Pp. 19–45 in S. Woodley, G. Francis, and J. Key, editors. Ecological integrity and the management of ecosystems. Lewis Publishers Inc., Chelsea, Michigan, USA.

King, A.W., A.R. Johnson, and R.V. O'Neill. 1991. Transmutation and functional representation of heterogeneous landscapes. Landscape Ecology 5:239–253.

King, A.W., W.R. Emanuel, and R.V. O'Neill. 1990. Linking mechanistic models of tree physiology with models of forest dynamics: problems of temporal scale. Pp. 241–248 in R.K. Dixon, R.S. Meldahl, G.A. Ruark, and W.G. Warren, editors. Process modeling of forest growth responses to forest stress. Timber Press, Portland, Oregon, USA.

Kuhn, T.S. 1970. The structure of scientific revolutions. Second edition. University of Chicago Press, Chicago, Illinois, USA.

Laplante, P. 1994. Fractal mania. Windcrest/McGraw Hill Publishing Co., New York, New York, USA.

Lewes, G.H. 1874–1875. Problems of life and mind. Longmans Green, London, UK.

Levin, S.A. 1992. The problem of pattern and scale in ecology. Ecology 73(6):1943–1967.

Lidicker, W.Z., Jr. 1988. The synergistic effects of reductionist and holistic approaches in animal ecology. Oikos 53:278–281.

Lidicker, W.Z., Jr. 1995. The landscape concept: something old, something new. Pp. 3–19 in W.Z. Lidicker Jr., editor. Landscape approaches in mammalian ecology. University of Minnesota Press, Minneapolis, Minnesota, USA.

McGarigal, K., and B.J. Marks. 1995. Fragstats: Spatial pattern analysis program for quantifying landscape structure. U.S. Forest Service Pancific Northwest Research Station General Technical Report PNW-GTR-351. Portland, Oregon, USA.

Mandelbaum, M. 1971. History, man, and reason. Johns Hopkins University Press, Baltimore, Maryland, USA.

Mauer, B.A. 1994. Geographical population analysis: tools for the analysis of biodiversity. Blackwell Science Inc., Cambridge, Massachusetts, USA.

May, R.M. 1994. The effects of spatial scale on ecological questions and answers. Pp. 1–17 in P.J. Edwards, R.M. May, and N.R. Webb, editors. Large-scale ecology and conservation biology. (35th Symposium of the British Ecological Society, with the Society for Conservation Biology) Blackwell Scientific Publications, London, UK.

Mayr, E. 1982. The growth of biological thought. Belknap Press of Harvard University Press, Cambridge, Massachusetts, USA.

Mayr, E. 1988. The limits of reductionism. Nature 331:475.

Medawar, P. 1973. A geometric model of reduction and emergence. Pp. 57–63 in F.J. Ayala and T.G. Dobzhansky, editors. Studies in the philosophy of biology: reductionism and related problems. University of California Press, Berkeley, California, USA.

Milne, B.T., A.R. Johnson, T.H. Keitt, C.A. Hatfield, J. David, and P.T. Hraber. 1996. Detection of critical densities associated with Piñon-Juniper woodland ecotones. Ecology 77:805–821.

Morgan, C.L. 1923. Emergent evolution. Williams and Norgate, London, UK.

Morgan, C.L. 1933. The emergence of novelty. Macmillan, New York, New York, USA.

Nagel, E. 1961. The structure of science: problems in the logic of scientific explanation. Harcourt, Brace and world Inc., New York, New York, USA.

Naveh, Z., and A.S. Lieberman. 1994. Landscape ecology: theory and application. Springer-Verlag, New York, New York, USA.

Odum, E.P. 1971. Fundamentals of ecology. W.B. Saunders Company, Philadelphia, Pennsylvania, USA.

O'Neill, R.V. 1979. Transmutation across hierarchical levels. Pp. 59–78 in G.S. Innis and R.V. O'Neill, editors. Systems analysis of ecosystems. Statistical Ecology No. 9. International Cooperative Publishing House, Fairland, Maryland, USA.

O'Neill, R.V., D.L. DeAngelis, T.F.H. Allen, and J.B. Waide. 1986. A hierachical concept of ecosystems. Monographs in Population Biology 23. Princeton University Press, Princeton, New Jersey, USA.

O'Neill, R.V., A.R. Johnson, and A.W. King. 1989. A hierarchical framework for the analysis of scale. Landscape Ecology 3:193–205.

Peak, D. 1994. Chaos under control: the art and science of complexity. W.H. Freeman and Company, New York, New York, USA.

Pearson, S.M. 1993. The spatial extent and relative influence of landscape level factors on wintering bird populations. Landscape Ecology 8:3–18.

Pickett, S.T.A., J. Kolasa, and C.G. Jones. 1994. Ecological understanding. Academic Press, San Diego, California, USA.

Platt, J.R. 1964. Strong Inference. Science 146:347–354.

Popper, K.R. 1968. The logic of scientific discovery. Harper and Row Publishers, New York, New York, USA.

Price, M.V. 1986. Structure of desert rodent communities: a critical review of questions and approaches. American Zoologist 26:39–49.

Ricklefs, R.E. 1973. Ecology. Chiron Press, Newton, Massachusetts, USA.

Ritchie, M.E. In press. Scale-dependent foraging and patch choice in fractal environments. Evolutionary Ecology.

Roughgarden, J., Y. Iwasa, and C. Baxter. 1985. Demographic theory for an open marine population with space-limited recruitment. Ecology 66:54–67.

Roughgarden, J., S.D. Gaines, and S.W. Pacala. 1987. Supply-side ecology: the role of physical transport processes. Pp. 481–518 in J.H.R. Gee and P.S. Giller, editors. Organization of communities past and present. Blackwell Scientific Publications, Oxford, UK.

Rosen, R. 1969. Hierarchical organization in automata theoretic models of biological systems. Pp. 181–199 in L.L. Whyte, A.G. Wilson, and D. Wilson, editors. Hierarchical structures. American Elsevier Publication Company, Inc., New York, New York, USA.

Salt, G.W. 1979. A comment on the use of the term *Emergent Properties*. American Naturalist 113:145–148.

Schoener, T.S. 1986. Mechanistic approaches to community ecology: a new reductionism. American Zoologist 26:81–106.

Sherburne, S.S., and J.A. Bissonette. 1993. Squirrel middens influence marten (*Martes americana*) use of subnivean access points. American Midland Naturalist 129:204–207.

Sherburne, S.S., and J.A. Bissonette. 1994. Marten subnivean access point use: response to subnivean prey levels. Journal of Wildlife Management 58:400–405.

Synder, J.E., and J.A. Bissonette. 1987. Marten use of clear-cuts and residual forest stands in western Newfoundland. Canadian Journal of Zoology 65:169–174.

Smuts, J.C. 1926. Holism and evolution. Macmillan, New York, New York, USA.

Thompson, P. 1989. The structure of biological theories. State University of New York Press, Albany, New York, USA.

Turner, M.G. 1989. Landscape ecology: the effect of pattern and process. Annual Review of Ecology and Systematics 20:171–197.

Turner, M.G., R.V. O'Neill, R.H. Gardner, and B.T. Milne. 1989. Effects of changing spatial scale on the analysis of landscape pattern. Landscape Ecology 3:153–162.

Turner, M.G., V.H. Dale, and R.H. Gardner. 1989. Predicting across scales: theory development and testing. Landscape Ecology 3:245–252.

Urban, D.L., R.V. O'Neill, and H.H. Shugart, Jr. 1987. Landscape ecology. Bioscience 37:119–127.

von Neumann, J. 1996. Theory of self-reproducing automata. (edited and completed by A.W. Burks after von Neumann's death in 1957). University of Illinois Press, Urbana, Illinois, USA.

Welsh, A.H., A. Townsend Peterson, and S.A. Altmann. 1988. The fallacy of averages. American Naturalist 132:277–288.

Weinberg, S. 1987. Newtonianism, reductionism and the art of congressional testimony. Nature 330:433–437.

White, P.S., and S.T.A. Pickett. 1985. Natural disturbance and patch dynamics: an introduction. Pp. 3–13 in S.T.A. Pickett, and P.S. White, editors. The ecology of natural disturbance and patch dynamics. Academic Press, Inc., San Diego, California, USA.

Whittaker, R.H. 1956. Vegetation of the Great Smoky Mountains. Ecological Monographs 221:1–44.

Whittaker, R.H., and W.A. Niering. 1965. Vegetation of the Santa Catalina Mountains, Arizona: a gradient analysis of the south slope. Ecology 46:429–452.

Wiens, J.A. 1989. Spatial scaling in ecology. Functional Ecology 3:385–397.

Wiens, J.A., N.C. Stenseth, B. Van Horne, and R.A. Ims. 1993. Ecological mechanisms and landscape ecology. Oikos 66:369–380.

Wikander, R. 1985. Parsimony and testability: a reply to Dunbar. Canadian Journal of Zoology 63:728–732.

Wilson, E.O. 1994. Naturalist. Island Press/Shearwater Books, Covelo, California, USA.

Wimsatt, W.C. 1976. Reductive explanation: a functional account. Pp. 671–710 in R.S. Cohen, C.A. Hooker, A.C. Michalos, and J.W. Van Eura, editors. Philosophy of Science Association (Proceedings of the 1974 biennial meeting). D. Reidel, Publishing, Dordrecht, Holland.

With, K.A., and T.O. Crist. 1995. Critical thresholds in species' responses to landscape structure. Ecology 76:2446–2459.

Wolfram, S. 1984. Universality and complexity in cellular automata. Physica D 10:1–35.

Wuensche, A., and M. Lesser. 1992. The global dynamics of cellular automata. Reference Volume 1, Santa Fe Institute Studies in the Sciences of Complexity. Addison-Wesley Publishing Company, Reading, Massachusetts, USA.

# 2
# Applications of Fractal Geometry in Wildlife Biology

Bruce T. Milne

## 2.1 Introduction

Unlike simple physical systems, such as frictionless pendula and leaking buckets, ecosystems possess complexities that have limited biologists' ability to describe, predict, and manage natural resources. Complexity can be addressed with tools to assess spatial pattern over vast expanses (e.g., geographic information systems and remote sensing), to uncover persistent interactions over space and time (Cressie 1991, Deutsch and Journel 1992), and to discover simplicity in the face of chaotic changes (Tilman and Wedin 1991, Solé et al. 1992, Kauffman 1993, Plotnick and McKinney 1993, Peak, Chapter 3). Techniques to unravel spatial and temporal complexity involve purposeful manipulation of the scale of observation to discover how phenomena change steadily, and predictably, with scale.

Ecology and wildlife biology share an awareness that natural complexity varies predictably with scale, in both space and time (Urban et al. 1987, Senft et al. 1987, Delcourt and Delcourt 1988). For example, hierarchical complexity can be discerned by measuring nature at a variety of scales (Allen and Starr 1982). By observing mortality due to old age and mortality due to predation, the observer jumps from the population level of organization to the community level (Allen and Hoekstra 1992). Reviews about scale (Wiens 1989, Levin 1992, Schneider 1994) and synthetic evaluations of complexity (Holling 1992) suggest that observations made at multiple scales are key to unraveling natural complexity.

Bruce T. Milne is Associate Professor of Biology in the Department of Biology, University of New Mexico. He is also principal investigator of the Sevilleta Long Term Ecological Research project. His research focuses on the role of spatial complexity in landscapes. Since 1985 he has explored the use of fractal geometry for the characterization of landscape patterns and has applied percolation theory in studies of beetle diffusion and ecotone structure. Recent interest in the biological controls of ecotones has led to investigations of soil, water, and energy balance models.

The word *scale* has many meanings, but in general it refers to the resolution at which patterns are measured, perceived, or represented. It is helpful to break the notion of scale into several components, namely, grain, extent, lag, and window length (Milne 1991a). These four components can be manipulated to reveal different properties of a study area or a species distribution. Grain is the smallest resolvable unit of study, for example a $1 \times 1$ m quadrat. On a digital map, a grain is one "cell" or "pixel," which represents the vegetation type present, the number of animals present, or some abiotic condition of the landscape. Grain generally determines the lower limit of what can be studied. Extent is the width of a study area, the boundary of a map, a species range, or the duration of a study. Extent determines the limit to how many grains can be observed and the limit of the largest patterns that can be observed (see Magnuson 1990).

Analyses of complexity involve comparisons of grains separated by a given distance. For example, we could easily compare all the grains that are 60 m apart; the distance 60 is called a lag distance. The results of comparisons vary predictably with lag (Burrough 1986, Cressie 1991, Rossi et al. 1992), and the dependence provides a very useful characterization of spatial complexity. For example, one could walk along recording elevation every 60 paces. A graph of the elevation at one location versus that at the next would portray the correlation between elevations measured at a lag of 60 paces. Estimating the correlations at 5, 10, 30, and 120 pace lags would reveal how the correlation changes with scale, thereby providing a characterization of terrain roughness.

Rather than considering one point at a time, a window (e.g., a quadrat or plot) surrounds a set of grains from which a statistical property, such as a mean, is measured. As for lags, statistical properties can be studied as a function of window size to reveal regularities in complex patterns.

Regardless of how scale is treated technically, many characteristics of landscapes that pertain to wildlife and habitat management vary with scale, such as vegetation (Palmer 1988), animal density (Peters 1983), patch geometry (Krummel et al. 1987) or resource availability (Milne 1992a). Changes in pattern with scale reflect transitions between the controlling influence of one environmental factor over another (Johnson et al. 1992). There are two ways in which factors operate across a range of scales, i.e., grain sizes, lags, or window sizes. First, a constraint such as a streambed may interact with something else, such as stream discharge, in a consistent way across a range of scales, as from 1st- to 10th-order streams. Second, the controlling factors may change abruptly, as when interspecific competition outweighs intraspecific competition (Yamamura 1976) to create a discontinuity in a pattern on the landscape, such as at an ecotone boundary (Milne et al. 1996). It is important for management purposes to: (1) identify what the major controlling factors are (e.g., forage availability versus climate) and (2) to quantify the "domain" of scales (Wiens 1989) over which a manageable factor operates, e.g., transmission distances of a disease. Tools described below detect such domains and characterize the scales over which a single factor operates (see Milne 1992b).

The main purpose of this chapter is to illustrate the fundamental notions of fractal geometry (Mandelbrot 1982) as they apply to wildlife biology. By definition, fractal geometry requires a multiscale approach to studies of nature. It is useful to ignore the puzzling details of fractal geometry, which are available in many reviews (e.g., Feder 1988, Peitgen and Saupe 1988, Milne 1991a) until an intuitive understanding is had of the issues, problems, and insights to be gained by the use of fractal geometry. Indeed, the beauty and usefulness of fractal geometry are closely tied to the intuition about natural complexity that is captured by rather simple quantitative approaches. Specific applications to wildlife biology included here are:

1. an estimate of bald-eagle nest density along a coastline
2. estimation of red-cockaded woodpecker habitat area for the southern United States
3. models of the fluxes of water, nitrogen, and reproductive energy costs for herbivorous mammals living on fractal landscapes
4. modification of the classic species–area relation to accommodate the fractal geometry of islands
5. aggregation of fine-scale maps to coarse resolution as in GAP analysis
6. analyses of pelican population changes in relation to climate fluctuations
7. visualization of habitat edges at multiple scales.

## 2.1.1 Allometry: Biological Scale Dependence

Species effectively operate at different scales by virtue of physiological and morphometric characteristics that vary predictably with body mass. Allometry describes how physical, morphological, physiological, and behavioral traits vary with body size both in animals (Peters 1983) and plants (Niklas 1994). Allometry affects reproductive success via clutch size, foraging rates, and traits related to locomotion, thermal regulation, and demography. In general, mammals respire, grow, and reproduce as the 3/4 power of body mass while physiological rates per unit mass vary as the $-\frac{1}{4}$ power of mass (Peters 1983). Allometrically controlled defecation rates and nutritional requirements pertain directly to ecosystem function because small animals process energy and wastes at vastly different rates than do large animals. Allometry in animals is tied closely to energetic requirements that have been adjusted evolutionarily to trade off the benefits of gathering energy against the rate at which energy can be converted to offspring (Brown et al. 1993). Tradeoffs between small size, which confers an advantage during reproduction, and large size, with associated low costs of transport to find food, indicate that the optimal mammalian body size is 100 g (Brown et al. 1993). For purposes of this chapter, I consider animals >100 g and thereby assume that relevant traits increase or decrease steadily with body mass.

Over broad areas, species interact with continental margins, mountain ranges, landscapes, and patches (Senft et al. 1987) in very different ways depending upon strong interactions between allometry, habitat requirements, and the geographic distribution of habitat. For instance, foragers select individual plants depending on palatability or caloric supply and simultaneously select among patches of plants, landscapes, and regions depending on food availability (Senft et al. 1987), safety from predation, competitive regimes, mate availability, and climatic factors (Root 1988). Source-sink (Pulliam and Danielson 1991), neighborhood (Clements 1905), and proximity effects (Milne et al. 1989) contribute to a species distribution in somewhat indirect ways. Thus, the myriad ultimate factors that regulate a species distribution interact, some locally and others over broad areas.

The fragmented, lumpy (Holling 1992), or irregular distributions of species abundances (Erickson 1945, Brown 1995) reflect locations where conditions satisfy a species' requirements at all scales. When several key factors regualte a species distribution, the spatial complexity of some factors will be important at fine scales but perhaps not at broad scales. For example, imagine a species that requires: (1) forage with a certain minimum protein content, (2) habitat with few parasites, and (3) a specified minimum chance of finding mates (Allee 1931). Each of these factors is distributed spatially in some pattern (Figure 2.1, top). Locations with suitable combinations of food, parasites, and mates can be mapped (Figure 2.1, bottom). If the suitable locations are resolved very finely, then variation produces a relatively disjoint set of habitat locales (Figure 2.1, bottom, scale = 1). The high resolution habitat map can be aggregated to coarser resolution (Milne and Johnson 1993). More coarsely resolved habitat maps lump "suitable" with "unsuitable" locations (relative to the finest scale), leading to errors of omission and commission (Figure 2.1, bottom; e.g., scale = 8, 32). At scale = 1, the intermittent occurrences are governed strongly by protein that fluctuates wildly (compare Figure 2.1 bottom for scale 1 with the protein curve). At scale = 32, mate availability is most important because sufficient protein and tolerable levels of parasites occur within each grain that contains mates. These observations are both: (1) at the root of explaining why virtually anything measured in nature exhibits greater variation at finer scales (Levin 1992) and (2) pertinent to a class of fractals (Mandelbrot 1982) called multifractals (Feder 1988), which are exceedingly rich in their application to ecology and wildlife biology. Quantitative approaches to the study of animal interactions with the landscape require knowledge of how the key factors vary spatially.

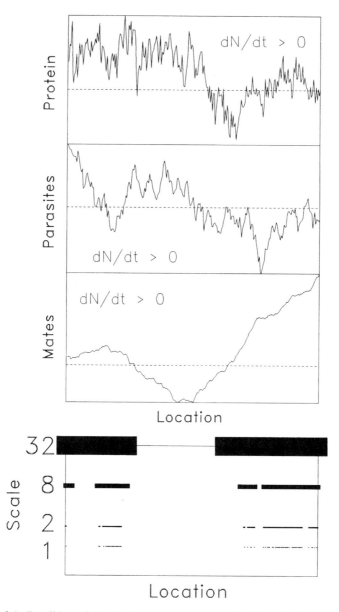

FIGURE 2.1. Possible origin of fragmented species habitat distributions. (Top) Amounts of protein in forage, parasites, and potential mates vary with location across a landscape. Each factor varies with its own fractal pattern. Thresholds (dotted lines) exist that determine whether the species can have positive growth rates (dN/dt > 0) above or below the threshold. For example, population growth is positive where a minimum protein content occurs. (Bottom) Locations where the population can grow, represented at four scales. Thickness of the horizontal bars is proportional to the spatial resolution (grain size) at which suitable habitat was assessed. At high resolution (scale = 1) suitable locations are fragmented. At low resolution the species apparently can live over a more extensive region.

## 2.2 The Notion of Scale Dependence

As a graduate student in 1982, I studied species diversity on the coast of Maine, USA (Milne and Forman 1986). The study was designed to test Simpson's (1964) peninsula diversity hypothesis that the number of species decreases with distance from the mainland. In search of independent replicate peninsulas, I spent many frustrating days trying to overcome what I eventually learned was the hallmark of a fractal: every small peninsula was but a knob on the side of a yet larger one (see the map in Figure 2.1 in Milne and Forman 1986, or virtually any map of a jagged coastline). Thus, no peninsula was "independent" of others, and the assumption of independent replicates (Hurlbert 1984) could not be satisfied. I learned that a profound stumbling block to an otherwise pedestrian study is a fertile topic in itself.

Indeed, coastlines remain the archetypal fractal because they illustrate the concepts of *self-similarity* and *scale dependence*. Self-similar patterns can be magnified to reveal more of the same pattern at increasingly finer scales. The classic example of scale dependence is Richardson's (1960) analysis of the coast of Britain. By asking "How long is the coast of Britain?" we discover that the length depends upon the resolution at which it is measured (Mandelbrot 1982), i.e., is scale dependent. At first this seems paradoxical, as we expect that a single length must exist. However, a simple exercise illustrates notions of both *scale dependence* and *self-similarity*.

### 2.2.1 Scale-Dependent Eagle Habitat

The coast of Admiralty Island is the nesting habitat for hundreds of bald eagles (Robards and Hodges 1976). Two practical questions arise: (1) how many eagle nests can fit along the coast of the island, and (2) are there as many eagles currently nesting there as might be supported? To answer these questions we need first to estimate the length of the coastline. Heeding the lesson from Richardson (1960), it is worth considering that the length of the coastline will vary directly with the scale at which it is measured. The classic way to measure the coastline at multiple scales is to apply a caliper set at some arbitrary gap width and to step along the coastline as rendered on a map or aerial photograph. Count the number of steps required to circumnavigate the perimeter. The counts are repeated for other gap widths, ideally spanning two orders of magnitude, say 1 to 100 km. A graph of the number of steps versus the caliper width will show a declining function. Since the length (km) is desired, rather than the number of steps, the counts are multiplied by the respective gap length.

The counts of steps $C(L)$ measured with a caliper of length $L$ follow a power law:

$$C(L) = bL^{-D_c} \qquad (1)$$

where $D_c$ is a fractal dimension estimated by the caliper method, and $b$ is a constant; the symbol $C(L)$ is subscript notation, not multiplication.

In geometry, a dimension specifies how to relate a small part of something to the whole. For example, there are 8 cells along one edge of a checkerboard and $8^2 = 64$ cells in all. The 8 is squared because the board is a two-dimensional plane. Similarly, the number of inches in a foot is $12 = 12^1$; the implicit exponent 1 indicates that a foot is a linear measure. The dimension 3 is used to relate the edge of a cube to its volume. Fractal objects, such as coastlines, generally have noninteger dimensions such as 1.28, indicating that the bends in the coastline tend to fill the plane in which it is embedded; the coastline does not occupy the entire two-dimensional plane (Figure 2.2). Rough terrain can be described as a bent and folded sheet that tends to fill the three-dimensional space surrounding the surface (Figure 2.3). Thus, terrain exhibits dimensions between 2 and 3 (Turcotte 1992). Fractal dimensions relate to the jaggedness of curves or the roughness of surfaces.

At first, the smallness of dimensions and, worse yet, the small difference between a dimension of, say, 1.2 and 1.3, may lead one to wonder how they could have much effect on wildlife. However, the dimensions are exponents in scaling relations like eq. 1, and thus a small change in the exponent can have a big impact on that which is measured, e.g., the coastline length.

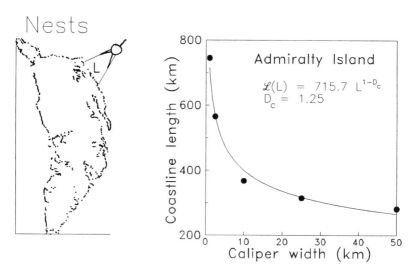

FIGURE 2.2. Distribution of bald-eagle nests (after Robards and Hodges 1976) and fractal geometry of Admiralty Island. The caliper dimension $D_c$ is related to the caliper width $L$ used to estimate coastline length $\mathscr{L}(L)$.

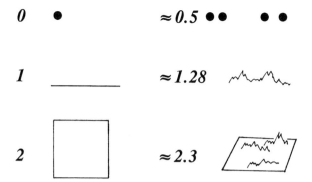

FIGURE 2.3. Euclidean and fractal dimensions of representative patterns. Euclidean sets such as points, lines, and planes have integer dimensions 0, 1, and 2, respectively. Fractals include sets of points along a line, a curve with fluctuations at all scales ($D \approx 1.28$), or rough terrain that resembles a bent and folded sheet with $D \approx 2.3$.

Patterns with noninteger dimensions are called fractals. Written as a power of $L$, measurements of fractals (e.g., eq. 1) reflect the idea that there is no single scale that best describes a fractal. Rather, an explicit reference to the scale is needed to infer the length, amount, density, or roughness of a fractal. Here is the most important point when using fractal geometry to solve problems in wildlife biology: the animal is the caliper. For example, on average, the area of an animal's home range (i.e., its caliper) increases with body mass because larger animals need more food and have to cover more ground to find it. Thus, a 0.1 kg animal needs much less area than a 100 kg animal. If we consider the populations of these two species we can imagine that: (1) small animals reach higher densities on a given fractal landscape (Morse et al. 1985), and (2) if the species happen to occupy coastlines, the small species will perceive a longer coastline than the large species. Moreover, let us assume that the two species use the same resource, such as grass. Then places where the resource is relatively dense for the small species will be different from those where the grass is dense for large species (Milne 1992a, Milne et al. 1992). The scale dependence of suitable habitat has profound implications for the coexistence of species.

Returning to the eagle example, counts of steps of length $L$ are converted to actual length of coastline $\mathscr{L}(L)$ by multiplying counts by the length of the step:

$$\mathscr{L}(L) = C(L)L = bL^{-D_c}L = bL^{1-D_c} \qquad (2)$$

By logarithmic transformation, eq. 2 can be written as the equation of a line, and linear regression can be used to estimate the slope, which is $1 - D_c$.

Specifically, we can write $\log \mathscr{L}(L) = \log b + 1 - D_c \log L$ which is in the form $y = b + mx$, the equation for a line. In practice, one takes the logarithm of $\mathscr{L}(L)$ and uses linear regression to relate it to the logarithm of $L$. The predicted slope $m$ from the regression equals $1 - D_c$ and $D_c = 1 - m$.

Not all patterns in nature have strong fractal properties. The appropriateness of a fractal representation is assessed by inspecting a log-log plot of the measurement versus $L$ to ensure that the relation appears linear, i.e., that the dimension is constant over 1 to 2 orders magnitude length scale. More careful inspections of the power law assumption should be made by graphing the residuals from the regression versus the predicted values and versus $L$. The residuals should be small and without curvature or trends. Loehle and Li (1996) recommend estimating the dimension piecemeal for each change in $L$ that is no more effective than careful inspection of residuals.

Thus, the coastline of Admiralty Island has a caliper dimension of 1.25 (Figure 2.2) corresponding to a jaggedness that is typical of coastlines. By including the constant $b = 715.7$ we obtain the scaling relation $\mathscr{L}(L) = 715.7L^{-0.25}$. The steepness of the scaling relation at small caliper lengths (Figure 2.2) indicates the great sensitivity of estimates of coastline length to the arbitrary choice of map scale. Analysis of the fractal geometry of the coastline indicates exactly how much the length varies with scale.

To estimate the number of eagle nests that fit along the coastline, I substituted the average inter-nest distance of 0.782 km (Robards and Hodges 1976) for $L$ in the scaling relation to estimate the effective length of coastline available to eagles, which is $715.7L^{-0.25} = 761.1$ km. Dividing 761.1 by the internest distance suggests that there are $761.1/0.782 = 973$ nests along the coast. Robards and Hodges (1976) observed 80 fewer nests, suggesting three possibilities: (1) they used an arbitrary, unspecified scale to estimate coastline length (Pennycuick and Kline 1986), thereby biasing the estimate of inter-nest distance; (2) the variance of inter-nest distance is important, if, for example, convolutions in the coastline alter social behavior, spacing, and density in ways that are not well represented by the mean inter-nest distance; or (3) the eagle population was depressed below carrying capacity during the pesticide-laden decade of the 1970s.

Thus, the regular decrease in coastline length with caliper width (Figure 2.2) illustrated the notion of *scale dependence*. Self-similarity was represented by the constant fractal dimension, or slope, of the curve, which indicated a very consistent relation between small crenulations and large undulations in the coastline. The scaling relation was used to estimate the effective coastline length for eagles and would have produced a different estimate for species that use a different caliper length. The dimension per se was not used as an "index" of habitat.

## 2.2.2 Box Dimension and the Red-Cockaded Woodpecker

Coastlines are quasi 1-dimensional structures that require the caliper method to estimate the dimension. In contrast, habitats that are better described as patches require different methods to measure scale dependence and self-similarity. As for Admiralty Island, one application is to estimate how much habitat is available to species that encounter the landscape at scales other than those used by cartographers. The box dimension (Mandelbrot 1982, Voss 1988, Milne 1991a) provides a scaling relation appropriate for binary maps of patches. Expectations for the box dimension are that it will equal 0 when a habitat map is composed of a single point (Figure 2.3) and will range up to a value of 2 when the habitat occurs everywhere at all scales. The pattern analysis literature generally poses a trichotomy of pattern types, namely random, regular, and clumped (Pielou 1977). Interestingly, both random and regular patterns share a box dimension equal to 2 (for boxes greater than the average inter-patch distance in regular patterns, such as checkerboards). In both cases the number of grains increases strictly as the square of the window length. By this criterion both patterns are best thought of as homogeneous or lacking in contagion.

To illustrate measurement of the box dimension, I began with a map of the red-cockaded woodpecker's potential habitat, represented by the southern pine forests of the United States (Evans and Zhu 1993) mapped at 1-km grain size (Figure 2.4). I found the box dimension of the potential habitat by first counting the number of 1-km-wide grains that contained southern pine forest. Next the map was overlaid with a grid of 2-km-wide boxes in which I counted the boxes that contained any number of 1-km grains of forest. The boxes were then increased to 4-km length, and so on up to 128 km to provide two orders of magnitude range in box size. Logarithms were taken of both the number of boxes $N(L)$ that contained forest and the corresponding box length $L$. Thus, the relation between the two was loglinear, $\log N(L) = \log k - D_b \log L$, which is equivalent to the fractal power law $N(L) = kL^{-Db}$ where $D_b$ is the box fractal dimension. I used linear regression to estimate the slope of the line. The slope provided a first approximation of the dimension (Figure 2.5).

Owing to the technical, mathematical definition of a fractal dimension (Mandelbrot 1982), the counts of occupied boxes should be done with the minimum number of boxes of length $L$ required to cover the habitat. Since the grid of boxes is rigid (i.e., boxes remain fixed in their positions relative to one another, much like cells in a checkerboard), the grid should be jostled to achieve the most efficient covering of the map. Another method, based on Mandelbrot measures (see below), automatically finds the minimum number of boxes and in general is the preferred method. However, it is both useful and simpler to learn the box counting method first.

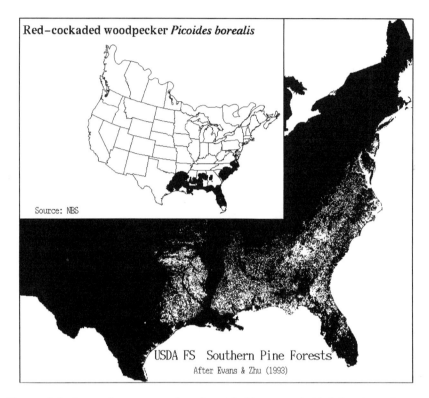

FIGURE 2.4. Approximate range (inset) and habitat map (white) for the red-cock-aded woodpecker, based on the Evans and Zhu (1993) forest cover map of southern pine forests.

Without the use of fractal geometry, there is danger that mapping habitat at an arbitrary scale could either underestimate or overestimate the habitat available to the birds. As in the coastline example, woodpeckers have their own caliper size with which they assess habitat availability. If woodpeckers maintain territories much less than $1\,km^2$, they would be using a "map" with smaller grains and the functional number (but not the area, see below) of habitat grains in the fractal environment could be more than was estimated by counting the number of 1-km-wide grains. This conclusion is based on the assumption that the habitat is fractal at scales less than 1 km.

The box counts can be transformed to area by multiplying both sides of the scaling relation by the box area, $L^2$. Thus, area scales with box size as $L^2 L^{-Db} = L^{2-Db}$, which is an increasing function of $L$ because $D^b$ is less than 2 (Figure 2.5). In this example, transformation to area magnifies undesirable deviations of the box counts away from the regression, and it is necessary to refine the characterization of the fractal geometry accordingly. Since it is legitimate to fit a fractal scaling relation only over the range of box sizes for which the estimates of area fit a power law, I fit the curve for area using length >10 km. Thus, the area dimension $D_a = 1.76$ (Figure 2.5).

Above 10 km (Figure 2.5), factors related to the geometry of terrain probably produce a reasonable fractal pattern by limiting the distribution of forests via gradients in soil moisture and solar radiation (Frank and Inouye 1994). At scales below 10 km considerably less habitat is available than would be expected from the fractal relation, possibly because of land use practices (Krummel et al. 1987) or seed dispersal limitation, which aggregate pine forests at short scales. Aggregation, or contagion, reduces the small fragments of forest that would be expected to persist if a fractal generating process were operating. Based on the argument that fractal habitat pattern enables species of different sizes to occupy different places within a landscape (Milne 1992a), the lack of fractal structure below 10 km implies that species of disparate body size may be forced to occupy the same, relatively large patches. Before reaching equilibrium, there could be an increase in species richness within a patch and also increased competition. Evidence of elevated richness due to contagion would come from

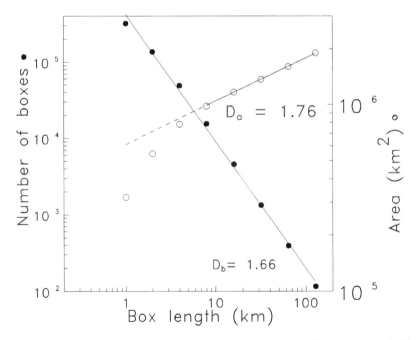

FIGURE 2.5. Scaling of the number of boxes and the corresponding area occupied by southern pine forests. The number of boxes of length $L$ occupied by pine forest scale with $L$ the $D_b$, the box fractal dimension. Transformation of box counts to areas revealed a strong departure from a fractal scaling law for box lengths less than 10 km (dashed line); the area dimension $D_a = 1.76$ was obtained from the regression of area with box lengths $\geq 10$ km.

studies of bird species richness in forest islands expressed per unit of patch area. Expressed per unit area, both richness and competition should be highest in landscapes with relatively few small patches, i.e., with area-scale relations that violate fractal scaling laws at fine scales. Support for the competition effect would argue for the deliberate creation or preservation of landscapes with a wide variety of patch sizes (Milne 1991b, Forman 1995). A very real trade-off in this design strategy relates to the greater success of nest parasites in small patches. However, costs due to parasitism might be outweighed by the benefits of the very large patches that are necessary to produce a fractal pattern that spans several orders of magnitude.

## 2.3  Mandelbrot Measures: Relations to the Landscape

The fractal properties of Admiralty Island and woodpecker habitat illustrate the statistical regularity and predictive power of fractal geometry. Equally useful insights come from mapping resources, habitats, or organisms at different scales. Such maps enable the statistical properties of fractals to be related directly to the ground (Milne 1992a), with potential applications in resource thinning and planting operations (Milne 1993). The statistical relations also enable assessments of the encounter rate between predators and prey (Mangel and Adler 1994). Geometrical treatments are useful for predicting species interactions through competition or predation, which may be altered considerably by landscape pattern (Spalinger and Hobbs 1992, Keitt and Johnson 1995).

It is useful to imagine two species, which by virtue of different body masses, occupy home ranges of different sizes. For example, home range area $H$ (km$^2$) of mammalian herbivores is predicted from body mass by the equation:

$$H = 0.032\, M^{0.998} \qquad (3)$$

where $M$ is body mass (Harestad and Bunnell 1979). The home range area of a 200-g plains pocket gopher (*Geomys bursarius*) is $0.032 \times 0.2^{0.998} = 0.0064\,\text{km}^2 = 6400\,\text{m}^2$, while that of a 2-kg black-tailed jackrabbit (*Lepus californicus*) is $63{,}911\,\text{m}^2$. It is convenient to take the square root of each area to obtain the "length" of the home ranges; $L = 80$ and $252\,\text{m}$ for the gopher and rabbit, respectively. The species integrate information about the landscape at different spatial scales that relate directly to foraging success (Milne et al. 1992).

Visualization of the two species' habitat perceptions can be made. To illustrate, I began with a digital map of grassland interspersed with woodlands (Milne et al. 1996). I assumed that the species live in grassland; more specific habitat maps could be made for each species. I then continuously passed a sliding window of length $L$ for each species over the map and counted the number of grassland grains in the window as the window slid across the map. Then, new maps were made showing the number of grassland grains in each window. The map of grain density for the rabbit (Figure 2.6B) seemed smoother than that of the gopher (Figure 2.6A) because the larger home range integrated over wider areas. Patches on the gopher map seemed more sharply defined because the small home range was nearly the size of grains on the map. This suggests that gophers will appear to be restricted more to the grassland habitat than are rabbits. This prediction

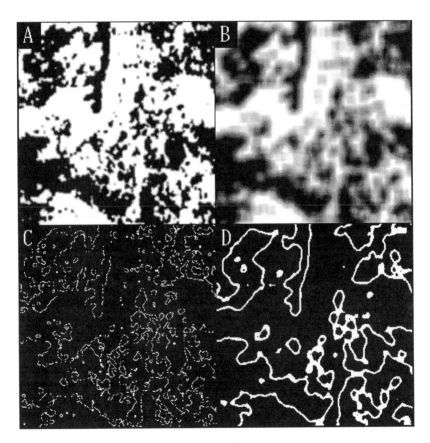

FIGURE 2.6. Multiscale density mapping. All panels describe analyses of a map of grassland from Sandia Park, New Mexico, USA. An aerial photograph taken in 1935 was scanned to produce image pixels representing $6.25\,m^2$ on the ground. Pixels were classified into two types, grassland and woodland. (A and B) Brightness is proportional to the number of grassland grains observed within (A) $3 \times 3$ and (B) $11 \times 11$ pixel sliding windows. (C and D) Locations (white) at which the number of grains within $3 \times 3$ (C) and $11 \times 11$ (D) windows ranged between 40 and 60% cover.

would be falsified if: (1) population densities were well below carrying capacity, and (2) if the nongrassland areas function as barriers, thereby reducing access to grassland within a window.

I incorporated the idea that some species prefer "edge" habitat by mapping only windows that contained 40 to 60% grassland cover (Figure 2.6C, D). The "edges" for rabbits were considerably smoother and at different locations than those for gophers. Here, I simply varied the window size and held the edge definition constant to reveal that changes in scale alter the amount of edge and locations of the edges (Milne 1992a), much as changes in caliper width alter estimates of coastline length.

As a technical aside, the counts of grassland cells, mapped at each location, form a geometric measure (Morgan 1988). When the measures are restricted to windows whose centers are occupied by the cover type of interest, in this case grassland, the measures are called Mandelbrot measures (Voss 1988). Since map boundaries prevent the centers of big windows from visiting the same set of locations as the centers of small windows, only locations common to both large and small windows were mapped (Figure 2.6).

## 2.3.1  Scaling of Mandelbrot Measures

One of the most useful aspects of fractal analyses based on Mandelbrot measures is a well-developed, and still developing, methodology to quantify the statistical behavior of subsets of fractal patterns. The methodology has been reviewed many times (Feder 1988, Voss 1988, Milne 1991a, 1992a, Solé and Manrubia 1995) and applied in fairly advanced ways for studies of the correlation between two or more fractal sets. e.g., as for studies of two or more species (Scheuring and Riedi 1994) and for a fractal thinning procedure designed to preserve tree density as perceived by a given target species (Milne 1993).

Here, I consider binary maps, composed of grains or pixels, that are either empty or occupied by habitat, e.g., Figures 2.4 and 2.6. As for the box dimension, the number of occupied cells $O(L)$ within a box of length $L$ increases as a power of $L$ according to the equation $O(L) = kL^{D}$, where $D$ is a fractal dimension of the set. In this case $D$ is specifically the "mass dimension" (Voss 1988) because it describes the "mass" or amount of habitat grains expected for a window of length $L$, which is centered on an occupied grain. This method differs from that of the box dimension by allowing overlapping windows to slide over the image rather than be fixed and nonoverlapping within a rigid grid. The sliding windows are intended to visit and enumerate the configurations of occupied grains on the map. For instance, having two occupied grains within a window is different from having 12 grains, and the sliding windows are intended to detect and characterize the frequency with which each mass of occupied cells occurs on a map. The motivation for this stems from the thermodynamic heritage of the technique (e.g., Grassberger and Procaccia 1983), which includes methods to specify all possible states of a system.

The dimension $D$ describes how fast the number of occupied cells $O(L)$ increases as the window size increases. Below, I illustrate scaling laws for studies of the simultaneous occurrence of two or more species and use the dimension to estimate how much habitat would be available jointly to species that operate at different scales. In studies of competition or disease transmission, the habitat the species have in common may be a predictor of the effect of one species on another. Similarly, the joint occurrence of a species and one or more environmental factors could be assessed at multiple scales.

Thus, several predictions can be made. Since habitat is the place where an organism lives, I assume that an animal's home range is centered on a grain of suitable habitat. Then the average number of grains available to an animal with a home range of length $L$ is

$$O(L) = kL^{D}. \tag{4}$$

Calculation of $D$ and $k$ is described in detail by Feder (1988), Voss (1988), Milne (1991a) and Milne et al. (1996). As for the box dimension, take logarithms of both sides to convert eq. 4 into an equation in the form $y = mx + b$ and use linear regression to estimate the slope $m = D$ and intercept $b$. Assuming natural logarithms were used, $k = e^{b}$.

Of course home ranges are not square and it is appropriate to consider better ways of applying $L$ to irregularly shaped home ranges. The simplest approach is to assume that the square root of home range area is a good approximation of the length when averaged over many home ranges which differ in shape. By computing $L$ for a species or a population based on many home ranges the nuances of any particular shape will be averaged out. A second approach is to begin with a map of each home range rendered as a cluster of grains; the cluster may have a highly irregular shape that includes gaps and jagged edges. The center of mass of the home range has coordinates equal to the mean latitude and mean longitude of grains in the cluster. Then, the "length" can be represented by twice the "radius of gyration" (Feder 1988, Cresswick et al. 1992), which is the square root of the average squared distance between each pair of grains in the cluster. The radius of gyration is a measure of length of highly irregular objects. Software for measuring the radius of gyration and other properties of clusters is available via the World Wide Web (http://sevilleta.unm.edu/~bmilne/khoros/ktool.html).

Returning to the use of Mandelbrot measures, the number of grains can be converted to percent cover $P(L)$, which is commonly used to characterize habitat, by dividing the number of grains by the area of the window and then multiplying by 100:

$$P(L) = 100\,O(L)\big/L^{2} = 100\,kL^{D-2} \tag{5}$$

To compute the expected number of habitat grains directly from body mass, first express $L$ in terms of mass via a transformation of home range area using eq. 3: $L = H^{\frac{1}{2}} = 0.032^{\frac{1}{2}}M^{0.998/2} = 0.179M^{0.499}$. Substitute this expression for $L$ into the expression for $O(L)$ (eq. 4) to get the number of grains of habitat $G(M)$ available to an animal of a given mass:

$$G(M) = k\left(0.179M^{0.499}\right)^{D} = k0.179^{D} M^{0.499D}. \tag{6}$$

Allometric relations for body mass and home range area include considerable variance. Equation 6, based on a statistical dependence between body mass and home range size, may be a suitable approximation in comparative studies across a wide range of body masses. For applications to a particular species in a particular study area it may be preferable to use an empirical estimate of home range area.

Equations 4 to 6 highlight one of the most useful applications of fractal geometry in wildlife biology, namely, to estimate or predict quantities such as cover and area of habitat available to species that operate at different scales. After all, a major challenge in landscape ecology is to transform maps generated by humans into maps that represent the landscape as it affects other species, few of which ever have a chance to see the whole landscape from above, much less measure the landscape's fractal geometry! In contrast to studies that attempt to use a fractal dimension as an index of landscape spatial complexity with little predictive power (e.g., O'Neill et al. 1988), eqs. 2 to 4 enable direct comparisons of the amounts of resources available to various species. The distinction between predicting the quantities versus using the dimension as a descriptive index relates to the role of the constant $k$, which is an essential ingredient for prediction (Milne 1992a).

## 2.3.2 Density Partitioning and Associations Between Species

Within a given map, some windows contain low densities of grains of a specific habitat (e.g., riparian), and other windows contain high densities. Remarkably, as the window size changes, the locations of low- versus high-density windows also change (Figure 2.7). Consequently, two wildlife species that require, say, 35% conifer cover but assess cover at different scales should settle in different places on the same landscape. Fractal analyses based on Mandelbrot measures (Mandelbrot 1982, Voss 1988) enable the investigator to "partition" the statistical information about a map according to relative densities of grains within windows of different sizes. I refer to a particular arrangement of occupied grains within a window as a configuration. Partitioning is somewhat analogous to finding various moments of a frequency distribution such as the mean, variance, and skewness. The analogy with statistical distributions breaks down however because: (1) fractal

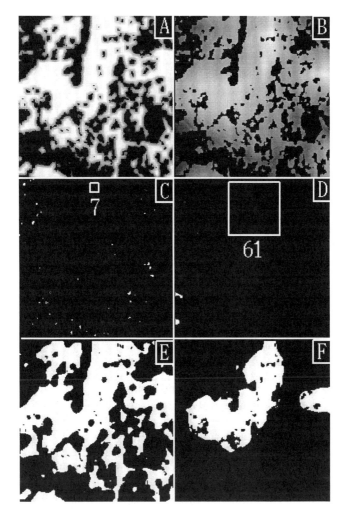

FIGURE 2.7. Density partitioning of a landscape at two scales. Original image was the same as for Figure 2.6. (A) Brightness is proportional to the number of grassland pixels (grains) present in $7 \times 7$ pixel windows; white regions represent the centers of windows that were saturated with 49 grassland grains. (B) Grain counts mapped in $61 \times 61$ pixel windows; both panels A and B were cropped to show only image locations for which the large window fit completely on the original image. Following Voss (1988), but unlike Figure 2.6, only windows whose center location included a grassland grain were used to form the counts. (C and D) Locations (white) of $7 \times 7$, and $61 \times 61$ pixel windows, respectively, for which $\leq 25\%$ of the window was occupied by grassland. White rectangles annotated with 7 or 61 describe the window sizes used in the respective columns of the figure. (E and F) Locations (white) of $7 \times 7$ and $61 \times 61$ pixel windows, respectively, with $\geq 75\%$ cover of grassland.

partitioning provides additional information about "negative moments," which describe the sparse configurations of grains (Milne 1991a), and (2) the scale dependence of each moment can be measured to learn how the geometry of sparse configurations differs from that of dense configurations (Feder 1988). Sparse configurations of grains can be thought of as "stepping stones" between higher concentrations of grains. Some landscapes contain more stepping stones than others, with implications for the connectivity between dense patches of habitat. Partitioning provides a suite of scaling exponents, one for each moment, and thus a more complete characterization of the pattern than can be had from any single dimension or exponent. In theory, the scaling exponents of all moments converge for infinite size maps but the finite extent of empirical maps makes the convergence a very special case.

In the management of pest outbreaks or of predator–prey interactions, it may be necessary to know the spatial association between host and parasite or predator and prey. In a nonfractal world, the associations can be described by a correlation coefficient between the abundances of one member of the pair and the other (Mangle and Adler 1994). However, if either or both players are fractally distributed, then the correlation between the two will change with scale (Scheuring and Riedi 1994).

It is reasonable to measure the association between a predator and a prey by focusing on the predator. Then, one can ask what the chance of finding a prey item is within a given search radius. Operationally, one centers a quadrat or window on the predator and counts the number of prey present. Replicate predators would provide a distribution of surrounding prey densities.

Measurements of the association between two species are made by scanning across a map of species 1 with a window of length $L$. Windows whose center location coincides with a grain containing species 1 are inspected. If any grains within the window are occupied by species 2, then the number of grains with species 1 is tallied to provide a measure of species 1 abundance, conditional on the presence of species 2 (Scheuring and Riedi 1994).

The analysis can be generalized for measurement of the fractal geometry of a given species contingent on the presence of any number of other species or environmental conditions. Specifically, the analysis involves counting the number of grains of species $j = 1$ that occur in windows of length $L$ occupied by at least one grain of species $j = 2, 3, \ldots S$, where $S$ is the number of species under study; grains for species $j \neq 1$ do not have to be in the center of the window. Each species occupies a set of points called $l_j$, which can be thought of as a map of all the grains that contain species $j$. For any given window size $L$ there is a number of windows $W(L,l_1 \ldots l_S)$ for which species 1 is present at the center location and species $2 \ldots S$ are present in the window. At each window location $i$, there are $n_i(l_1|l_2 \ldots l_S)$ grains of species 1 conditional ("|") on the other species. Since the counts need to be divided by the total mumber of windows that satisfy the condition, the total number of grains of species 1 that satisfy the condition is $n(l_1|l_2 \ldots l_s) = \Sigma_i n_i (l_1|l_2 \ldots l_s)$. The portion of the total at location $i$ is $p_i(l_1|l_2 \ldots l_s) = n_i(l_1|l_2 \ldots l_s)/n(l_1|l_2 \ldots l_s)$. The portions $p_i(l_1|l_2 \ldots l_s)$ have explicit spatial coordinates and as such constitute a geometric measure (Morgan 1988) on the landscape. Scheuring and Riedi (1994) compute scaling exponents from the portions $p_i (l_1|l_2 \ldots l_s)$ using an ambiguous series of equations.

I compute the scaling behavior of the measures by recasting the portions in terms that enable use of an alternative expression (Voss 1988, Milne 1991a, 1992a). Since a frequency distribution can be made of the grain counts $n_i(l_1|l_2 \ldots l_s)$, I follow Voss (1988) and first compute the probability $P(m,L)$ of observing $m = n_i(l_1|l_2 \ldots l_s)$ points in a window of length $L$. The values of $m$ are $\leq L^2$ so $\Sigma_{m=1} P(m,L) = 1$. The fractal moments are formed:

$$M(L)^q = \sum_{k=1}^{L^2} m^q P(m,L),  \tag{7}$$

which defines the $q$th fractal moment $M(L)^q$. Since raising $m$ to a high power effectively creates a large weight by which $P(m,L)$ is multiplied, positive values of $q$ produce moments that emphasize the contribution of densely filled windows, i.e., locations where many grains of species 1 occur with all the subsidiary species. In contrast, negative $q$ values downweight dense windows in favor of sparse windows. Thus, $q$ can be thought of as a tuning parameter that enables the moments to reflect selected partitions of the species distribution. As it turns out, some values of $m$ at a fine scale produce high values of $m^q P(m,L)$ and thereby contribute a lot to the size of the moment (eq. 7). At broader scales other values of $m$ contribute a lot but these values of $m$ generally occur at different locations than the important $m$ values at the fine scale. Since the moment at scale $L$ characterizes "what the animal encounters," we conclude that animals that operate at different scales are influenced by different locations.

The $q$th roots of the moments increase as a power of $L$ (Voss 1988, Milne 1991a, 1992a) according to:

$$\left\langle M(L)^q \right\rangle^{1/q} = kL^{Dq}  \tag{8}$$

for $q \neq 0$; for $q = 0$, $D_0$ is obtained from the slope of $T(L) \sim \ln L$ where $T(L)$ $= \Sigma_m \log m\, P(m,L)$. The brackets "$\langle\,\rangle$" indicate that an average has been taken. The result is an entire family of scaling exponents $D_q$. The notion that a fractal is composed of partitions, each of which has its own scaling exponent, led to the term *multifractal* (Feder 1988). Scaling relations (e.g., eq. 8) are valid only when the moments increase as a power of $L$. It is necessary to inspect doubly logarithmic graphs of the transformed $q$th root of the moments and $L$ to verify the assumption of multifratality (Scheuring and Riedi 1994).

Applied to the maps of gopher and rabbit edges (Figure 2.6), the multifractal analysis characterizes the association of the edge habitats of the species at all scales. Thus, on average (i.e., for $q = 1$, which corresponds to the mean number of occupied grains in a window) there will be $M(L) = 0.08 L^{1.86}$ cells of gopher edge wherever rabbit edge also occurs within windows of length $L$ (Figure 2.8). Compared to the scaling of gopher and rabbit edges by themselves, the joint occurrence of gopher with rabbit edge reflects the density of gopher edge when measured in small windows but takes on the density of rabbit edge in large windows (Figure 2.8). This transition reflects the fact that rabbit edges were defined at a broader scale and that the analysis of association was conditioned on the presence of rabbit edges.

Little research has addressed the processes that give rise to particular values of fractal dimensions in ecology (but see Sommerer and Ott [1993] for an example from physics). in general, deviations of the box or mass dimensions from values of 2.0 (i.e., random or regular patterns) indicate that a process that creates contagion, or nonrandom pattern, is operating. The multifractal analysis of cooccurrence has potential for identifying envi-

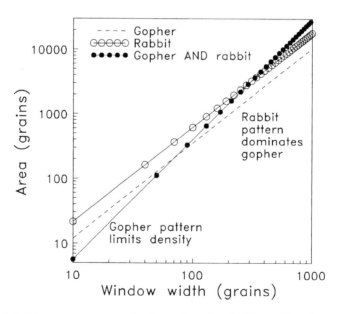

FIGURE 2.8. The cooccurrence of edges of gopher habitat with edges of rabbit habitat and the scale dependence of each edge type alone. By considering both edges simultaneously, a new slope emerges which, in this analysis, characterizes the dependence of gopher edges on the presence of rabbit edges.

ronmental factors or species associations that regulate a given species distribution, as indicated by the novel dimension that emerged from considering the cooccurrence of gopher and rabbit edges. Solé and Manrubia (1995) show that changes in the parameters of a simulation model (i.e., changes in the process rates) alter the dimensions. In this sense, the scaling behavior of any given ecological fractal could be considered diagnostic of some underlying physical or biological process that constrains the fractal to some finite subset of the landscape.

Scheuring and Riedi (1994) show how to use multifractal results to form standard $\chi^2$ contingency tables to evaluate cooccurrence at all scales simultaneously. Their innovation addresses concerns about finding "the" proper measurement scale for studies of species associations in spatially complex landscapes, at least for ranges of window sizes that satisfy the multifractal assumption. Simultaneous tests for association across a range of scales are possible.

## 2.4 Flux

Fluxes form the basis for many landscape and resource management decisions. Fluxes describe the production of forage, trace greenhouse gases such as methane, and reproductive effort expended on a landscape during the breeding season. Where nonrandom patterns or incomplete mixing of gasses occur, ecologists and wildlife biologists who extrapolate fine-scale field measurements to predict patterns and fluxes over broad regions need to incorporate the spatial heterogeneity of the landscape into the predictions. Some of the most difficult problems relate to estimating fluxes, or rates of production of matter and energy per unit area through time.

Substitution of the home range equation (eq. 3) into expressions for metabolic fluxes could be used in models of nutrient cycling rates at various scales. For example, mammals excrete $N$ at rates (g/s):

$$N(M) = 3.41 \times 10^{-6} M^{0.74} \tag{9}$$

(Stahl 1962). Flux, which is the mass of N per unit area per unit time, is found by dividing $N(M)$ by the area from which flux can occur, namely the number of habitat grains occupied by an animal of a given mass $G(M)$, (eq. 6). Thus, the flux from an animal of mass $M$ living on a landscape with dimension $D$ has units $gm^{-2}s^{-1}$:

$$F(M) = N(M)/G(M) = 3.41 \times 10^{-6} M^{0.74}/(k0.179^D M^{0.499 D})$$

$$= 3.41 \times 10^{-6} M^{0.74 - 0.499 D}/k0.179^D. \tag{10}$$

Equation 10 is simply a ratio of nitrogen loss rate $N(M)$ divided by the grains of habitat available to the animal within its home range $G(M)$. As before (eq. 4,5,6), $k$ is a constant derived by regression and is related to the geometry of gaps (Mandelbrot 1982), or equivalently the degree of contagion (Milne 1992a) within the landscape. Similar expressions can be derived for other emissions from animals such as methane, evaporative water (e.g., Crawford and Lasiewski 1968), and feces (Blueweiss et al. 1978). Following the logic of eq. 10, the flux of evaporative water (Crawford and Lasiewski 1968) is:

$$W(M) = 4.4 \times 10^{-4} M^{0.88-0.449D} / k0.179^{D}. \tag{11}$$

Comparisons of the per individual fluxes of $N$ and evaporative water for mammals ranging from 0.1 to 1000 kg revealed a startling interaction between the relative magnitudes of the fluxes and the fractal dimension of the landscape (Figure 2.9A). Small mammals (0.1 kg) generate relatively high $N$ flux, compared to water flux, for landscapes of all dimensions. The ratio of $N$ flux to water flux is lower for large mammals, reflecting the tremendous water demand of large mammals. The water conservation strategies of small mammals, combined with a preference for high $N$ foods, such as seeds, suggests that small mammals could explain the preponderance of variation in $N$ flux from one landscape to the next.

Interestingly, species of modest size (25 kg) display comparable fluxes for both $N$ and water in landscapes of any dimension. Such species would be perfect generalists, performing equally well in any landscape. It is intriguing to wonder whether people "selected" goats, sheep, and swine as domestic animals (rather than rodents or elephants) because of the merits of a 25 kg beast.

The compression of the curve for 25 kg animals (Figure 2.9A) reflects a mass-dependent reversal of fluxes in low- versus high-dimensional landscapes. For example, the 0.1 kg animal has a lower evaporative flux in a landscape where $D = 1.0$ than in a landscape with $D = 2.0$ (Figure 2.9A). In contrast, the highest water flux of 1000-kg species occurs at $D = 1.0$. Presumably the homogeneity of a $D = 2.0$ landscape, coupled with the lower transport costs of large animals, implies greater access to water and to cover for thermoregulation. Collectively, the ends of the curves bound a set of observable fluxes (assuming a constant value of $k$ that would probably vary among species with different habitat requirements, even within a given landscape).

I used an allometric expression for population density (Damuth 1981) to calculate the total flux of evaporative water and total reproductive cost (Peters 1983, p. 127, after Brody 1945) for 1600 km$^2$ landscapes. I multiplied the per-animal reproductive cost $R(M)$, measured in joules, by the number of females expected over the area (Figure 2.9B). As for individual fluxes, landscape heterogeneity interacts with body mass such that large animals have the highest per-landscape reproductive costs on the most homogeneous landscapes ($D = 2.0$), possibly due to intraspecific competition related to overlapping home ranges. The reverse is true for small animals. The body

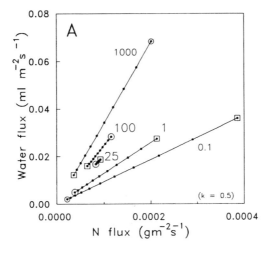

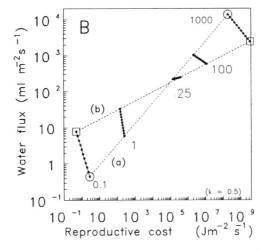

FIGURE 2.9. (A) Fluxes of nitrogen and evaporative water for individual animals. Each curve describes the fluxes for an animal of a given mass (kg). The 10 points on each curve correspond to fractal dimensions ranging from 1.0 to 2.0 in increments of 0.1. Circles and squares indicate landscape dimensions $D = 1.0$ and $D = 2.0$, respectively. (B) Flux of evaporative water and reproductive energy (expressed per $m^2$ of occupied landscape) from animals of various body masses and identical habitat requirements living in a $1600\,km^2$ landscape. Each curve (filled circles) describes the fluxes for a population of animals of a given mass (kg). Each point on a curve corresponds to a landscape fractal dimension, ranging from 1.0 to 2.0 in increments of 0.1. Open circles (curve a) and squares (curve b) indicate landscape dimensions of $D = 1.0$ and $D = 2.0$, respectively.

mass at which species are insensitive to landscape dimension is somewhat less than 25 kg. For landscapes with $D = 1$, water flux is related to reproductive cost $R(M)$ according to $W(M) = 0.25R(M)^{0.57}$ (Figure 2.9B, dashed curve (a)), while fluxes for $D = 2$ behave according to $W(M) = 9.41R(M)^{0.25}$ (Fig. 2.9B, dashed curve (b)). Changes in the exponents of these relations for $D = 1$ versus $D = 2$ indicate changes in the dominance of one or more implicit physiological factors, such as litter mass and total reproductive cost.

The dependence between body mass and reproductive cost argues strongly for the preservation of large tracts of land to support the reproduction of large species. Since the absolute costs to large animals are higher than costs to smaller species in landscapes of all dimensions $(1 < D < 2)$, it is untenable to argue that wise landscape design (i.e., design that manipulates $D$) can increase the fitness of large species relative to smaller species. However, changes in fitness for a given species may be achieved by engineering landscapes with constants $k$ and dimension $D$ that are more favorable to a species.

Interspecific differences in habitat requirements will be manifest in interspecific differences in $k$ within a given landscape. Thus, the possibility remains that body mass curves in Figure 2.9 could overlap for a given landscape, indicating that landscape design could reduce the reproductive costs of a large species to levels below those of a small one.

Some consideration should be made here about the validity of SI units (e.g., $gm^{-2}s^{-1}$), which have integer exponents. The implicit assumption behind integer exponents is that estimates of $N$ production and water loss are insensitive to fractal environmental heterogeneity, which may or may not have been present in the experiments that were used to estimate the rates. Integer exponents are probably valid if $N$ production was averaged over time and space scales for which autocorrelation did not exist, i.e., variation was homogeneous in time and space. Pennycuick and Kline (1986) proposed that units of measure be superscripted by the appropriate dimension to account for the effects of fractal patterns that are at odds with the integer dimensions of SI units. Assuming homogeneity during calibration, eqs. 10 and 11 suffice to incorporate spatial heterogeneity of the landscape.

## 2.5  Fractal Applications to GAP Analysis

The goal of GAP analysis is to identify areas where conservation efforts may have omitted critical habitats of threatened and endangered species. Economic considerations often limit map resolution to a 1-km grain size, necessarily eliminating some of the rarer cover types. For example, a 1-km grain mapped as open water could easily contain a 30-m-wide island of pine marten (*Murtes amerieavia*) habitat. Since the fractal geometries of cover classes differ, a GAP map made at a 1-km grain size sacrifices information about some classes more than others. Moreover, the risk that a given 1-km grain omits a particular class depends on the multifractal association between the mapped class and the omitted classes.

There are many ways to aggregate fine-scale maps to coarse scales. Some of the best understood methods are renormalization methods (Gould and Tobochnik 1988, Cresswick et al. 1992, Milne and Johnson 1993) which involve aggregation of $2 \times 2$ blocks of grains. A block is converted into a single grain if the set of occupied grains within it satisfy a rule chosen by the investigator. For example, the "percolation" rule (Gould and Tobochnik 1988) results in an occupied block if at least two occupied grains are adjacent vertically (Fig. 2.10A). In cases that involve multiple cover classes on random maps, the minority classes are lost upon aggregation because they make up less than 61% of the total cover on the map (Gould and Tobochnik 1988). Autocorrelation within actual landscapes results in similar effects at coverages not equal to 61%. Some rules eliminate classes below other critical densities (e.g., 50% for the majority rule, Turner et al. 1989). In contrast, the similarity dimension rule preserves cover density by maintaining a constant fractal dimension of a selected class (Milne and Johnson 1993 Figures 2.10, 2.11).

The effectiveness of the fractal preservation rule is apparent from comparison with the percolation rule (Figure 2.11). For example, repeated renormalization of a map with an initial coverage of $p = 0.1$ by the percolation rule takes it to $p = 0$ by the time the grain size equals 4 (Figure 2.11A),

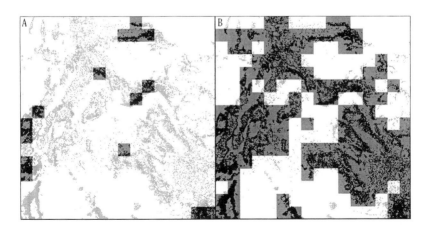

FIGURE 2.10. Renormalization of piñon-juniper woodland mapped originally at a 1-km grain size (Evans and Zhu 1993). Large blocks represent single grains at the renormalized scale. (A) Renormalization by the percolation rule. Since the woodlands cover <61% of the area, the renormalized map had a lower percent cover. Black within the blocks identifies locations containing woodland at the original scale. (B) Renormalization by the similarity dimension rule yielded a coarse-grained map that had high fidelity with the pattern of the original.

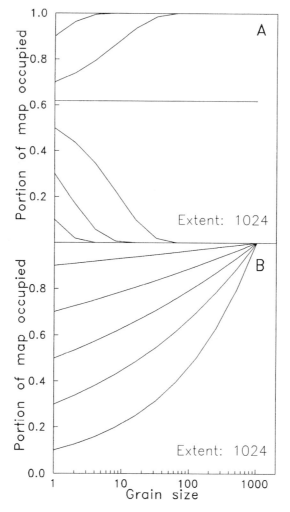

FIGURE 2.11. Rates of commission (positive slopes) and omission (negative slopes) for landscapes which have been renormalized by (A) the percolation rule and (B) the similarity dimension rule. Shallow slopes over 10-fold decreases in map resolution and positive slopes for all initial densities testify to the suitability of the fractal method.

which is a 16-fold reduction in map resolution. In contrast, preservation of the similarity dimension distorts the same map to $p \cong 0.15$ after the same reduction in resolution; a 100-fold reduction in resolution inflates $p$ to just 0.2 (Figure 2.11B). If particular classes are of special interest, then methods should be sought to minimize their loss upon aggregation. Renormalization trajectories, as in Figure 2.11, are a great help in choosing methods. Software and a tutorial for renormalization and other fractal analyses are available at: (http://sevilleta.unm.edu/~bmilne/frac/clara/clarat.html).

## 2.6 Species Area Relations Are Fractal

The classic species area curves, that describe the number of species per area, form the basis of major theories about the distribution and abundance of organisms (Preston 1962, MacArthur 1972, Rosenzweig 1995). The species area relation is an essential tool for estimating biodiversity. Interestingly, the species area relation can be expressed readily as a fractal scaling law by taking the square root of area to obtain a length scale.

In particular, the classic relation between the number of species $S$ and island area $A$ is:

$$S = cA^z \tag{12}$$

and includes the exponent $z$. Taking the square root of area gives

$$S = c\left(A^{1/2}\right)^z = cA^{z/2}, \tag{13}$$

which can be rewritten in terms of length:

$$S = cL^\gamma \tag{14}$$

for which $\gamma = z/2$. Since $S$ is an "affine function" (Barnsely 1988), it is inappropriate to call $\gamma$ a fractal dimension. It is, however, a scaling exponent, and a dimension could be obtained from studies of the autocorrelation or semivariance of $S$ (Milne 1991a). Consequently, empirical $z$ values may be explained at least partly as a consequence of life on fractal terrain, such as islands. For instance, if an observation window is centered on an island or continent, increasing window sizes encompass all of the island until they ultimately include some of the surrounding water where terrestrial species are absent. This is similar to a Scheuring and Riedi (1994) analysis of one species contingent on the presence of one or more other species. Here however, the "species" of interest is the set of all species (which are counted to give $S$) and the land is analogous to "other species." Since terrestrial species must occur on land, $S$ is implicitly conditioned on the presence of terra firma. With increasing window size, the absence of land beyond the island boundary is the constraint that limits $S$. Since the land boundary is fractal (Figure 2.2), its dimension should dominate the scaling exponent of the species area curve when the window size is greater than roughly half the island diameter and less than or equal to the width of the island. Thus, expressing the species area curve as a fractal scaling relation emphasizes the ultimate constraint of land mass on species richness. This, of course, is the intention of the classical species area curve, but the fractal geometry of the land is generally ignored (but see Scheuring 1991).

To incorporate the geographic complexity of islands, I compute the box fractal dimension of an island by making a map of it, overlaying a grid of boxes of length $L$, and then counting the number of grid cells that contain any part of the island. For box sizes smaller than the island width, the number of boxes that contain parts of the island $N(L)$ is:

$$N(L) = kL^{-Db} \tag{15}$$

where $D_b$ is the "box" dimension (Voss 1988). The exponent is negative because the number of boxes decreases as the box size increases. Multiplying $N(L)$ by the area of the box ($L^2$) gives the area at scale $L$: $A(L) = N(L)L^2 = kL^{2-Db}$. Since $D_b$ is <2, the apparent area increases with $L$. This expression for area can be substituted in eq. 12 to incorporate the geometry of the island:

$$S(L) = c\left(kL^{2-Db}\right)^z. \tag{16}$$

Throughout the history of the study of species–area relations, investigators have neglected to use eq. 15 to partial out the scale dependence of island area, which is equivalent to rewriting eq. 16 as

$$S(L) = ck^z L^{z(2-Db)} = ck^z A^{z-(zDb)/2}, \tag{17}$$

in which the dimension of the island is confounded by the species area exponent $z$. Equation 16 eliminates ambiguity about the scale at which the island area was measured and enables estimates of $S(L)$ from various islands to be controlled, or normalized, for $L$. Normalization entails expressing the area of each island at a common $L$. For example, given the scaling relations for two islands and areas computed at arbitrary scales (km$^2$),

$$\text{Island A: } A_A(L) = 0.51 L^{2-1.6}; \quad A_A(1) = 0.51\,\text{km}^2 \text{ for } L = 1$$
$$\text{Island B: } A_B(L) = 0.32\, L^{2-1.2}; \quad A_B(2) = 0.55\,\text{km}^2 \text{ for } L = 2,$$

it is possible to express the area of the small Island A at the same measurement scale as Island B; $A_A(2) = 0.51(2^{2-1.6}) = 0.67\,\text{km}^2$. Expressed at the same scale, Island A is larger than B, because the higher fractal dimension and constant conspire to provide more island area as the resolution is increased. Normalizing island area across all islands should disentangle the exponent in the classic species–area relation from the implicit exponent of eq. 14. If whole archipelagos are to be treated as study units, then the box dimension of each archipelago, coupled with species richness estimates for the archipelago, would enable tests of Scheuring's (1991) hypothesis that species richness is controlled by energy expenditures that scale with body mass raised to the 3/4 power. His accounting for surface roughness on islands could also be combined with the above suggestion that the box dimension is related to the classic species area curve.

It should not be surprising that species richness is disproportionately high in the Andes of South America where rough terrain creates islands of habitat with low box dimensions. In the extremes, when habitat is confined to jagged peaks that create patches with $D_b \sim 1$, eq. 16 becomes approximately $S(L) = c(kL)^z$, whereas smooth, round patches with $D \sim 2$ imply $S(L) = c(kL^0)^z = ck^z$, which should be less diverse. Under this hypothesis, rough terrain partitions habitat into smaller units and fosters diversification over evolutionary time. Connectivity or homogeneity, which are maximal when island area scales as $L^2$, would tend to enable competitive dominants to eliminate other species or would allow genetic introgression to disrupt locally adapted genotypes, thereby reducing evolutionary potential and diversity.

## 2.7  Scaling of Population Changes

Much has been written about the scale dependence of fractal changes through time, such as variation in river flows and sunspots (Hurst 1951, Mandelbrot and Wallis 1969, Feder 1988). Temporal fractals require different quantitative approaches than those used for coastlines and binary maps (Mandelbrot 1982, Voss 1988, Milne 1991a). Limited applications in ecology (Solé et al. 1992) provide one reliable method of characterizing fluctuations in population sizes across a range of time scales so as to identify periods of scale dependence and possible relations with fractal environmental variables.

For example, I hypothesized that the El Niño Southern Oscillation Phenomenon (ENSO), which is widely recognized as an historically important perturbation of foodwebs in the Pacific Ocean, would exhibit scaling behavior similar to fluctuations in pelican population sizes. The ENSO fluctuations are characterized by the Southern Oscillation Index (SOI), which is the pressure differential between Darwin, Australia and Tahiti. Persistent low values of the index presage an El Niño event that disrupts ocean currents, upwelling, and the delivery of nutrients to the foodchain. Presumably, collapse of the fish populations during El Niño cascades up the foodchain to affect pelican population sizes. If pelican fluctuations are indeed tied to ENSO, then both should exhibit similar scaling behavior.

Following Solé et al. (1992), I analyzed the durations of excursions that the SOI (Quinn et al. 1987) and pelican populations (Pauly and Tsukayama 1987) made away from their respective mean values (Figure 2.12). After subtracting the respective mean from each time series, the durations of excursions above or below the mean (Figure 2.12, bottom) were estimated by linear interpolation. I found power-law scaling for the number of excursions as a function of excursion length (Figure 2.13). As null hypotheses for the scaling behavior, I studied the frequency distributions of excursion lengths generated from white noise and from the running integral of white noise, also known as Brownian motion (Feller 1951). White noise is simply a series of numbers that are drawn in random order from a Gaussian distribution. A Brownian noise that varies less through time is made by summing a series of random numbers.

The frequencies of pelican and SOI excursion lengths exhibited similar slopes, which differed strongly from white noise and from Brownian motion. Thus, the ENSO phenomena were more correlated through time than a random series, but less correlated than a random walk. As expected, the similarity in the slopes of the SOI and pelican fluctuations support the hypothesis that the pelican population is driven by climate fluctuations.

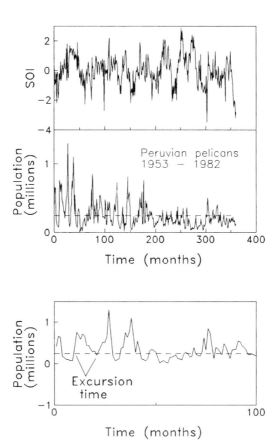

FIGURE 2.12. The monthly Southern Oscillation Index (SOI) and Peruvian Pacific Ocean pelican population sizes for the period 1953–1982 (data from Pauly and Tsukayama 1987). The bottom panel illustrates the excursion time, which is the time between intersections of a time series with its mean.

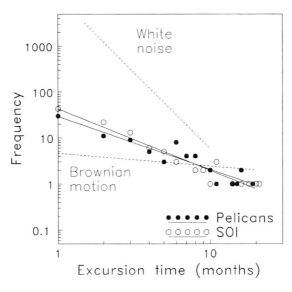

FIGURE 2.13. Scaling of the frequencies of excursion times for white noise, Brownian motion, the monthly Southern Oscillation Index (SOI), and pelican population sizes.

This method could be applied to other observations to test for the effects of constraints that govern the fluctuations.

## 2.8 Landscape Design

Natural resource management entails landscape planning and design. Although the use of fractals to make striking images of landscapes, terrain, and vegetation (Mandelbrot 1982, Barnsley 1988) provides visually convincing arguments that fractal geometry has uses in ecology and wildlife biology, a lack of theoretical basis for the ecological benefits of fractal designs has hampered their application. Milne (1991b) explained the use of iterated function systems (Barnsley 1988) to generate fractal designs that are aesthetically pleasing, naturalistic, and easy to generate. Thus, it is conceivable that patterns such as clearcuts, plantations, and parklands could be manipulated based on fractal designs with potential benefits to wildlife because the designs are self-similar, as are patterns in nature. Morse et al. (1985) suggested that fractal environments support many species by accommodating a wide range of body sizes, each of which perceives the environment at different scales. Theory predicts that fractal landscape patterns of different dimensions affect ecological processes (Fig. 2.9). Thus, landscapes designed to have a given dimension and the associated self-similarity across a range of scales could increase the reproductive success of a targeted species of reduce that of a pest.

## 2.9  Conclusions

A wealth of alternative fractal models exist to aid in the characterization of complex landscape patterns. Species abundances, habitat patterns, coastline geometries, fluxes, cartographic estimation, landscape design, and population fluctuations all are amenable to fractal analysis. In this chapter I focused on selected applications that may typify the types of problems encountered in wildlife biology. Existing allometric relations (Peters 1983) formed the basis for analyses of energy, water, and nutrient fluxes in landscapes with various dimensions. This application illustrates how new fractal relations can be generated for particular problems.

An underlying theme throughout this chapter is that fractal geometry is most effective when used to predict quantities such as habitat area, coastline length, and fluxes. Many papers in the field of landscape ecology have emphasized the measurement of one or several types of fractal dimension. These studies often stop short of prediction, having halted at description. There are, indeed, instances when a dimension forms a stringent index that, if equal to a theoretical value, is diagnostic of fundamental dynamical processes, e.g., Feder (1988), Sommerer and Ott (1993). However, use of fractal dimensions without explicit a priori expected values ranks as one of the weakest uses.

Fractal power laws are diagnostic of one or more processes that act in a consistent fashion over a range of scales (Krummel et al. 1987). However, such laws generally apply over a finite range of scales (Fig. 2.5), and other forms of scale dependence may apply, such as exponential relations. Care is needed, in the form of graphical inspection of the measured quantities as a function of scale, to evaluate the appropriateness of a fractal scaling law for a given application. Such scrutiny is absolutely necessary for multifractal applications (eq. 8) and for studies of co-occurrence, especially when characterizing the scale dependence of sparse occurrences, i.e., for $q < 0$.

Finally, fractal geometry provides highly effective tools to address the complexity of nature. The regular statistical behavior of complex patterns provides quantitative insights into what otherwise are bewildering irregularities that confound many of the traditional approaches in ecology, wildlife biology, and management.

## 2.10  Summary

Species use resources at different scales by virtue of allometric variation in home range area, metabolic rate, and movement speeds. Consequently, fractal distributions of habitat and resources imply that a given landscape represents effectively different resource patterns for each species. Assessments of resource availability at multiple scales are needed to predict the coexistence of species, to estimate effective habitat area, and to characterize the ways in which species respond to spatial complexity. The fundamentals for such assessments include the notion of scale dependence, the measurement of fractal dimensions, and visualizations of resource density at many scales. This chapter introduced a range of insights possible from multiscale analyses of resource patterns. Specific applications to wildlife biology included estimation of bald-eagle nest density along a coastline; estimation of red-cockaded woodpecker habitat area for the southern United States; models of the fluxes of water, nitrogen, and reproductive energy costs for herbivorous mammals living on fractal landscapes; modification of the classic species–area relation to accommodate the fractal geometry of islands; aggregation of fine-scale maps to coarse resolution as in GAP analysis; analyses of pelican population changes in relation to fluctuations in the El Niño Southern Oscillation phenomenon; and visualization of habitat edges at multiple scales. Fractal analyses in wildlife biology simultaneously provide statistical and spatial information with which to assess habitat availability and to guide management decisions.

*2.11 Acknowledgments.* Many thanks to John Bissonette for motivating this work and to Alan Johnson for suggesting the pelican study. The red-cockaded woodpecker range map was obtained from the National Biological Service via http://www.im.nbs.gov/bbs/htmra/h3950ra.html. Yeulong Yang provided a constructive critique of the manuscript. Support was provided by the NSF (grants BSR-9058136 and BSR-9107339) and the Electric Power Research Institute. Sevilleta LTER publication no. 94.

## *2.12 References*

Allee, W.C. 1931. Animal aggregations: A study in general sociology. University of Chicago Press, Chicago, Illinois, USA.

Allen, T.F.H., and T.B. Starr. 1982. Hierarchy. University of Chicago Press, Chicago, Illinois, USA.

Allen, T.F.H., and T.W. Hoekstra. 1992. Toward a unified ecology. Columbia University Press, New York, New York, USA.

Barnsley, M. 1988. Fractals everywhere. Academic Press, New York, New York, USA.

Blueweiss, L., H. Fox, V. Kudzma, D. Nakashima, R. Peters, and S. Sams. 1978. Relationships between body size and some life history parameters. Oecologia (Berlin) 37:257–272.

Brody, S. 1945. Bioenergetics and growth. Reinhold Publishing Company, Baltimore, Maryland, USA.

Brown, J.H. 1995 Macroecology. University of Chicago Press, Chicago, Illinois, USA.

Brown, J.H., P.A. Marquet, and M.L. Taper. 1993. Evolution of body size: consequences of an energetic definition of fitness. American Naturalist 142:573–584.

Burrough, P.A. 1986. Principles of geographical information systems for land resources. Assessment. Clarendon Press, Oxford, UK.

Clements, F.E. 1905. Research methods in ecology. The University Publishing Company, Lincoln, Nebraska USA.

Crawford, E.C. Jr., and R.C. Lasiewski. 1968. Oxygen consumption and respiratory evaporation of the emu and rhea. Condor 70:333–339.

Cressie, N.A.C. 1991. Statistics for spatial data. J. Wiley & Sons Inc., New York, New York, USA.

Cresswick, R.J., H.A. Farach, and C.P. Poole, Jr. 1992. Introduction to renormalization group methods in physics. J. Wiley & Sons Inc., New York, New, Yrok, USA.

Damuth, J. 1981 Population size and body size in mammals. Nature 290:699–700.

Delcourt, H.R., and P.A. Delcourt. 1988. Quaternary landscape ecology: relevant scales in space and time. Landscape Ecology 2:23–44.

Deutsch, C.V., and A.G. Journel. 1992. GSLIB: Geostatistical software library and user's guide. Oxford University Press, Oxford, UK.

Erickson, R.O. 1945. The Clematis fremontii var. Riehlii population in the Ozarks. Annals of the Missouri Botanical Garden 32:413–460.

Evans, D.L., and Z. Zhu. 1993. AVHRR for forest mapping: national applications and global implications. p. 76–79 in A. Lewis, editor. Looking to the future with an eye on the past: proceedings of the 1993 ACMS/ASPRS. American Society of Photogrammetry and Remote Sensing.

Feder, J. 1988. Fractals. Plenum Press, New York, New York, USA.

Feller, W. 1951. The asymptotic distribution of the range of sums of independent random variables. Annals of Mathematical Statistics 22:427.

Forman, R.T.T. 1995. Land mosaics. Cambridge University Press. Cambridge, UK.

Frank, D.A., and R.S. Inouye. 1994. Temporal variation in actual evapotranspiration of terrestrial ecosystems: patterns and ecological processes. Journal of Biogeography 21:401–411.

Gould, H., and J. Tobochnik. 1988. An introduction to computer simulation methods: applications to physical systems. Part 2. Addison-Wesley Publishing Company, Reading, Massachusetts, USA.

Grassberger, P., and I. Procaccia. 1983. Measuring the strangeness of strange attractors. Physica D 9:189–208.

Gupta, V.K., and E.D. Waymire. 1989 Statistical self-similarity in river networks parameterized by elevation. Water Resources Research 25:463–476.

Harestad, A.S., and F.L. Bunnell. 1979. Home range and body weight: A reevaluation. Ecology 60:389–402.

Holling, C.S. 1992. Cross-scale morphology, geometry and dynamics of ecosystems. Ecological Monographs 62:447–502.

Hurlbert, S.H. 1984. Pseudoreplication and the design of ecological field experiments. Ecological Monographs 54:187–211.

Hurst, H.E. 1951. Long-term storage capacity of reservoirs. Transactions of the American Society Civil Engineers 116:770–808.

Johnson, A.R., B.T. Milne, and J.A. Wiens. 1992. Diffusion in fractal landscapes: Simulations and experimental studies of Tenebrionid beetle movements. Ecology 73:1968–1983.

Kauffman, S.A. 1993. The origins of order. Oxford University Press, Oxford, UK.

Keitt, T.H., and A.R. Johnson. 1995. Spatial heterogeneity and anomalous kinetics: Emergent patterns in diffusion-limited predator–prey interactions. Journal of Theoretical Biology 172:127–139.

Krummel, J.R., R.H. Gardner, G. Sugihara, R.V. O'Neill, and P.R. Coleman. 1987. Landscape pattern in a disturbed environment. Oikos 48:321–324.

Levin, S.A. 1992. The problem of pattern and scale in ecology. Ecology 73:1942–1968.

Loehle, C., and B.-L. Li. 1996. Statistical properties of ecological and geologic fractals. Ecological Modelling 85:271–284.

MacArthur, R.H. 1972. Geographical ecology. Princeton University Press, Princeton, New Jersey, USA.

Magnuson, J.J. 1990. Long-term ecological research and the invisible present. BioScience 40:495–501.

Mandelbrot, B.B. 1982. The fractal geometry of nature. W.H. Freeman and Co., New York, New York, USA.

Mandelbrot, B.B., and J.R. Wallis. 1969. Some long-run properties of geophysical records. Water Resources Research 5:321–340.

Mangel, M., and F.R. Adler. 1994. Construction of multidimensional clustered patterns. Ecology 75:1289–1298.

Milne, B.T. 1991a. Lessons from applying fractal models to landscape patterns. Pp. 199–235 in M.G. Turner, and R.H. Gardner, editors. Quantitative methods in landscape ecology. Springer-Verlag, New York, New York, USA.

Milne, B.T. 1991b. The utility of fractal geometry in landscape design. Landscape and Urban Planning 21:81–90.

Milne, B.T. 1992a. Spatial aggregation and neutral models in fractal landscapes. American Naturalist 139:32–57.

Milne, B.T. 1992b. Indications of landscape condition at many scales. Pp. 883–895 in D.H. McKenzie, D.E. Hyatt, and V.J. McDonald, editors. Ecological indicators. Vol. 2. Elsevier Applied Science, London, UK and New York, New York, USA.

Milne, B.T. 1993. Pattern analysis for landscape evaluation and characterization. Pp. 131–143 in M.E. Jensen, and P.S. Bourgeron, compilers. Ecosystem management: principles and applications. United States Department of Agriculture, Forest Service Pacific Northwest Research Station General Technical Report PNW-GTR-318. Portland, Oregon, USA.

Milne, B.T., and R.T.T. Forman. 1986. Peninsulas in Maine: woody plant diversity, distance, and environmental patterns. Ecology 67:967–974.

Milne, B.T., K. Johnston, and R.T.T. Forman. 1989. Scale-dependent proximity of wildlife habitat in a spatially-neutral Bayesian model. Landscape Ecology 2:101–110.

Milne, B.T., M.G. Turner, J.A. Wiens, and A.R. Johnson. 1992. Interactions between the fractal geometry of landscapes and allometric herbivory. Theoretical Population Biology 41:337–353.

Milne, B.T., and A.R. Johnson. 1993. Renormalization relations for scale transformation in ecology. Pp. 109–128 in R.H. Gardner, editor. Some mathematical questions in biology: predicting spatial effects in ecological systems. American Mathematical Society, Providence, Rhode Island, USA.

Milne, B.T., A.R. Johnson, T.H. Keitt, C.A. Hatfield, J.L. David, and P. Hraber. 1996. Detection of critical densities associated with piñon-juniper woodland ecotones. Ecology 77:805–821.

Morgan, F. 1988. Geometric measure theory. Academic Press. New York, New York, USA.

Morse, D.R., J.H. Lawton, M.M. Dodson, and M.H. Williamson. 1985. Fractal dimension of vegetation and the distribution of arthropod body lengths. Nature 314:731–734.

Niklas, K.J. 1994. Plant allometry: the scaling of form and process. University of Chicago Press, Chicago, Illinois, USA.

O'Neill, R.V., J.R. Krummel, R.H. Gardner, G. Sugihara, B. Jackson, D.L. DeAngelis, B.T. Milne, M.G. Turner, B. Zygmunt, S.W. Christensen, V.H. Dale, and R.L. Grahm. 1988. Indices of landscape pattern. Landscape Ecology 1:153–162.

Palmer, M.W. 1988. Fractal geometry: a tool for describing spatial patterns of plant communities. Vegetatio 75:91–102.

Pauly, D., and I. Tsukayama, editors. 1987. The Peruvian anchoveta and its upwelling ecosystem: three decades of change. Instituto de Mar del Peru, Callao, Peru.

Peitgen, H.-O., and D. Saupe. 1988. The science of fractal images. Springer-Verlag, New York, New York, USA

Pennycuick, C.J., and Kline, N.C. 1986. Units of measurement for fractal extent, applied to the coastal distribution of bald eagle nests in the Aleutian Islands, Alaska. Oecologia 68:254–258.

Peters, R.H. 1983. The ecological implications of body size. Cambridge University Press, Cambridge, UK.

Pielou, E.C. 1977. Mathematical ecology. J. Wiley & Sons, New York, New York, USA.

Plotnick, R.E., and M.L. McKinney. 1993. Ecosystem organization and extinction dynamics. Palios 8:202–212.

Preston, F.A. 1962. The canonical distribution of commonness and rarity, Part I. Ecology 43:185–215, 431–32.

Pulliam, H.R., and B.J. Danielson. 1991. Sources, sinks, and habitat selection: a landscape perspective in population dynamics. American Naturalist 137:S50–S60.

Quinn, W.H., V.T. Neal, and S.E. Antunez de Mayolo. 1987. El Niño over the past four and a half centuries. Journal Geophysical Research 92:14449–14461.

Richardson, L.F. 1960. Statistics of deadly quarrels. E.Q. Wright and C.C. Lienau, editors. Boxwood Press, Pacific Grove, California, USA.

Robards, F.C., and J.I. Hodges. 1976. Observations from 2760 bald eagle nests in southeast Alaska: progress report 1969–1976. United States Department of the Interior, Fish and Wildlife Service, Eagle Management Study. Juneau, Alaska, USA.

Root, T. 1988. Energy constraints on avian distributions and abundances. Ecology 69:330–339.

Rosenzweig, M.L. 1995. Species diversity in space and time. Cambridge University Press, Cambridge, Massachusetts, USA.

Rossi, R.E., D.J. Mulla, A.G. Journel, and E.H. Franz. 1992. Geostatistical tools for modelling and interpreting ecological dependence. Ecological Monographs 62:277–314.

Scheuring, I. 1991. The fractal nature of vegetation and the species-area relation. Theoretical Population Biology 39:170–177.

Scheuring, I., and R.H. Riedi. 1994. Application of multifractals to the analysis of vegetation pattern. Journal of Vegetation Science 5:489–495.

Schneider, D.C. 1994. Quantitative ecology: spatial and temporal scaling. Academic Press, San Diego, California, USA.

Senft, R.L., M.B. Coughenour, D.W. Bailey, L.R. Rittenhouse, O.E. Sala, and D.M. Swift. 1987. Large herbivore foraging and ecological hierarchies. BioScience 37:789–799.

Simpson, G.G. 1964. Species density of North American recent mammals. Systematic Zoology 13:57–73.

Solé, R.V., D. López, M. Ginovart, and J. Valls. 1992. Self-organized criticality in Monte Carlo simulated ecosystems. Physics Letters A 172:56–61.

Solé, R.V., and S.C. Manrubia. 1995. Are rainforests self-organized in a critical state? Journal of Theoretical Biology 173:31–40.

Sommerer, J.C., and E. Ott. 1993. Particles floating on a moving fluid: a dynamically comprehensible physical fractal. Science 259:335–339.

Spalinger, D.E., and N.T. Hobbs. 1992. Mechanisms of foraging in mammalian herbivores: new models of functional response. American Naturalist 140:325–348.

Stahl, W.R. 1962. Similarity and dimensional methods in biology. Science 137:205–212.

Tilman, D., and D. Wedin. 1991. Oscillations and chaos in the dynamics of a perennial grass. Nature 353:653–655.

Turcotte, D.L. 1992. Fractals and chaos in geology and geophysics. Cambridge University Press, Cambridge, UK.

Turner, M.G., R.V. O'Neill, R.H. Gardner, and B.T. Milne. 1989. Effects of changing spatial scale on the analysis of landscape pattern. Landscape Ecology 3:153–162.

Urban, D.L., R.V. O'Neill, and H.H. Shugart. 1987. Landscape ecology. BioScience 37:119–127.

Voss, R.F. 1988. Fractals in nature: from characterization to simulation. Pp. 21–70 in H.-O. Peitgen and D. Saupe, editors. The science of fractal images. Springer-Verlag, New York, New York, USA.

Wiens, J.A. 1989. Spatial scaling in ecology. Functional Ecology 3:383–397.

Yamamura, N. 1976. A mathematical approach to spatial distribution and temporal succession in plant communities. Bulletin of Mathematical Biology 38:517–526.

# 3
# Taming Chaos in the Wild:
# A Model-free Technique for
# Wildlife Population Control

DAVID PEAK

## 3.1 Introduction: Fundamental Dilemmas
## of Wildlife Management

In an ideal world, a wildlife manager would have access both to the detailed histories of a well-defined set of interconnected populations and to an accurate deterministic model of the ecological dynamics of those populations. The manager would use the model to generate forecasts for population behavior with different assumptions about climatic conditions and land development and, as a result, would construct sound conservation policies that would ensure the continued robustness of the managed species while simultaneously optimizing human economic interests.

The real world is different. Here, a wildlife manager is confronted with having to construct resource management policies employing, at best, incomplete data and a largely implausible model of the underlying dynamics, chosen primarily for its tractability.

For clarity, it is useful to cast the ideal world setting described above in the following formal description. In the ideal world, there are $N$ relevant populations, interconnected in some complex ecological web. The number $N$ may include spatial information: for example, mule deer in spatial patch 1 and mule deer in spatial patch 2 may be considered as two distinct populations. These populations are counted at discrete times, $t_n$, and change from census to census in a way that can be expressed as

$$\mathbf{P}\left(t_{n+1}\right) = \mathbf{F}\left(\mathbf{P}\left(t_n\right), \beta\right). \tag{1}$$

---

David Peak is Professor and Assistant Department Head of Physics at Utah State University. He has taught a wide variety of courses at four quite different colleges and universities, including courses for both undergraduates and graduates on the principles and applications of fractal geometry and chaotic dynamics. His research interests include the study of the mechanical properties of complex materials, and the detection and control of chaos in physical, biological, and social systems.

In eq. 1, $\mathbf{P}(t_n)$ represents the set $\{P_1(t_n), P_2(t_n), \ldots, P_N(t_n)\}$ of actual populations of relevance at counting time $t_n$ (for example, at the end of the $n$th breeding season of one of the species), $\mathbf{F}$ is a set of $N$ deterministic rules for generating the set of populations at each successive counting time, and $\beta$ is a set of parameters (such as intrinsic birth and death rates, or imposed relocation or harvest rates) that control how the populations change in time.

Given such a formal system, one expects to be able to make accurate predictions (at least over the short term; when the populations change chaotically, long-term prediction is impossible—see below) for a range of scenarios that entail different parameter values. For historical reasons, any formal description of change, such as eq. 1, is called dynamics, and the associated rule of change, $\mathbf{F}$, is called the dynamic. More precisely, eq. 1 is said to define an "$N$-dimensional dynamical system."

In real-world situations, population data are almost always acquired by some sort of sampling algorithm. That is, rarely is the entire population of a given species counted at any one time. Instead of having available the desired time series of actual populations, $\mathbf{P}(t_1), \mathbf{P}(t_2), \ldots$, one has only a time series of population estimates, $\mathbf{S}(t_1), \mathbf{S}(t_2), \ldots$ Naturally, one seeks a sampling procedure that merely introduces a rescaling of the populations without bias, that is, $S_i(t_n) = a_i P_i(t_n)$, for each species $i$ in the system. In this event, the dynamics of the estimated populations—though modified in form from eq. 1—will be every bit as deterministically predictive as the dynamics of the actual populations.

Unfortunately, sampling measurements are often plagued with undesirable characteristics. A partial list of these includes:

- $N$ is wrong: The number of sampled populations may well misrepresent the number of relevant populations; that is, $\mathbf{S}(t_n)$ is a set $\{S_1(t_n), S_2(t_n), \ldots, S_N(t_n)\}$, but, $N$ is either less than the actual number of relevant populations or, when irrelevant populations are included, larger than the number of relevant populations.
- The level of aggregation is wrong: As stated previously, the population number $N$ may include spatial information. In writing eq. 1, the deterministic dynamics is tacitly assumed to be operating on some well-defined spatial scale and that the various population values are either restricted to spatial patches of appropriate scale or averaged over a number of such patches (a metapopulation). Sampling, of course, inevitably requires aggregating data on a scale established by the algorithm. There is no guarantee that the level of aggregation associated with the sample is the level that is appropriate to the dynamics. Because sampled and real scales may be incommensurate, variations observed in a sampled population may not be representative of the variations described in eq. 1.
- There is measurement error: Almost always, the sampling technique will be infected with noise, leading to population estimates of unknown accuracy.

And, after all of this is said and done, even if reliable population estimates can be obtained, one rarely (if ever) knows the dynamical rules, **F**, or the relevant control parameters, $\beta$, needed for reliable forecasting.

Faced with such daunting realities, it is tempting to conclude that careful quantitative analysis is largely irrelevant to the business of wildlife management and that its pursuit is a waste of time and resources. What I will argue in the following sections, however, is that in at least some instances, careful quantitative analysis may well be rewarded with an implementable strategy for managing populations in the wild to achieve a desirable range of population sizes.

In the next section I will give a brief introduction to the phenomenon of deterministic chaos. Following that, I will outline a technique for controlling chaos in systems for which no underlying theoretical model is known. I will then turn to the more relevant question: Can similar model-free techniques be employed in systems that display large amplitude fluctuations that are not chaotic, but instead are stimulated by random inputs? I will relate this question to variations in populations in the wild, showing how stocking and harvesting can suppress large-scale fluctuations. Finally, I will extract a few cautionary lessons.

## 3.2  Rudiments of Deterministic Chaos

Almost all real wildlife population data are characterized by the kind of large fluctuations shown in Figure 3.1. Here, fur pelt counts are used as a sampling tool for estimating wolverine population aggregated over the entire province of British Columbia (Novak et al. 1987). What causes such variable behavior? A few possibilities immediately come to mind. First, it is possible that the sampling technique is unreliable and inherently noisy; the fluctuations might be due to chance in the measuring process, for example, bad or good luck in trapping, random errors in counts, and so forth, and not reflect actual population variation. On the other hand, the sampling tool may be perfectly adequate, and the fluctuations in the samples may be caused by actual fluctuations in the population. Such variations can have stochastic as well as deterministic origins.

Let us examine an example of deterministic randomness. A simple population dynamics model, similar to that introduced by Moran (1950) to describe insect populations and by Ricker (1954) to describe fish populations, will suffice for the purposes of illustration. In the model there is a single relevant species with a population, $P(t_n)$. Each member of the population has an optimum average energy requirement $\varepsilon$. The landscape, or waterscape, in which the population resides can supply an average total energy $E$, to this species. The ratio $C = E/\varepsilon$ is the maximum population that the landscape can support at optimal conditions. In other words, $C$ is the maximum theoretical population size for which no competition for resources would be experienced by any of its members.

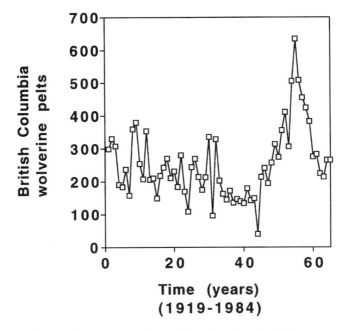

FIGURE 3.1. A plot of wolverine pelt yields in British Columbia for the years 1919–1984 shows irregular variations, similar to what are found in typical wildlife population data.

Now, the landscape can be imagined to be partitioned into $C$ equal cells, each of which represents the average home range a member of the population can forage over without competing with another member for resources. The population is assumed to obey the density dependent rule of change

$$P(t_{n+1}) = P(t_n) - d\Delta t P(t_n) + Cb\Delta t f[P(t_n)]. \tag{2}$$

In eq. 2, $\Delta t = t_{n+1} - t_n$ is the sampling interval; $d\Delta t$ represents the probability of death of a single individual in $\Delta t$; $b\Delta t$ represents the average number of offspring produced per individual in $\Delta t$, in the absence of competition for resources; and $f[P(t_n)]$ is a density-dependent factor describing the probability of surviving birth in each of the $C$ cells. In the model employed in this discussion, $f(P)$ is taken to be the probability that a cell is occupied by exactly one member of the population, and that, in turn, is assumed to be adequately approximated by the Poisson distribution,

$$f(P) = \frac{P}{C} \exp\left(-\frac{P}{C}\right).$$

This assumed form produces the intuitively appealing result that when $P$ is small compared to $C$, the number of births from $t_n$ to $t_{n+1}$ is just proportional to $P(t_n)$, but as the population grows, competition among the members for resources causes the effective birth rate to decline. Though this model is highly simplistic, variants of it have been used to describe qualitative features of various real populations (see, for example, McCullough 1979, Logan and Allen 1992, Power and Power 1995).

Note that eq. 2 admits a nonzero equilibrium solution, $P_{eq}$, that can be found by replacing each $P$ on both sides of the equation with $P_{eq}$ and solving. The result is $P_{eq} = C \ln(\frac{b}{d})$. Of course, this result makes sense as a population only if the result is positive, and that requires that $b$ be greater than $d$. Much of the literature of population biology is a discussion of the evaluation of equilibrium and a search for reasons why populations have not settled into it. It is to this latter issue that I now turn.

I want to discuss the nature of chaos and its implications for managers. I beg the reader's indulgence, then, to allow me to make a few mathematical simplifications. In particular, I will assume that the generations of the species described by eq. 2 are nonoverlapping. This is a poor assumption, of course, for many animal species (though it may be appropriate for some insects and fish). I take the sampling interval to be one generation, or, equivalently, one breeding season, and write the scaled population, $P(t_n)/C$ as $x_n$. With these assumptions, eq. 2 becomes

$$x_{n+1} = \beta x_n \exp(-x_n), \tag{3}$$

where $\beta$ is the ratio of birth rate to death rate, $\frac{b}{d}$. In this paper, I refer to eq. 3 as the Ricker Equation. Note that the relation more traditionally referred to as the Ricker Equation in theoretical ecology (May and Oster 1976) can be obtained from eq. 3 by a trivial rescaling of the population.

In any case, iteration of eq. 3 produces a time series, a succession of $x$ values, one for each new generation. Because, in eq. 3 the next generation, $x_{n+1}$, depends explicitly only on the previous generation, $x_n$, and not on any generation before that, such as $x_{n-1}$, $x_{n-2}$, and so forth, the Ricker Equation is said to be a one-dimensional dynamical system.

The Ricker Equation has several special solutions; among them are $x_n = 0$ and $x_n = \ln(\beta)$. Substitution of these values into eq. 3 shows that $x_{n+1} = x_n$. The values 0 and $\ln(\beta)$ therefore will repeat again and again, generation after generation. They are called fixed points of the dynamics (of eq. 3). The value $\ln(\beta)$ corresponds to the nonzero equilibrium population discussed previously.

When $\beta < 1$, $\ln(\beta)$ is negative and is said to be a spurious solution. In the case where $\beta < 1$, the other fixed point, 0, is said to be a stable fixed point, or a fixed point attractor. The reason for this terminology is that any starting value $x_0 > 0$ will evolve to 0 as time goes on—that is, 0 attracts any starting value $x_0$ to it. Clearly, a population in which each member is not at least replaced in each generation will become extinct in a finite number of generations.

On the other hand, when $\beta > 1$, 0 is no longer stable; it is no longer an attractor of the dynamics. In this case, any $x_0 > 0$ is carried away from 0.

When births exceed deaths in a generation, the population grows. For a range of βs greater than 1, the second fixed point, ln(β), is an attractor. For these birth rates, the dynamics is said to evolve to a stable, nonzero equilibrium value for the population. Examples of extinction and of a nonzero fixed point attractor are shown in Figure 3.2.

The dynamics of the Ricker Equation becomes more complicated and interesting for sufficiently high fecundities. For example, when β is in the range 7.39 < β < 12.42, the fixed point ln(β), like 0, is also unstable. That is, when the species' intrinsic birth rate is sufficiently high, the population will not stabilize at a single fixed value. Instead, for this range of β, the steady state values of $x_n$ are observed to repeat every other generation (Figure 3.3). The attractor of the dynamics is now said to be a 2-cycle. The value of β (namely, $e^2 = 7.38905 \ldots$) that demarks the separation of the steady states whose attractor is a fixed point from those whose attractor is a 2-cycle is called a bifurcation value.

The dynamics bifurcates again and again as β is increased beyond 12.42— first near β = 12.43, where the steady state becomes a 4-cycle, then again near β = 14.22, where the steady state becomes an 8-cycle, and so on. In each of these cases, the attractor is a finite set of discrete x-values bounded between some $x_{min}$ and some $x_{max}$.

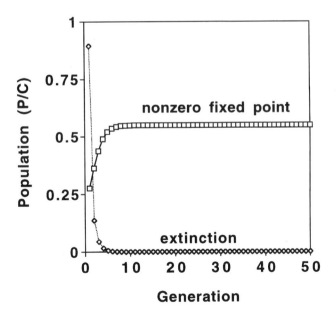

FIGURE 3.2. Shown are time series derived from the Ricker Equation (eq. 3 in the text) with two different intrinsic birth parameters, β. The asymptotic, or attractor, populations are: for β = 0.5, extinction—a stable equilibrium value of 0; for β = 1.7, a stable, nonzero equilibrium population.

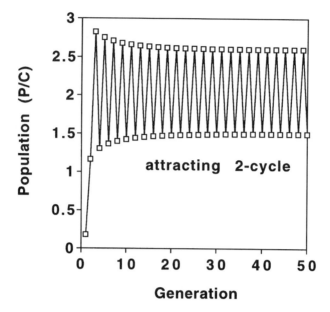

FIGURE 3.3. A time series derived from the Ricker Equation with $\beta = 8.17$ reveals an attractor population that cycles every other generation between two values (about 1.32 and 2.88).

Eventually, near $\beta = 14.77$, something new and interesting occurs. The steady state still appears to be periodic, at least on a rough scale, and characterized by an attractor. That is, there is a well-defined interval ($x_{min}$, $x_{max}$) into which almost any initial population $x_0$ is drawn; and, once $x_n$ enters that interval, all successive $x$s stay within it. But, there is actually a well-defined value of $\beta$ near 14.77—we shall call it $\beta^*$—at which something qualitatively different appears in the long-time behavior of the dynamics. Just below $\beta^*$, two starting values $x_0$ and $x'_0$ that are initially very close to each other iterate to successive values that eventually become even closer. That is, if $|x_0 - x'_0| < \varepsilon$, then $|x_n - x'_n| < \varepsilon$ for all $n$ sufficiently large. This is shown in Figure 3.4. Such behavior is consistent with a periodic attractor. That is, values starting near the attractor converge to it.

On the other hand, for $\beta$ just a bit larger than $\beta^*$, two very close starting values stay close only for a finite time. Eventually, $x_n$ and $x'_n$ can get quite far apart, though both remain within the attracting interval of allowed values (Figure 3.5). This eventual divergence of initially close states is called sensitive dependence on initial conditions. Moreover, a very careful scrutiny of the time series of $x$-values produced by the Ricker Equation when $\beta$ is slightly larger than $\beta^*$ reveals these values never repeat! (Careful scrutiny means, for example, that a Fourier analysis of a large amount of late-time data shows a single peak for $\beta < \beta^*$, but that for $b$ slightly larger than $\beta^*$

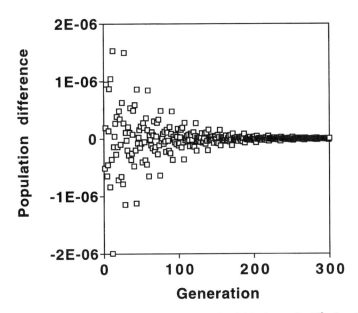

FIGURE 3.4. Two Ricker populations that start off within 1 part in $10^6$ of each other get progressively even closer, when $\beta$ is a little less than $\beta^*$ (about 14.77). This behavior is consistent with both starting populations being attracted to the same periodic attractor.

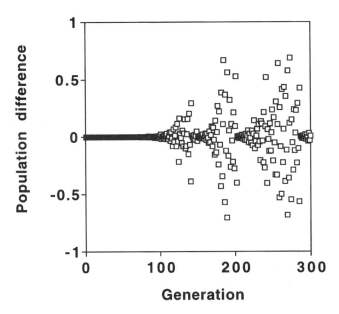

FIGURE 3.5. When $\beta$ is a little greater than $\beta^*$, two populations that start off within 1 part in $10^6$ of each other stay close only for a finite time. In this example, the populations become very different after about 100 generations. This strange behavior is called sensitive dependence on initial conditions, and is a characteristic symptom of deterministic chaos.

many broad peaks appear, similar to what one would see if the population time series were a long-period cycle that was slightly blurred with random noise.) The attractor in this case consists of clusters of $x$-values that are no longer isolated and discrete as they are in the periodic case.

Thus, the steady state of the dynamics for the case where $\beta$ is slightly larger than $\beta^*$ is bounded and aperiodic; the associated attractor, because of its sensitivity to initial values, is said to be strange. Such bounded, aperiodic behavior is called deterministic chaos (Li and Yorke 1975). The degree of apparent randomness in the population time series can be expanded by increasing $\beta$ further. For example, Figure 3.6 shows a sequence of wildly chaotic population values generated by using the Ricker Equation with $\beta = 18.17$. Despite their uniformly random appearance, these data are perfectly deterministic: chance plays no role their generation. A more complete introduction to these and many related ideas can be found in Peak and Frame (1994).

## 3.3 Controlling Chaos with Small Perturbations

Because of sensitive dependence on initial conditions and the inevitability of measurement uncertainty, long-term prediction for chaotic dynamics is impossible. Any uncertainty in the specification of the starting state of a chaotic process will eventually grow so large that states predicted $n$ time-steps into the future, for $n$ sufficiently large, will have only occasional and accidental agreement with actual state values. Nonetheless, because chaos is deterministic, short-term prediction is possible and so is control—even when an underlying model is not known. A model-free method for controlling chaos employing only small, infrequent changes in some system parameter was outlined by Ott, Grebogi, and Yorke (OGY) (1990). Since then, there has been an explosion of interest in demonstrating how the OGY method can be used to control actual laboratory systems and in developing new, more refined algorithms for chaos control (Ott et al. 1994; Ditto et al. 1995). To date, OGY control (or some variant of it) has been exhibited in mechanical (Ditto et al. 1990; Starrett and Tagg 1995), fluid (Singer et al. 1991), electronic (Hunt 1991), laser (Gills et al. 1992), chemical (Petrov et al. 1993), and biological (Garfinkel et al. 1992; Schiff et al. 1994) systems.

The OGY strategy for controlling chaos is based on two fundamental properties of a strange attractor: an infinite number of unstable periodic behaviors lie arbitrarily close to values on the attractor, and the dynamics of a chaotic process comes recurrently close to any value on the attractor.

It is useful to think of chaos as a long sequence of failed attempts at periodic behavior. For example, in the Ricker Equation with $\beta = 18.17$, a parameter where most initial $x$-values lead to wildly chaotic behavior, an initial value $x_0 = 2.89977188 \ldots$ produces a sequence that repeats every time; that is, 2.89977188 . . . is a fixed point. Similarly, the initial value $x_0$ = 0.46887416 . . . generates the 2-cycle sequence 5.33066961 . . . , 0.46887416 . . . , 5.33066961 . . . , and so on. (Note that " . . . " implies a nonrepeating decimal.) Other starting values spawn 3-cycles, 4-cycles, and on and on. None of these periodic sequences is stable, however. Thus, if the starting value 2.90000000 is chosen, a departure from 2.89977188 . . . by less than 0.008%, the first nine generations are all approximately 2.9, but the tenth is more nearly 2.8, the eleventh is about 3.0, the twelfth 2.6, the thirteenth 3.4, the fourteenth 2.0, the fifteenth 4.8, the sixteenth 0.7, and so on. In short order, sensitive dependence on initial conditions takes the time series to parts of the underlying attractor that are far from the unstable fixed point value.

On the other hand, if one waits long enough, any unstable periodic behavior that is permitted by the dynamics will reoccur. In Figure 3.6, the groups of time series data labeled A, B, C, D, and E recurrently come close to the unstable fixed point, 2.89977188. . . .

The OGY strategy for controlling chaos has two parts: (1) examine time series data as they are collected for evidence of allowed periodicities; (2) when the data revisit a previously identified, desired periodic behavior, perturbatively alter the dynamics producing the data to keep the data from running away from the target values. Depending on what type of system

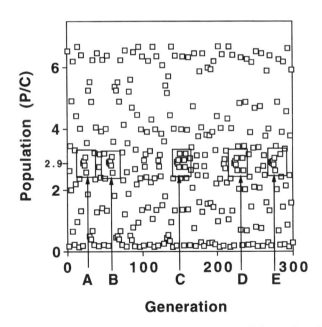

FIGURE 3.6. A Ricker time series with $\beta = 18.17$ appears wildly random. The epochs labeled A through E are evidence of recurrent returns to the vicinity of the unstable fixed point given by $\ln(\beta) = 2.9$.

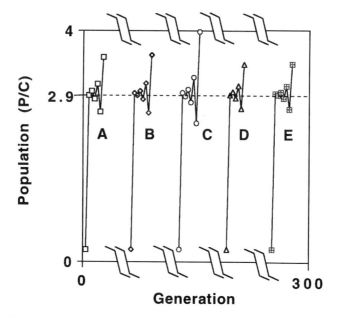

FIGURE 3.7. The recurrent returns to the fixed point seen in Figure 3.6 are magnified. In each case the time series shows a similar pattern: approach to the fixed point value from well below it, then gradual escape via oscillations of growing amplitude.

over which control is to be exercised, the latter part may be pretty tricky, but it is reasonably straightforward for population management purposes. As I discuss below, no model of the dynamics is needed to effect OGY-type, perturbative control of chaos: the data alone teach one how to do it.

The simplest, most prominent, and often, from a practical standpoint, the most useful unstable periodic behavior to look for in a chaotic time series is an unstable fixed point. The approximate population recurrences already noted in Figure 3.6 are magnified in Figure 3.7. Each recurrence has a similar precursor and each undergoes a similar pattern in escaping the fixed point—one time above the fixed point, the next below (and farther away), the next above (and farther still), and so on. This repetitive sequence of events is a clear signature of a "saddle point" (a fixed point that is approached along one direction and escaped along another), and its identification is key to implementing perturbative control of chaos (see Appendix to this chapter). While Figures 3.6 and 3.7 are produced by using the Ricker Equation, similar patterns can be expected in chaotic data from other sources.

The OGY-inspired, perturbative control algorithm is best understood graphically. To do so, it is necessary to define what is meant by a *first return map*. Suppose $x_1, x_2, x_3, \ldots, x_N$ is a time series. Form the sequence of pairs $(x_1, x_2), (x_2, x_3), (x_3, x_4), \ldots, (x_{N-1}, x_N)$; draw cartesian axes with $x_n$ along the horizontal and $x_{n+1}$ along the vertical and plot all the pairs formed from the data. The result is a first return map.

By making a first return map, one is "embedding" one-dimensional data (i.e., just a string of numbers) in two dimensions. Embedding time series data in higher dimensions sometimes leads to a clarification of the source of the data. For example, suppose the time series of interest is a string of population numbers generated by the Ricker Equation in the wildly chaotic regime seen in Figure 3.6. Inspection of a string of such chaotic values is not very instructive: they hop around in a willy-nilly way that appears as good as random. If the data were random and totally uncorrelated, plotting them in two dimensions would produce a structureless blob, as uninformative as the original string. But, chaotic data from the Ricker Equation are anything but random. A first return map for Ricker data with $\beta = 18.17$ is shown in Figure 3.8. Immediately one sees the Ricker attractor as it appears embedded in two dimensions and infers from it that a simple, deterministic rule connects successive values in the time series. This result is a specific instance of a more general theorem due to Takens (1981) that says that it is possible to completely reconstruct the dynamics of any chaotic process solely by embedding one-dimensional data in a sufficiently high-dimensional embedding space.

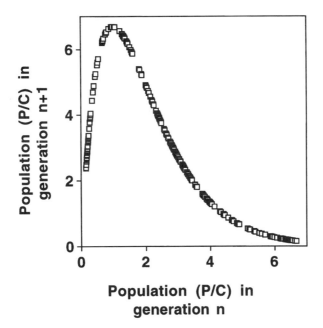

FIGURE 3.8. A first return map of the data shown in Figure 3.6 demonstrates that, despite appearing random, successive population values actually have a simple, deterministic relationship.

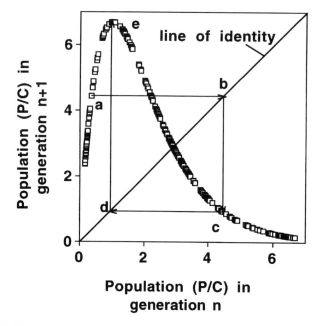

FIGURE 3.9. The succession of points on a first return map is determined graphically by drawing a horizontal line from the first point to the line of identity, then a vertical line from the line of identity to the next point. Thus, point a is followed immediately by point c (via point b), and that, in turn, is followed immediately by point e (via point d).

Irrespective of where the data come from (deterministic rule or random source), the points plotted on the first return map have an orderly sequence. For all $k$, $(x_{k-1}, x_k)$ immediately precedes $(x_k, x_{k+1})$: the second entry in one data pair is the first entry in the next. Thus, if one starts at an arbitrary $(x_{k-1}, x_k)$, the next data pair, $(x_k, x_{k+1})$, can be identified graphically in the following way. Start by adding to the first return map a "line of identity," that is, a line along which $x_{n+1} = x_n$. Now, from $(x_{k-1}, x_k)$ (point a in Figure 3.9, for example) draw a horizontal line (a line along which all points have vertical component equal to $x_k$) that intersects the line of identity (at point b); this point of intersection will have coordinates $(x_k, x_k)$. From the inter-section point draw a vertical line (a line along which all points have horizon-tal component equal to $x_k$); $(x_k, x_{k+1})$ corresponds to the data pair that lies on that vertical line (at point c). The data pair $(x_{k+1}, x_{k+2})$ is subsequently found by drawing a horizontal line from $(x_k, x_{k+1})$ to the line of identity (intersecting with it at point d), then (from d) a vertical line to the data set (at point e). This graphical process can be repeated again and again until all data pairs are hit.

Now, for a chaotic time series, the line of identity on the first return map will have a point $(x_{fixed}, x_{fixed})$ $[= (\ln(\beta), \ln(\beta))$, for the Ricker Equation, for example] that is often surrounded by data pairs with a characteristic geometric pattern inherited from the generic, characteristic pattern shown in Figure 3.7: approach to the fixed point pair along an approximately horizontal direction, then escape along an approximate line with a steep slope (that is, a slope with magnitude greater than 1). Such a pattern is illustrated in Figure 3.10. Note that, in order to show as much detail as possible in this figure, the scalings of the horizontal and vertical axes are different.

The reason, incidentally, that escape from the fixed point is along a line in the first return map, is that for any deterministic process the dynamics near the fixed point is approximately linear. For concreteness, concentrate on the behavior of the Ricker Equation near its nonzero fixed point. Let $x_n = \ln(\beta) + \varepsilon_n$ and $x_{n+1} = \ln(\beta) + \varepsilon_{n+1}$ where both $\varepsilon_n$ are $\varepsilon_{n+1}$ are assumed to be much less than 1. Now, eq. 3 defines how $x_n$ and $x_{n+1}$ are related, so if $\ln(\beta) + \varepsilon_n$ is substituted for $x_n$ into the right hand side, the exponential is

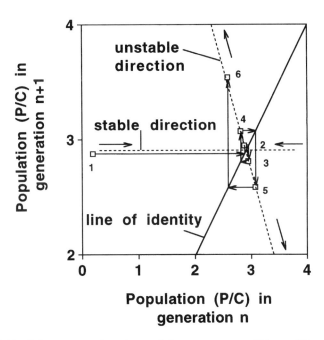

FIGURE 3.10. A first return plot of one of the recurrences of Figure 3.7: approach to the fixed point is along a line called the stable direction; escape from the fixed point is along a second line called the unstable direction. The intersection of the stable and unstable directions defines the position of the fixed point. If the starting point were exactly on the stable direction, the next, and every subsequent, point would be at the fixed point.

expanded as a power series, and only constant terms and terms proportional to ε are retained, the result for $\varepsilon_{n+1}$ is

$$\varepsilon_{n+1} = \left(1 - \ln(\beta)\right)\varepsilon_n.$$

The set of points $(\ln(\beta) + \varepsilon_n, \ln(\beta) + (1 - \ln(\beta))\varepsilon_n)$ lies on a straight line passing through $(\ln(\beta), \ln(\beta))$, with a slope equal to $1 - \ln(\beta)$. For a nonzero fixed point to exist, $\beta$ has to be greater than 1, so $\ln(\beta)$ has to be greater than 0. As long as $\ln(\beta)$ is less than 2, the magnitude of the slope $1 - \ln(\beta)$ is less than 1. For those $\beta$s, the magnitude of $\varepsilon_{n+1}$ is less than the magnitude $\varepsilon_n$ and deviations from the fixed point shrink. In other words, the fixed point is stable for $\ln(\beta) < 2$. But, for chaotic Ricker data, $\beta > 14.77$, so $\ln(\beta) > 2.7$, and therefore $1 - \ln(\beta)$ is more negative than $-1$. Thus, if $\varepsilon_n$ is positive, $\varepsilon_{n+1}$ will be negative, and vice versa. Furthermore, the magnitude of $\varepsilon_{n+1}$ will be greater than the magnitude of $\varepsilon_n$, since the magnitude of $(1 - \ln(\beta))$ is greater than 1. Escape from $(\ln(\beta), \ln(\beta))$, then, is a kind of hip-hop with each successive distance from the fixed point pair being larger than the previous.

The direction of approach to the fixed point pair is called the stable direction. The direction of escape is called the unstable direction (Figure 3.10). The stable and unstable directions intersect exactly at the fixed point on the first return map. The two-part perturbative control strategy of OGY refers directly to the stable and unstable directions of a fixed point:

1. Start with a training regime that consists of plotting time series data as a first return map and identifying the fixed point, as well as the fixed point's stable and unstable directions. Though in the Ricker example presented here the origin of the data is known, that is not necessary: the training part is purely algorithmic and requires no deeper knowledge.

2. Once this information is gathered, control can be implemented. One waits until the time series gets within some predefined tolerance level of the fixed point. The properties of chaos ensure that this will happen sometime. Then the previously acquired knowledge of the unstable direction is used to predict what the time series will do after each new value is added. The dynamics is approximately linear as long it remains close to the fixed point, so prediction simply consists of linear forecasting. If the prediction indicates that the time series will depart from the fixed point by more than the tolerance level, an intervention is required. The intervention consists of placing the state of the system on the stable direction of the desired fixed point by altering the natural dynamics.

For the purposes of wildlife management, alteration of the natural dynamics entails harvesting or relocation from the spatial domain of interest and/or stocking, that is, relocating into the domain of interest. Suppose $x_k$ is the population of interest in the current generation, and that $|x_k - x_{\text{fixed}}|$ is less than a preselected tolerance level. At the same time, suppose that $x_{k+1}^{\text{pred}}$, the linear forecast for the population in generation $k + 1$, is outside of the tolerance level. An intervention strategy employing harvesting or stocking is required that takes the population from $x_k$ to an intermediate value, $x_{k+\frac{1}{2}}$, then from $x_{k+\frac{1}{2}}$ to $x_{k+1}$.

Two possible interventions are illustrated in Figure 3.11. The point $(x_{k-1}, x_k)$ (point a in the figure) is forecast, using knowledge of the unstable direction (a → b → c), to go to $(x_k, x_{k+1}{}^{\text{pred}})$ (point c). In the example shown, $x_{k+1}{}^{\text{pred}}$ is too far below the desired fixed point value. The point $(x_k, x_{k+1}{}^{\text{pred}})$ is outside of the "square of tolerance" defined by [$x_{\text{fixed}} \pm$ tolerance level, $x_{\text{fixed}} \pm$ tolerance level]. Intervention can occur before the next breeding season (case 1, in Figure 3.11) or after it (case 2, in Figure 3.11).

Intervention before the breeding season requires a nonintuitive act. Remember, the population is predicted to fall too low in generation $k + 1$. What is required in case 1 is to *reduce* the population size before breeding can occur to decrease competitive pressure. This is accomplished by setting a harvest level at $\Delta x_k = (x_k - x_{\text{fixed}})$. After the harvest intervention (a → d), the population will theoretically be $x_{k+\frac{1}{2}} = x_k - \Delta x_k = x_{\text{fixed}}$ (point d is $(x_{k-1}, x_{k+\frac{1}{2}})$). Natural breeding is then allowed to occur (d → e), after which $x_{k+1} = x_{\text{fixed}}$, in principle.

On the other hand, intervention after the breeding season (case 2) requires a more conventional act, namely, infusion of new members into the population. The predicted number to be added needed equals $\Delta x_k = (x_{\text{fixed}} - x_{k+1}{}^{\text{pred}})$. Natural breeding is allowed to occur before the intervention (a → b → d) producing, as a result, a population $x_{k+\frac{1}{2}}$. If $x_{k+\frac{1}{2}}$ actually equals $x_{k+\frac{1}{2}}{}^{\text{pred}}$ (that is, point d = point c), then the population after intervention (d → e) will be $x_{k+1} = x_{k+\frac{1}{2}} + \Delta x_k = x_{\text{fixed}}$ (point e), and subsequent populations, in principle, will then remain at $x_{\text{fixed}}$.

The strategies required for the case where $x_{k+1}{}^{\text{pred}}$ is too high follow directly; there, one could intervene by stocking, in order to exacerbate the competition for available resources, before the breeding season or by harvesting after it.

Of course, if $x_{\text{fixed}}$ were known exactly and the populations could be adjusted exactly, only a single intervention would be required. This situation is impossible to achieve in practice, however. When the fixed point is determined empirically (and, thus, with finite precision), the intervention will never succeed in exactly hitting $x_{\text{fixed}}$. Also, there is no guarantee that the natural breeding process will actually produce the expected population. Thus, continual and systematic—not occasional and haphazard—monitoring and intervening will be required to maintain control.

I will now turn to an example of controlling Ricker population data in which the intervention is restricted to removal of members from the population. It is important to keep in mind, once again, that I am presenting here a highly simplified population dynamics model for the purpose of extracting the essence of chaos and its potential control. A more realistic model might well include multiple interacting species, overlapping generations, and har-

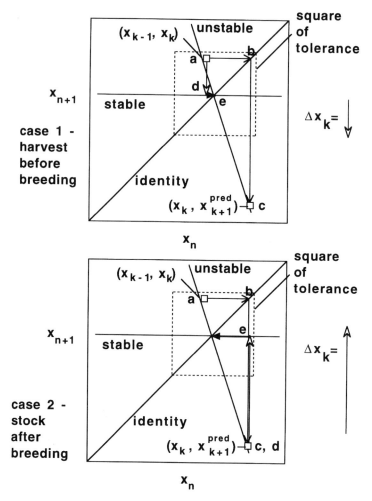

FIGURE 3.11. Two scenarios for controlling chaotic population fluctuations are demonstrated. In both cases the population at time $k$ is close to the desired fixed point, but is predicted at time $k + 1$ to be lower than can be tolerated. The prediction is shown graphically on a first return plot by the sequences a → b → c. In case 1, the controlling intervention is harvesting prior to the breeding season. The harvest is designed to take the population to the fixed value (point d), after which the natural dynamics keeps it there (point e). In case 2, the controlling intervention is by stocking after the breeding season. The strategy is to raise the population after the breeding season (from d) to the desired fixed point value (e).

vest rates that reflect management and economic considerations. The harvesting intervention I will now describe should be understood as being in addition to whatever harvesting is already included in the uncontrolled population dynamics.

Figure 3.12 shows an example of controlling the chaos of a Ricker population (with β = 18.17), using the method outlined above. Because the intervention is restricted to harvesting, in order to successfully place the system on the stable direction of the desired fixed point (in this case, the value, $\frac{P}{C} = \ln(\beta) = 2.9$), the timing of the intervention has to be varied—sometimes intervention is done before the breeding season, sometimes after. The figure shows a typical time series before and after control is implemented. For the first 25 or so generations, the whereabouts of the dynamics' fixed point along with its stable and unstable directions is learned directly from the unfolding time series. Once the fixed point is established, control is implemented whenever the dynamics returns to within a preestablished threshold of the fixed point (10%, in the example). This occurs around generation 40. After that, intervention prior or subsequent to breeding, as dictated by the season-by-season forecast, leads to control of the population to within about ±5% of the fixed point value. In this example, interventional harvesting is only done when necessary—about once every 4 to 5 seasons—and requires removal of only a few percent the population. If the same simulation is run many times with randomly chosen

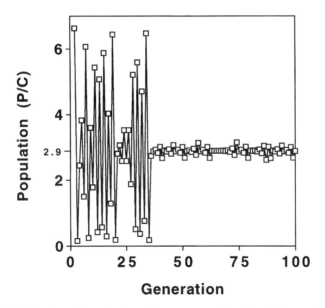

FIGURE 3.12. A series of population values generated by the Ricker model with β = 18.17 is controlled by harvesting. Initially evidence of an unstable fixed point is sought. The existence of the fixed point, along with its stable and unstable directions, is established after about 25 generations. A harvesting intervention is implemented at the next close pass to the fixed point, which occurs at about generation number 40. Subsequently, the population is maintained near the desired fixed point by occasional, small harvests.

starting states, it always takes 20 to 30 generations, on average, to learn where the fixed point is and what the associated stable and unstable directions are. Control can usually be implemented immediately once these quantities are known; that is, one does not have to wait an additional 20 to 30 generations after establishing the position of the fixed point.

It is important to reiterate that the control illustrated in Figure 3.12 is perturbative—interventional harvesting for the purpose of control is applied only when the population forecast requires it and the harvest yield is typically small—and that the controlling strategy is model-free, e.g., it is derived directly from the continuously updated population data. It is also important to note that control by this method cannot be directed to just any state—it works only for those special unstable periodic states that the dynamics admits. As such, perturbative control is radically different from standard control scenarios that are population selective and continuously applied, but that also require some underlying model to implement successfully.

## 3.4  A More Realistic Situation

In the preceding section I showed how the properties of chaos can be exploited to effect control of a population without knowing anything about the underlying dynamics. The example discussed was presented for pedagogical purposes; it is highly unrealistic. There are no known examples of real populations that obey such simple, one-dimensional dynamics as the Ricker Equation. Chaos in low-dimensional models typically requires extraordinary parameter values. For example, in the Ricker model the intrinsic birth rate has to exceed 14.7 offspring/individual for chaos to exist. In addition, first return maps for real population data are never as simple as what is seen in Figure 3.8. In a real situation, the next population value is never observed to depend on the current one only.

In an $N$-dimensional dynamical system, the next entry in a time series of values of one component of the system can be predicted, in principle, if the $N$ previous values are known. In a random process, no finite set of data suffices to predict the next entry. Thus, a random process is effectively infinite-dimensional. (Distinguishing between a high-dimensional deterministic process and a truly random one is one of the most interesting and challenging pursuits of contemporary statistical analysis.) From the standpoint of prediction and control, real population dynamics are, at best, examples of high-dimensional determinism.

So, is the technique described in the preceding section of any practical use? The answer is, it may be. Here is why. Suppose that one has available an estimate, $S(t_n)$, of a single population that is actually an integral part of an $N$-population dynamical system. If the high-dimensional dynamics is chaotic, it typically will have a set of (perhaps many) unstable fixed points.

In the vicinity of those fixed points the dynamics will be approximately linear. As long as the sampling algorithm that generates the estimated population is reasonably free of pathologies, the sampled population $S$ will be near a fixed value, $S_{fixed}$, when the $N$-dimensional population set is near a fixed point, and the dynamics of $S$ there will also be approximately linear.

Thus, even if the real dynamics involves many more populations than the one being sampled, the real dynamics will be episodically low dimensional. Hence, identification of $S_{fixed}$ and the associated stable and unstable directions around $(S_{fixed}, S_{fixed})$ on a first return map may be possible. And, despite the fact that $S$ refers to only one of the populations, attempts to control this single population during a low-dimensional episode of the system may be sufficient to control the entire high-dimensional system, though, of course, such control is hardly a foregone conclusion. Control of high-dimensional processes using only low-dimensional information has been demonstrated in at least two laboratory settings (Garfinkel et al. 1992; Schiff et al. 1994), giving one hope that the phenomenon may be fairly wide spread.

As an example of the perturbative control of a single population immersed in a high-dimensional population network, consider the following. Suppose, as in the previous examples, that the population of interest obeys eq. 3, but now instead of the intrinsic birth rate being constant, the birth rate is assumed to vary from generation to generation:

$$x_{n+1} = \beta_n x_n \exp(-x_n) \tag{4}$$

In eq. 4, $\beta_n$ is taken to be of the form $\beta(1 + \xi_n)$, where, as before, $\beta$ is the intrinsic birth parameter $\frac{b}{d}$. All of the interactions of the population of interest with its high-dimensional environment are supposed to be adequately approximated via the time dependence of the as yet unspecified variable $\xi_n$. In the absence of detailed knowledge of the hierarchy of dynamical relationships affecting the behavior of $x_n$, this complex of interactions is modeled by assuming that $\xi_n$ is a random variable drawn uniformly from some interval $[-\xi_0, \xi_0]$, where $\xi_0$ is the amplitude of the stochastic driving force in the dynamical system, eq. 4. Thus, in this model, the time-course of the population of interest is altered by coupling to very high dimensional processes.

For specificity, suppose that $\beta$ is 8.166. . . . In the absence of stochastic driving, a time series manufactured from the Ricker Equation with such a parameter value would be a 2-cycle, alternating between $x = 1.32 \ldots$ and $x = 2.87. \ldots$ It is useful to explore this case because of its similarity to oscillations in predator–prey models, where the predator and prey are isolated from other populations and where their environment is strictly unchanging. The Ricker Equation has an unstable fixed point for this parameter value at $x = 2.1$. Without stochastic driving, the existence of that unstable fixed point would never be evident in the 2-cycle time series.

On the other hand, when random fluctuations are injected into the effective birth parameter, $\beta_n$, the otherwise stable 2-cycle begins to show large scale fluctuations. The epoch labeled a in Figure 3.13, for example, depicts a long time series of *x*-values produced by the modified Ricker Equation, eq. 4, with a birth parameter that fluctuates around 8.166 . . . by up to ±5%. For much of the time the resulting behavior appears to be a noisy 2-cycle with bands centered at about $x = 1.3$ and 2.9. Occasionally, the dynamics undergoes excursions to populations midway between these bands, that is, to values that encompass the unstable fixed point.

A first return map of the populations (Figure 3.14) plotted from the data in epoch a of Figure 3.13 produces a broad smudge of points, suggesting a high-dimensional origin. Clearly, these points are not a result of chaos in the one-dimensional dynamics of eq. 4, because a first return map for such data would be similar to the smooth curve of Figure 3.8. On the other hand, the observed population swings are not strictly random either, because from them the existence of a recurrent fixed point can be extracted using the linear-approach-linear-escape signature that characterizes such a fixed point. In fact, the empirically inferred fixed point is about where it should be for a population time series obeying eq. 4 with $\ln(\beta) = 2.1$ and with no forcing. The fluctuations seen in Figure 3.14, therefore, are neither chaotic nor purely stochastic: they lie somewhere in between.

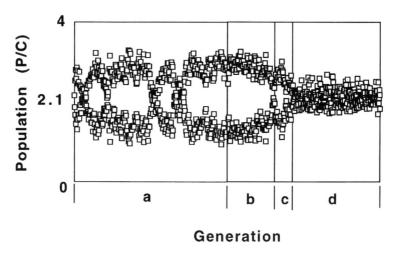

**Generation**

FIGURE 3.13. The time series shown in epoch a is a Ricker population with birth parameter that fluctuates randomly in the range $\beta = 7.77$ to 8.57 (that is, $8.17 \pm 5\%$). Without the random fluctuation, the time series would be the stable 2-cycle shown in Figure 3.3; the associated dynamics would have an unstable fixed point at $\frac{P}{C} = 2.1$. Search for evidence of the unstable fixed point occurs during epoch b. Once the unstable fixed point along with its stable and unstable directions are established, recurrence of a population near the fixed point is awaited (epoch c), after which control of the population is implemented (epoch d).

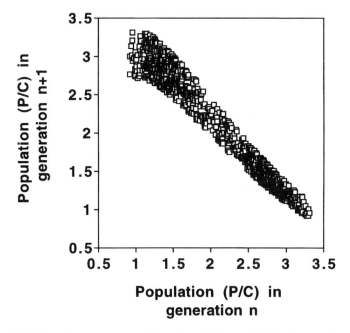

FIGURE 3.14. The first return plot of the data of Figure 3.13 during epoch a shows a broad blob of points suggesting high-dimensional (possibly random) dynamics.

During epoch b in Figure 3.13, the population data are monitored continuously for evidence of an unstable fixed point. Once that evidence is collected (at the end of b, in the example) control can be attempted whenever the population is sufficiently close to the empirically derived fixed point. The waiting period for recurrence is epoch c in Figure 3.13. Finally, upon such a recurrence, perturbative control can be implemented. Interventional harvesting produces the time series seen in epoch d.

The histograms shown in Figure 3.15 emphasize the profound consequences of perturbative population control. Before control is implemented, the distribution of population values (histogram a) is consonant with expectations prompted by Figure 3.13: the distribution is bimodal and very broad. On the other hand, once control is implemented, the population distribution contracts to a single, fairly narrow peak. Among other things, this illustration suggests that appropriately applied interventions can suppress phenomena like predator–prey oscillations and direct unruly populations to special fixed values.

In Figure 3.15, two control strategies are portrayed: histogram b results from intervention by harvesting only, while histogram c results from intervention by stocking only. Curiously, the stocking intervention leads to a sharper distribution than does the harvesting intervention. As will be argued below, the latter result is general: for controlling purposes, stocking always beats harvesting.

While hardly as dramatically clean as the control of a single, chaotic Ricker population (Figure 3.12), the control demonstrated in this example with stochastic inputs is still fairly impressive. It is reasonable to expect that a number of conservation and economic benefits would flow from such taming of wild population swings.

## 3.5 Some Lessons

### 3.5.1 Are Fixed Points Real?

The examples presented in this paper may be suggestive, but one wonders whether real population data are anything like the data examined in the illustrative examples. In particular, do real population data have fixed points with stable and unstable directions? The answer, frequently, is yes. Figure 3.16, for example, shows the same wolverine pelt data as in Figure 3.1. The highlighted points reveal a typical fixed-point signature: approach along a stable direction, escape along an unstable direction. Fixed-point

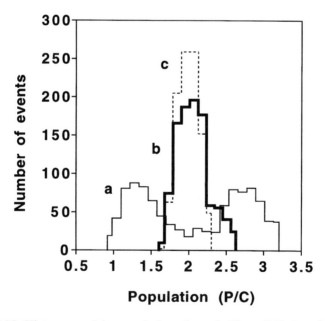

FIGURE 3.15. Histograms of the populations shown in Figure 3.13 show (a) a decidedly broad bimodal distribution in the uncontrolled state and (b) a narrow unimodal distribution after harvesting control is implemented. The histogram labeled c is associated with stocking control and is more sharply peaked than b.

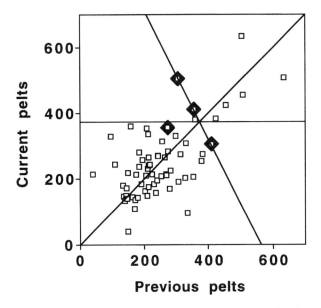

FIGURE 3.16. The bolded points in a first return plot for the wolverine pelt data of Figure 3.1 suggests the existence of an unstable fixed point.

analysis for other population data sets often produce similar positive results.

Does the existence of a fixed point with associated stable and unstable directions guarantee that the data are generated by deterministic dynamics? Not at all! (Witkowski et al. 1995). Figure 3.17 shows a first return map of Gaussian random noise containing an excellent example of a "fixed point." Of course, this pattern of data points is a chance occurrence. In a much larger data set produced by the same random process, numerous additional examples of "fixed points" would be observed, but, in all likelihood, they would all be quite different with quite different stable and unstable directions. Their existence would be a probabilistic fluke. In contrast, a genuine fixed point recurs again and again at the same value with the same stable and unstable structures. To convince oneself that a given fixed point is real, many repetitions of the same fixed-point-approach-escape pattern should be observed. Unfortunately, wildlife data sets are rarely sufficiently long to achieve this.

A practical demonstration of the reality of an observed fixed point is to implement control: take data in real time; when a fixed point pattern emerges, immediately try to institute control. If there is a positive effect, then the fixed point is probably genuine. If attempts at control fail, the fixed point may be an accident. For example, all attempts to control time series generated by the process that produced the Gaussian data set of Figure 3.17 fail. There is no determinism undergirding the data of Figure 3.17; random events are uncontrollable, at least via the strategy outlined in this paper.

One might wish that evidence for fixed points could be educed from just a few seasons of observation. Regrettably, such is not typically the case. If the associated dynamics has very high dimension, only rarely may fixed points be visited. On the other hand, for population dynamics that are fundamentally low-dimensional but driven with perturbative interactions and therefore appear to be high-dimensional, as in the example above, fixed-point information can recur with reasonable frequency. In any event, an interesting open research question is, how might the acquisition of fixed point information be accelerated?

## 3.5.2 You Cannot Maintain Control
## If You Do Not Keep at It

Attempts to control wildlife populations by perturbative intervention have a chance of working only if the state of the system remains sufficiently close to the target fixed point. If the state is allowed to wander too far away,

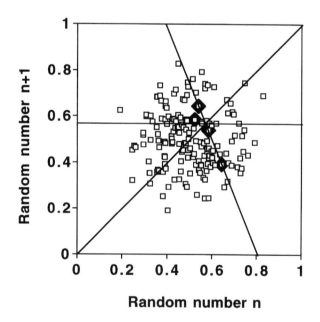

FIGURE 3.17. The bolded points in a first return plot of data drawn from a Gaussian random process also suggests the existence of an unstable fixed point. This fixed point is a statistical accident and will almost never recur with the same stable and unstable directions.

perturbative control will not work except by coincidence, and attempts to do so, in fact, may produce behavior that is even more undesirable than if the system were left alone. In the situation outlined in the previous section, each new sampling time incorporates some unpredictable element, so one cannot make even rough predictions beyond the next breeding cycle. Thus, insofar as the model of the previous section captures important aspects of reality, one can conclude from it that harvest and stock levels set as a result of forecasts that span several breeding seasons, as opposed to a single season, will have no effect in actually controlling wildlife population sizes.

### 3.5.3 You May Not Be Able to Control the Prey by Controlling the Predator

In many model predator–prey systems, the predator population in a given season is strongly influenced by the number of prey killed over the previous season. In such models, the predator is always one season out of phase with the environmental variations that affect the breeding conditions of the prey. While direct control of the prey population under these conditions, as in the last example above, is possible, such a time lag makes it impossible to indirectly control the prey population by trying to control the predator population. In fact, it is impossible to achieve any control of the predator at all under these conditions.

Incidentally, note that harvesting the predator can lead to disastrous results. Figure 3.18 shows results obtained from a simple predator–prey model where the prey birth rate fluctuates randomly and where the predator is controlled by harvesting. Because of bad luck, specifically, an unfortunate fluctuation in the prey's fecundity in the previous season, the harvested predator is driven into extinction.

### 3.5.4 Stocking Is Always More Stable Than Harvesting for Perturbative Control

Figure 3.19 illustrates an interesting aspect of perturbative control of an unstable fixed point. What is depicted is a blow-up of a portion of a first return map near a fixed point. The fixed point's stable and unstable directions are displayed, as is the line of identity (the 45° line). The population in generation $k$, at point a, is within the tolerance level of the fixed point, that is, it is within the dotted box that defines what points on the first return map are tolerably close to the fixed point. Using the unstable direction at generation $k$, the population in generation $k + 1$ is predicted to go to point c (via the graphical iteration a → b → c) that is greater than the fixed point by an amount that exceeds the tolerance level. Two interventions are possible: stock prior to the breeding season or harvest after it.

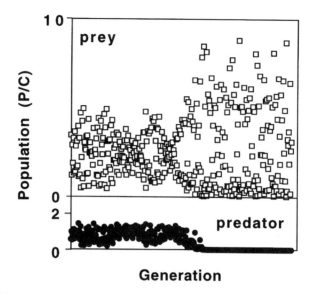

FIGURE 3.18. Attempts to control a predator–prey model system by harvesting the predator harbor two important lessons: the prey cannot be controlled by trying to control the predator, and a large predator harvest during a season in which the prey has an abnormally low birth rate can extinct the predator.

To implement the stocking intervention, the population is increased be-fore the breeding season by an amount that places the population on the stable direction of the fixed point vertically above it in the first return map. The open arrow from a to d shows that intervention. If the dynamics is unchanged from generation $k$ to generation $k + 1$, the population in genera-tion $k + 1$ will be the fixed point value (that is, d $\rightarrow$ e on the graph). But, suppose a fluctuation causes the birth rate to be higher than it was in generation $k$. Then the actual population in generation $k + 1$ is obtained graphically by the sequence d $\rightarrow$ e $\rightarrow$ f.

The harvesting intervention requires using the forecast to set a harvest level. This is the amount by which the population would have to be reduced to bring the population from the forecasted value (c) after the breeding season to a point on the stable direction vertically below it (g). In this strategy the population is first allowed to change according to the breeding rules in effect at time $k + 1$—to point h on the first return map (via a $\rightarrow$ b $\rightarrow$ h). If the birth rate has fluctuated between time step k and time step $k + 1$, h will not be the predicted value c. After the breeding season the harvest is applied. As a result, the population is reduced from point h to point i, which, again, because of the fluctuation, will not be on the stable direction at the desired point g.

The figure demonstrates that for this scenario, the stocking end point f is closer to the desired fixed point than is the harvesting end point i. For this

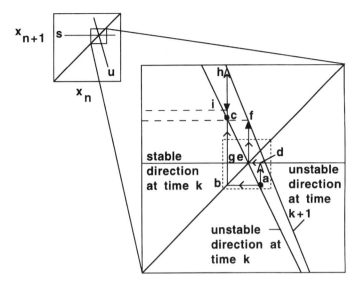

FIGURE 3.19. Stocking and harvesting produce different results when the population dynamics includes fluctuations. Here, the population at time k is predicted to be too large in time $k + 1$ ($a \to b \to c$). In this example, an unexpectedly high birth rate would actually take the population to h in the absence of an intervention. Stocking before the breeding season takes the population to point f (via $a \to d \to e \to f$). Harvesting after the breeding season, using the predicted necessary harvest, results in the population at i (via $a \to b \to h \to i$). The point f is closer to the desired fixed point than is i, so, in this case, harvesting is worse than stocking. Similar arguments lead to the conclusion that stocking is always a more stable controlling strategy than is harvesting.

case, at least, stocking is a better strategy than harvesting. In fact, it is a straightforward exercise to show that the same outcome prevails for all possible combination of circumstances—the predicted value is either too high or too low, the unstable direction fluctuation is either higher than or lower than expected, and the stable direction has either positive or negative slope. Remarkably, control via perturbative stocking is always more effective than control by perturbative harvesting.

## 3.6 Summary

Deterministic chaos can arise in nonlinear processes, of which population dynamics is one example. To the uninitiated observer chaos appears to be random, but it has nothing to do with chance: it admits at least short range forecasts and control. Perturbative control of low-dimensional chaos is possible by extracting characteristic signatures from the erratically varying, chaotic data without reference to an underlying model of the dynamics. Though low-dimensional chaos is not observed in real population data, high-dimensional chaos as well as some noise induced fluctuations are episodically low-dimensional. These high-dimensional situations may be controlled during low-dimensional episodes by the same methods that allow control of low-dimensional chaos. As a consequence, control of wildlife populations by continued vigilance to population time series and by the utilization of small harvesting or stocking interventions applied at the right moments may be possible.

## 3.7 References

Ditto, W.L., S.N. Rauseo, and M.L. Spano. 1990. Experimental control of chaos. Physical Review Letters 65:3211–3214.

Ditto, W.L., M.L. Spano, and J.F. Lindner. 1995. Techniques for the control of chaos. Physica D 86:198–211.

Garfinkel, A., M.L. Spano, W.L. Ditto, and J.N. Weiss. 1992. Controlling cardiac chaos. Science 257:1230–1235.

Gills, Z., C. Iwata, R. Roy, I.B. Schwartz, and I. Triandaf. 1992. Tracking unstable steady states: extending the stability regime of a multimode laser system. Physical Review Letters 69:3169–3172.

Hunt, E.R. 1991. Stabilizing high-period orbits in a chaotic system: the diode resonator. Physical Review Letters 67:1953–1955.

Li, T., and J.A. Yorke. 1975. Period three implies chaos. American Mathematics Monthly 82:985–992.

Logan, J.A., and J.C. Allen. 1992. Nonlinear dynamics and chaos in insect populations. Annual Review of Entomology 37:455–477.

May, R.M., and G.F. Oster. 1976. Bifurcations and dynamic complexity in simple ecological models. American Naturalist 110:573–599.

McCullough, D.R. 1979. The George Reserve deer herd. University of Michigan Press, Ann Arbor, Michigan, USA.

Moran, P.A.P. 1950. Some remarks on animal population dynamics. Biometrics 6:250–258.

Novak, M., M. Obbard, J.G. Jones, R. Newman, A. Booth, A.J. Satterthwaite, and G. Linscombe. 1987. Furbearer harvests in North America, 1600–1984. Ontario Trappers Association, Toronto, Ontario, Canada.

Ott, E., C. Grebogi, and J.A. Yorke. 1990. Controlling chaos. Physical Review Letters 64:1196–1199.

Ott, E., T. Sauer, and J.A. Yorke, editors. 1994. Coping with chaos. John Wiley and Sons, New York, New York, USA.

Petrov, V., V. Gaspar, J. Massere, and K. Showalter. 1993. Controlling chaos in the Belousov-Zhabotinsky reaction. Nature 361:240–243.

Peak, D., and M. Frame. 1994. Chaos under control: the art and science of complexity. W.H. Freeman, New York, New York, USA.

Power, M., and A.G. Power. 1995. A modeling framework for analyzing anthropogenic stresses on brook trout (*Salvenius fontinalis*) populations. Ecological Modeling 80:171–185.

Ricker, W.E. 1954. Stock and recruitment. Journal of the Fisheries Research Board of Canada 11:559–623.

Schiff, S.J., K. Jerger, D.H. Duong, T. Chang, M.L. Spano, and W.L. Ditto. 1994. Controlling chaos in the brain. Nature 370:615–620.

Singer, J., Y.-Z. Wang, and H.H. Bau. 1991. Controlling a chaotic system. Physical Review Letters 66:1123–1125.

Starrett, J., and R. Tagg. 1995. Control of a chaotic parametrically driven pendulum. Physical Review Letters 74:1974–1977.

Takens, F. 1981. Detecting strange attractors in turbulence. Lecture Notes in Mathematics 898:366–381.

Witkowski, F.X., K.M. Kavanaugh, P.A. Penloske, R. Plonsey, M.L. Spano, W.L. Ditto, and D.T. Kaplan. 1995. Evidence for determinism in ventricular fibrillation. Physical Review Letters 75:1230–1233.

## 3.8 Appendix
## Algorithm for Finding Flip Saddle Fixed Points and Stable and Unstable Directions

The most important property of a time series that is approaching an unstable fixed point is that successive values will be nearly the same for a short interval. So, the first clue to look for in the analysis of any time series is a sequence of values that are reasonably close. But, these close values also have to display the characteristic hip-hop escape depicted in Figure 3.10.

The formal test can be expressed like this. Take a sequence of 5 values $x_1$, $x_2$, $x_3$, $x_4$, and $x_5$ in which $x_1$ and $x_2$ are not close (according to some preselected criterion), but $x_3$ is close to $x_2$, $x_4$ is close to $x_2$, and $x_5$ is close to $x_3$. Then make the following tests: if $x_2$ is less (greater) than $x_3$ then $x_3$ must be greater (less) than $x_4$, and $x_4$ must be less (greater) than $x_5$; moreover, $|x_5 - x_3| > |x_4 - x_2| > |x_3 - x_2|$.

If these conditions are all met, then calculate:

$$x_{21} = x_2 - x_1$$
$$x_{32} = x_3 - x_2$$
$$x_{43} = x_4 - x_3$$
$$x_{54} = x_5 - x_4$$
$$D = x_{32}^2 - x_{43}x_{21}$$

$$c = \frac{x_{54}x_{32} - x_{43}^2}{D}$$

$$d = \frac{x_{43}x_{32} - x_{54}x_{21}}{D}$$

$$\text{root1} = \frac{d + \left(d^2 + 4c\right)^{1/2}}{2}$$

$$\text{root2} = \frac{d - \left(d^2 + 4c\right)^{1/2}}{2}$$

$$x_{\text{fixed}} = \frac{cx_1 + dx_2 - x_3}{c + d - 1}$$

$$\lambda_u = \min\left(\text{root1}, \text{root2}\right)$$

$$\lambda_s = \max\left(\text{root1}, \text{root2}\right).$$

The quantities $x_{\text{fixed}}$, $\lambda_u$, and $\lambda_s$ are the values of the fixed point and slopes of the unstable and stable directions respectively.

# 4
# Patch Dynamics: The Transformation of Landscape Structure and Function

S.T.A. PICKETT and KEVIN H. ROGERS

## 4.1 What Is Patch Dynamics?

One of the most important developments in modern ecology is the recognition that heterogeneity, or spatial pattern, is a key part of the structure and functioning of nature. No person working to understand the natural world or to manage natural resources can afford to neglect the fact that habitats, physical environments, resources, organisms, and other ecological objects have a complex spatial diversity (Bell et al. 1991, Caldwell and Pearcy 1994). However, ecologists and wildlife biologists have recognized spatial heterogeneity primarily at coarser scales. Coarse scale heterogeneity appears in such patterns as elevational gradients of plant and animal distribution, the contrast between different climatic zones with latitude, and contrasts between such community types as bogs and uplands, grassland, and forest (McIntosh 1991).

---

S.T.A. Pickett is a Scientist at the Institute of Ecosystem Studies (IES). His research interests have spanned a wide spectrum of interests from focus on plant population ecology to landscape disturbance dynamics, and from old growth forest to urban effects on ecosystems. In addition to technical papers he has written on the methodology of ecology and the application of contemporary ecology to conservation and management. He has contributed to international courses on landscape ecology at the Ben Gurion University in Israel and the University of Chile, as well as graduate courses at IES combining basic ecological information with application.

Kevin H. Rogers is Director of the Centre for Water in the Environment and a professor in the Department of Botany at the University of the Witwatersrand, Johannesburg. He is also leader of Research for the Kruger National Park Rivers Research Programme. He teaches a range of general and freshwater ecology courses at undergraduate and graduate levels. His main research interests are in integrating fluvial geomorphology, hydrology/hydraulics, and vegetation processes in rivers and wetlands to develop spatially explicit, predictive models to support decision making in conservation. He finds that landscape ecology provides an excellent framework for linking studies across disciplines that traditionally have operated at very different scales.

In spite of an appreciation for heterogeneity on some scales and for some ecological phenomena, most ecological theories assumed that the systems they dealt with had a uniform structure and that the mechanisms that structured them were also uniform in space (Kingsland 1985). Therefore, most scientists, planners, and managers sought out or tried to maintain homogeneity and ignored the fact that system characteristics change naturally over time. An advance in thinking was to recognize that even when systems could be considered to be uniform, they existed within a matrix of other systems that differed in structure and function (Wiens 1995). The realization that the edge between contrasting systems was a habitat of importance to certain wildlife species was a further important step in accepting the diversity of the physical environment as important to natural populations (Harris 1988). Spatial context therefore enhances our ability to understand and work with patchy landscapes and their elements. This paper will explain the nature and dynamics of patchy landscapes on various scales and present the relevance of heterogeneity and spatial context to management.

In order to grasp the nature of patchiness as a fundamental tool for understanding and managing nature, we must first define and amplify the concept. We will then summarize the current state of knowledge concerning patchiness and its changes, and follow with an analysis of its utility to managers. Finally, we will assess the frontiers of patch dynamics and opportunities for improving the basic knowledge and the application in natural resources management.

## 4.1.1  Patches and Patch Mosaics

A *patch* is simply a recognizable area on the surface of the Earth that contrasts with adjacent areas and has definable boundaries (Kotliar and Wiens 1990). The list of natural and human-made patches that are important to ecology and wildlife management is almost endless. The cluster of shrubs in a desert grassland, the forest woodlot surrounded by farm fields and pastures, the alpine lake bordered by areas of conifers and bare rock, or the cluster of tasty seedlings emerging from a fertile spot in a savanna are all examples of patches. Some of these patches look like islands in a nearly continuous sea of contrasting habitat, just like real islands in the ocean, while others are set in a complex array of other patches, some of which are quite similar in size and shape. This second kind of patchiness is more akin to a quilt made of patches of fabric (Box 4.1).

Patches can be formed by many kinds of agents, both human and natural (Pickett and White 1985). Natural mechanisms of patch formation include both abiotic and biotic agents. The presence, form, and metabolism of organisms, especially plants, often create patches. For instance, savanna trees generate patches that differ from the surrounding canopy of grasses in soil organic matter, nutrient content, water holding capacity, and other

---

**Box 4.1. Basic Terms and Concepts in Patch Dynamics**


Patch   A definable area on the Earth's surface whose structure or composition differs from adjacent areas. When the third dimension is considered, a patch body is recognized. Patches may appear in terrestrial and aquatic habitats.

Mosaic   An array of patches. Some or all may be permanent, although most real mosaics have at least some dynamic component, such that individual patches and consequently the composition and spatial arrangement of the entire mosaic change through time.

Heterogeneity   The general condition of spatial contrast and variability of which patchiness is a special case in which discrete boundaries can be perceived.

Landscape   An area in space that is heterogeneous. Particular kinds of landscapes can be defined on the basis of the kinds and configurations of patches present and having in common the agents and processes that create or maintain patches. Strictly, a landscape can be recognized at any spatial scale.

Element   A component of a landscape. A synonym for *patch* in landscape ecology.

Scale   A relationship between two measures. Scale can be characterized by the extent of the measure and the resolutions of the units of measure. A coarse-scale pattern or phenomenon is larger or less resolved than some reference measure, while a fine-scale pattern or phenomenon is smaller than a reference.

---

important features. Burrowing or wallowing animals may create patches in wetlands, grasslands, savannas, deserts, or forests. Beaver create wetland patches along a woodland stream and open areas of forest canopy in the adjacent upland forest. Fires, tornadoes, hurricanes, and thunderstorm downdrafts can all create patches that differ from adjacent unaffected areas. Floods rearrange the substratum in streams, creating new sand or gravel bars in particular locations while removing others. Some natural agents of patch formation are quite long lasting and are related to the underlying geological structure of the land. Acid or basic soil patches, springs and seeps, watering holes, cliffs, and the like are examples of features that may be essentially permanent features of the landscape on ecological time scales. Patches may arise anew within an area or may be left when an area is extensively changed. For instance, a forest may establish in a former agricultural field, or people may reduce a forest to an isolated patch by constructing housing. How patches are created is important because it determines the structure, resources, shape and size, and location of patches relative to other patches and ot the populations that are affected by them. Forman and Godron (1986) summarize patch types as disturbance patches, remnant patches, regenerated patches, environmental resource patches, and planted or constructed patches that are introduced into a landscape by people. The examples cited above show just how common patches are and suggests the wide variety of events and processes that cause patchiness at various spatial scales.

Some examples of how patch type is related to wildlife species can expand the basic concepts introduced so far. A remnant patch of old forest in a cultivated landscape may harbor the only hawks in a large part of an eastern U.S. county. An occasionally burned patch may permit the persistence of the Kirtland's warbler in sandy areas in Michigan. In the pinelands of New Jersey, some disturbance is needed to maintain the open habitats of certain wildflowers (Little 1979). Many bird species require large areas of intact old growth forest for their persistence, while populations of some browsing mammals are well known to be stimulated by forest openings or edges. Termites concentrate nutrients and water in their large mounds in many tropical and subtropical environments. In turn, such termite-generated patches serve as habitat for mammals such as warthogs or porcupines in southern African savannas. Prairie potholes, depressions in mesic or seasonally wet grasslands function as habitat for a sequence of waterfowl as the water level in the potholes draws down through the year. Sodic soil patches found at specific locations as a result of hydrological pathways in savanna, act as crucial salt licks for mammals and indeed for some invertebrates.

The examples should not mislead us to focus only on individual patches. In fact, it is often the mixture of patches that is really the total resource for wildlife and biodiversity (Gilbert 1980). Species often feed, breed, and shelter in different patches. Such a requirement for different patch types results from the fact that patches in which shelter is maximized due to closed structure are often those that do not have high productivity, or patches in which courting display is not obscured do not provide hiding places for broods. Therefore, organisms may often require a commodious mixture of patches or different patch types at different stages in their lives or seasons of the year, rather than a single, optimal patch type. This situation suggests that ecologists and managers should focus on the entire mosaic of patches in a landscape (Angelstam 1992).

The final feature of patch dynamics that makes it such an exciting development in ecology is that the concept can be applied at many spatial scales (Wiens 1989, Wu and Levin 1994). To use the entities often recognized by ecologists as "levels of organization" on different scales, patches can apply to individual organisms, populations, behavioral or feeding guilds, communities, ecosystems, or landscapes (Kolasa and Pickett 1991). In each of these ecological realms, spatial pattern can exist and can change through time. Patches on one scale can be composed of smaller scale patches that influence how the original patch functions. Alternatively, patches can be parts of larger scale assemblages of patches. Thus, patches can be parts of hierarchies. Both changes within individual units and changes in the whole assemblage of units can influence resources and ecological services and amenities.

So far, we have indicated that patches are an important feature of all sorts of landscapes at various scales, that such patchiness has important functions for wildlife species, that patches originate from a vast array of abiotic and biotic processes and conditions, and that the whole array of patches may be important to wildlife and, hence, to managers. This overview is only part of the story, because it has focused on patch creation and has tacitly taken patches to be static. In the next section, we add the important time dimension to the consideration of patches and patchiness.

## 4.1.2 Patch Dynamics

### 4.1.2.1 Changes Within Patches

One of the most important insights about the nature of patches is that they are changeable (Watt 1947, White 1979). Remnant patches may change through time as old individuals of a species die and are not replaced (Burgess and Sharpe 1981), as in the case of the old oaks in prairie-grove remnants in central Illinois. Patches in which the natural disturbance regime or the human land-use or management regime has drastically changed, may undergo community succession, as when grazing animals undergo drastic population declines or extirpations or when agricultural fields are abandoned and succession occurs. Succession, or ecosystem and community dynamics, may take place in almost any sort of patch, regardless of its origin, as organisms interact, influence one another and the physical environment, and new organisms invade.

Many features of patches change as the result of such dynamics (Niering 1987, Breen et al. 1988). For instance, during postagricultural succession, species composition, canopy layering, soil features, and resource availability all change through time. Such changes affect individual patches, and managers must be sensitive to the changes in the patches for which they are responsible, because the target of management may increase or decrease due to changes in the patch caused by succession or disturbance. But the situation is even more complex than attending to change in individual patches.

### 4.1.2.2 Change of a Mosaic

When considering a set of patches covering some area, a complex pattern likely appears. While some patches are old and relatively unchanging, others may be freshly disturbed by natural events or created anew by humans. Still others may be in different successional stages (Pickett 1976), or perhaps in differing stages of deterioration following some stress or isolation from critical supplies of resources or colonizing animals. Therefore, not only is it possible to view a landscape as a patchwork quilt, but, unlike the household artifacts that lend their name to the natural patchwork, the natural and human landscapes out of doors are a changing patchwork (Watt 1947, Bormann and Likens 1979).

This is the essence of patch dynamics. Patches exist in a landscape. Some of them are essentially permanent in ecological time—for example, those based on the underlying rock, slopes, and hydrology of an area—but other patch types are more mutable. Some of these changeable patch types are ephemeral, such as a seasonal spring pool in California grasslands or the small gap, created by the fall of a single tree, that is destined to fill in from the edges (Hibbs 1982). Other patches change slowly, like the highly eroded abandoned field in the coastal plain, or a large strip-mined area. Different attributes of patches can change at different rates. For example, natural succession may cover a disturbed area quickly, but return to a species composition and structure similar to that present before the disturbance only after several hundred years.

The lesson to draw from all this variety of patch types and modes of patch change is that a single landscape can contain examples of all these kinds of patches and very often will contain patches that are in different stages of development or deterioration at any one time. Patch dynamics is important not only because it alerts us to the changes in individual patches, but also because it indicates that the entire matrix of patches will be dynamic and shifting. It is these complex and inclusive arrays of patches that are the true focus of our planning, management, restoration and conservation efforts (Alverson et al. 1994).

## 4.2 What Is the Status of Knowledge of Patch Dynamics?

This section of the chapter seeks to extract generalizations from many of the examples and ideas used to introduce the study of patch dynamics, so that wildlife managers can apply that knowledge to their particular situations.

### 4.2.1 Mosaics and Their Effects

How do the composition and dynamics of patches in an area affect the biological diversity and ecosystem functioning of the area?

A mosaic of patches can occur across a forest floor measuring but a fraction of a hectare. Such a mosaic may be composed of pits and mounds of various ages, created by the uprooting of trees (Beatty 1984). Downed logs and branches in various stages of decay are patches in this mosaic, as are accumulations of litter in shallow depressions. Other patches may have been created by the burrowing of a fox or the bedding of deer. Clusters of clonal wildflowers or shrubs also form patches. There may be gaps in the canopy, formed by insect outbreak, or windthrow of trees. More permanent patches may be caused by seeps of groundwater at the break in a slope or outcrops of rock. Hence, the mosaic is a complex of patches of different sizes, shapes, volumes, origins, juxtapositions, and resource states, and the particular combination of structural characteristics and process interaction determines biological diversity and ecosystem functioning.

The forest mosaic described above is itself part of a larger forest pattern. At the coarser scale, the detailed forest floor mosaic appears as a part of a birch-sugar maple forest stand. At this larger scale, the forest matrix as a whole may also contain beech–hemlock patches, riparian corridors, and sparse pine stands on extensive rock outcrops. These examples have a general lesson that is applicable to any system: (1) Patchiness occurs on all scales, and fine-scale patchiness is nested within coarse-scale patchiness. The insights in the previous section suggest that (2) such patch mosaics are made up of elements that change at different rates. Examples include the slow filling of a treefall pit versus the fast filling of a small canopy gap, and the rapid canopy turnover due to fire or ice damage on the exposed rock outcrops versus the slow succession in the hemlock patches.

The effects of nested and dynamic patterns of heterogeneity are great. First, the patchiness existing in the physical environment, and the modification caused by organism growth and metabolism is a major source of diversity. Biological diversity responds to environmental heterogeneity (Huston 1994). Individual patches support different species, or individual species require multiple patch types. In either case, patchiness contributes to biodiversity (Pickett 1996).

The second major effect of patchiness is to generate and control flow of materials, energy, organisms, and information through the environment (Wiens et al. 1985, Breen et al. 1988, Wiens 1992). Nutrients cycle along hydrological pathways or are transported by wind or organisms. In some cases, transport is from areas of higher concentration to areas of lower concentration, as when mycorrhizae tap rich pockets of phosphorus in the soil and transport that nutrient to a plant, often a long distance away (Allen 1991). In other cases, organisms can concentrate nutrients from a wide area, as in the deposition of dung in a nesting or denning area within a large foraging range. The patchy distribution of microorganisms is one of the most significant generators of patchiness in ecosystem functioning (Waring and Schlesinger 1985). An example is the mediation of soil nitrogen flow to stream waters by riparian vegetation (Groffman et al. 1993). Information, in the form of environmental regulators or genetic material, may flow through mosaics as well (Breen et al. 1988). The penetration of wind into a patch of trees controls the physiology and behavior of some organisms in the patch. The movement of pollen or dispersing animals into a patch may change the genetic structure of the populations in the patch and affect their evolution.

## 4.2.2 Scale

Patches and their dynamics can apply to many scales, and specific patterns and dynamics can characterize particular scales. However, there are important details of how scaling of patch dynamics occurs and what it means for ecological systems.

In order to scale the concept of *patch* appropriately, it is important to realize that patches have three-dimensional structure (Breen et al. 1988, Pickett and Cadenasso 1995). Although patches are represented most often as two-dimensional areas on maps, in the ecological world such patches represent highly textured features of ecological systems. For example, forests are notoriously layered, and the typical temperate deciduous forest can have five idealized layers: canopy and subcanopy trees, understory trees, shrubs, herbs, and mosses. These layers are not always present or all continuous, and it may be possible and useful to divide some layers more finely, such as dividing the herb layer into seedling-sized trees, herbs, mosses, and lichens. Even lower statured ecosystems not thought to have much texture—perhaps because people are too large to walk within them—do in fact have different layers. For example, oldfields dominated by broad-leaved herbs can have a dense, continuous overstory as well as a layer of intermediate-sized flowering plants overtopping a layer of rosettes and stoloniferous forbs, and a ground layer of mosses, lichens, and litter. Small mammals, such as voles and mice, respond differently to such oldfield patches (see Milne, Chapter 2). Deserts have shrub patches that comprise a woody overstory and a layer of herbaceous annuals. The intervening open spaces exhibit some mixture of layers of bulb-bearing perennials, rainy-season annuals, and a soil crust formed of algae, cyanobacteria, mosses, lichens, and their chemical exudates. Whether a particular area in a desert serves as a source or sink of runoff water, or of nutrients, or of seeds, depends on its three-dimensional structure and which layers are present and dominant (Shachak and Brand 1991). Wetlands, too, are three dimensional (Breen et al. 1988) in terms of water depth, substratum topography, chemical gradients in water and substratum, and architectural complexity of the vegetation, among others. As an example of aquatic systems, river communities also have three-dimensional structure formed by the substratum, channel form, debris, organisms, and current velocity and turbulence. Attached algal communities can exhibit the same increasing vertical complexity expected in many forest successions, albeit at a much finer scale.

The three dimensional structure of patches can be generalized by using the term *patch bodies* (R. Naiman, personal communication, 1996). There are three basic components to the three dimensional structure of any patch, be it aquatic or terrestrial: below substratum, below canopy, and above canopy. Each one of these three general zones can be further subdivided depending on the particulars of the system at hand. Such a classification emphasizes the role of boundaries and fluxes among and within patch bodies (Breen et al. 1988). It is also important to recognize the role of organisms as engineers of three dimensional patch structure (Jones et al. 1994). Not only do their roots and shoots form the physical superstructure of patches, but their debris, chemical gradients, and behaviours also contribute to patch architecture.

One situation that makes clear the significance of three-dimensional structure of patches is the role of forest edges. Although there is little empirical information on functional responses to forest edges, theoretical predictions exploit the changing structure of edges from the time a new edge is created by carving an isolated forest from a larger block of forest (Angelstam 1992, Forman and Moore 1992, Pickett and Cadenasso 1995). The structure and depth of a forest edge can affect the suitability of the forest interior to habitat specialists, the predation risk of some bird species, the predator pressure on seeds and seedlings, and the architecture of woody seedlings and shrubs.

How organisms use patches can be explored by examining the cost–benefit ratio for staying versus leaving a patch. For organisms, costs can be cast in terms of risk of mortality, energy for maintenance, and time for movement. Benefits can be cast in terms of survival, growth, and reproduction (Wiens 1995). Other ecological phenomena can be analyzed in terms of costs and benefits as well. Costs and benefits of nonorganismal phenomena may be in terms of water or nutrient yield or retention, sequestration or mitigation of contaminants, and susceptibility to disturbance. Such relationships are highly scale dependent (Turner et al. 1989). The array of patch types (the extent) available for an organism or ecological process to spread over and the shapes, sizes, and arrangements of the patches can all potentially affect the costs and benefits. This sort of analysis gives managers a way to assess a planned or ongoing management effort. This approach might be more expensive or time consuming to implement, but it is a promising area for research aimed at making it practical.

Because patchiness can be addressed on various scales, there is no one scale alone that is appropriate for assessment and use of patchiness. Indeed, patchiness at one scale may be related to patchiness on coarser or finer scales. Patchiness on one scale of resolution can be aggregated into coarser scale patches, or divided into finer scale patches, depending on the needs of the scientists or managers examining the system.

## 4.2.3 Approaches to the Study of Patchiness

Because patchiness is so ubiquitous and important in nature, there are a number of approaches to the phenomenon. Each illustrates important aspects of patchiness, and the series indicates how the understanding of patchiness has changed through time. We must note that throughout most of its history, ecology has sought out uniform systems or assumed uniformity in its theories and models. Only recently has ecology begun to take heterogeneity and patchiness seriously, and to see its wide-ranging implications and importance (Kolasa and Pickett 1991). We will divide the approaches to patchiness into three types: (1) the island biogeographic or fixed template approach, (2) the shifting mosaic approach, and (3) the landscape or human land use approach (Figure 4.1).

### 4.2.3.1 Island Approach

One of the first ways that patchiness was introduced to the majority of ecologists was in the theory of island biogeography. The most notable introduction of this approach was in the book by MacArthur and Wilson (1967). Ecologists apparently were unprepared to accept an earlier precursor, according to Wu and Vankat (1995). But once available in a clear, generalizable, and theoretically compelling form, islands became a key tool

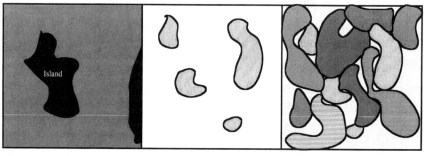

A ISLAND                    B PATCH DYNAMIC                    C LANDSCAPE

FIGURE 4.1. Approaches to patchiness. (A) The island biogeographic approach, indicating an oceanic island or other insular habitat surrounded by a hostile matrix. A mainland is indicated at the right hand edge of the map. Species are supplied to the island by the mainland by long-distance dispersal. (B) The patch dynamic approach, indicating patches of a particular type, e.g., treefall gaps in a forest canopy, that is considered homogeneous for purposes of studying the ecology of the gaps. Such an approach might be appropriate for species that disperse among gaps carried by specialist avian dispersers. (C) The landscape approach, in which the uniform matrix in a patch dynamic approach is resolved into a variety of contrasting patches that have different effects on the process of interest. Such an approach might be appropriate for studying overland dispersal in which the dispersal agents respond differentially to different patch types.

for understanding not only the oceanic islands for which the theory was originally developed, but even terrestrial and aquatic systems.

The advantages of the island approach are that it invites ecologists to see their systems as open to the flux of organisms (and by extension, other materials and influences) from outside, and it emphasizes that periodic local extinction can be an important component of ecological systems[1]. Disadvantages of the strict island approach are that it assumes a fixed template of patches contrasting strongly with an essentially uniform matrix. In addition, the strict island approach focuses on populations of species, and certain key interactions that may occur during the colonization and extinction processes are hidden from direct view. Competition, establishment dynamics, predation, and extinction dynamics were not generally exposed in the theory, which focused instead on the simple turnover of species through time (Abbott 1980). Attention was on the net invasion and net loss of species. Empirical studies led ecologists to realize that ecological history and heterogeneity were also given short shrift in the pure form of the theory. For instance, former connections to larger land masses are a historical effect on land bridge islands, and elevation and altitudinal complexity affect species richness on other islands (Brown 1986, Shafer 1990).

### 4.2.3.2 Shifting Mosaic Approach

A more refined approach to patch dynamics, compared to the coarser approach of classic island biogeography, is that of the shifting mosaic (Watt 1947). The phrase is attributable to Bormann and Likens (1979) and emphasizes that patches are bounded by other patches that may be more or less similar (as opposed to highly contrasting and often hostile habitats as in the case of oceanic islands) and that the mosaic of patches itself may change through time. An example of a shifting mosaic is the changing pattern of burned forest, forest of various successional ages after fire, and the shapes and sizes of various patch types and ages through time in boreal forest of the Boundary Waters Canoe Area (Heinselman 1973).

Such a mosaic is obviously important to wildlife management, because the availability of browse and herbage for animals, snags and damaged trees for cavity nesters, and more subtle habitat requirements all change across the spatial mosaic, and through time, as the mosaic changes (Foster 1980, Franklin 1993, Grumbine 1994). The patches at the landscape scale seen from a low-flying aircraft are different vegetation communities and are home or larder to different animal populations and assemblages. The shifting mosaic approach thus unifies population ecology (how plants and ani-

---

[1] Note that the term *ecological system* is used for ecological aggregations ranging from populations, through communities and ecosystems, to landscapes. It may refer to either a concrete instance of one of these entities or the conceptual model that represents it.

mals coexist locally) and community ecology (how those assemblages change through time), with the landscape ecology (how communities interact in space). Explanations of community structure and coexistence are therefore much improved by taking into account the landscape perspective. For example, many models and concepts of community organization based strictly on local resource partitioning and niche division are inadequate and could be improved by recognizing that organisms track resources that flow through dynamic mosaics, thus dispersing to less competitive or otherwise more suitable patches in a landscape.

Similarly, populations often are seen to be divided among similar but separated patches, with the subpopulations in some patches going extinct, while other vacant patches are being colonized. If these compensatory processes are in balance, the subdivided population is said to exhibit true "metapopulation" dynamics (Levins 1970, Harrison 1994). But even if the two processes are not in strict balance, a system of populations of a species can respond as a shifting mosaic. In either case, it is appropriate and valuable to determine whether an individual population is behaving as one population or an interacting group of populations of the same species. For instance, a population in a specific patch may in fact be kept viable by immigration from distant patches in which the species reproduces better. Alternatively, a low-quality patch may be a drain on the overall success of a species in a landscape (Pulliam 1988; Box 4.2). Again, the shifting mosaic approach is an improvement over the island model for metapopulations

---

## Box 4.2. Concepts about Populations in Mosaic Systems

Metapopluation    Strictly, a system of populations of a given species in a landscape linked by balanced rates of extinction and colonization. More loosely, the term is used for groups of populations of a species some of which go extinct while others are established, but the entire system may not be in equilibrium.

Source    In a landscape, a population or an area that supplies colonists to other patches.

Sink    In a landscape, a population or site that attracts colonists, while not supplying migrants to other sites or populations.

Rescue effect    The maintenance of a population or patch by periodic supply of colonists from outside. The target population is not sustainable alone.

because the subtleties of contrast in species response to the intervening matrix can be taken into account, rather than dividing the environment into fatal (matrix) versus welcoming (island) states.

### 4.2.3.3 Landscape Approach

There is one final approach to patchiness that makes the subject of greater utility to managers—the inclusion of human-occupied or human-created patches (Turner et al. 1990, McDonnell and Pickett 1993). Ecologists and some wildlife managers have focused on "wild" nature, eschewing areas where humans have much obvious presence. Yet the reality of the industrialized and petroleum-subsidized world we inhabit is that anthropogenic land-cover types, such as urbanized, settled, agricultural, and transportation corridors, and the land uses associated with them, are widespread. Similarly, many ecological studies have ignored the historic or prehistoric influence of humans on the structure of ecological systems. For instance, anomalous areas of vegetation in savanna mark the locations of prehistoric settlements. Human-generated and maintained patches are important for contemporary landscape ecology (Forman and Godron 1986), and are increasingly incorporated into the rest of ecology. Thus, ecologists must no longer exclude human-caused patches from the mosaics they study. Nor can the human-generated or human-controlled fluxes from outside the study area be ignored.

## 4.2.4 Methods

Currently, the methods for examining patch dynamics have two main components. First is detection of patches and second is determination of change in patches (Dunn et al. 1991, O'Neill et al. 1991, Turner et al. 1991). Although different specific methods are often linked to a particular scale, the principles are scale independent. Patch detection relies either on contrasting the "texture" of one area with another or on detecting a boundary on a linear transect. Boundary detection is perhaps more straightforward and more easily accomplished. It requires a small sampling transect relative to the size of the neighboring patches (Figure 4.2). Discrete segments of the transect are sampled in sequence, and a rapid change in the variable(s) measured is used to document a boundary. Mean values, statistical variance around the mean, or similarity of composition of some collection of species or other attributes of the environment can be assessed. One popular method for detecting similarity of features along a transect is the moving-window technique, in which a set distance is selected to sample along the transect (see Milne, this volume). That distance is the sampling "window," which is moved along the transect. Boundaries are indicated by a high degree of contrast between adjacent "openings" of the window.

| Patch Hierarchy | Geomorphic classification | Management scales | Prediction scales | Patch and Mosaic Types | Biotic Response |
|---|---|---|---|---|---|
| Scale 1<br><br>km² × 1000 | **Mpumalanga Watershed**<br>5 perennial rivers. | Water resources planning scale. | Water resources modelling. | River specific fauna, flora, landscape, tourism potential, water quality and quantity issues. | Catchment vegetation and landuse change due to intersecting rainfall, geology and evaporation gradients. |
| Scale 2<br><br>km × 100+ | **Sabie River and Catchment** | Water resources planning and management scale. | | Least developed river (upstream development imminent to cater for post-apartheid community upliftment). | Downstream fish response to increasing temperature. Vegetation response to decreasing rainfall. |
| Scale 3<br><br>km × 10-40 | **Kruger National Park Zone**<br>4 river zones each defined by geology and regional slope differences. | Conservation policy set and success judged. Interaction between conservation and water resources. | | Broad characteristics repeatable between rivers. | Large scale pollution problems. |
| Scale 4<br><br>km × 5-20 | **Macro - Reach**<br>9 macro reaches defined by discharge, sediment input and slope. | Conservation manager's first subjective scale for decision making. | Hydrological and sediment yield modelling. | Non-repeating within-river mosaics. That is, each macro-reach is different. | |
| Scale 5<br><br>km × 1-10 | **Reach**<br>Contains any or all five channel types. | River conservation goals set and representative mosaics monitored. | Long term vegetation response to geomorphic change. | Secondary within-river repeating mosaic. | Differential plant "community" distribution between patch bodies. |
| Scale 6<br><br>m × 100-km × 5 | **Channel Type**<br>Particular combinations of morphological units. | | **Primary scale for predicting geomorphic change in response to upstream development.** | Primary within-river repeating mosaic. | Differential plant species distribution between patch bodies. |
| Scale 7<br><br>m × 10-100 | **Morphological Unit**<br>Position on cross section provides further classification as active, seasonal or ephemeral. | Species management scale e.g. Hippopotamus or invasive plant species. | Cross-section hydrodynamic modelling. Fish community response to hydraulic/hydrological change. | **Patch body**<br>3D geomorphic unit with above (e.g. forest canopy) and below ground (e.g. hyporheic) habitat. | Differential plant species distribution within patch body. Grain for established plant response. |
| Scale 8<br><br>m × 0.1-10 | **Micro - site**<br>Defined by micro-topography and elevation on the unit. | | Prediction of plant regeneration response. | **Micro-patch body**<br>With above and below ground micro-habitat. | Grain for plant establishment response. Invertebrate habitat, anaerobic leaf pack, bird nest site, cover for fish fry, etc. |

Determination of entire patches and mosaics can be accomplished best through remote sensing. Media range from aerial photography to digital thematic mapper (TM) and SPOT images from satellite. Limitations of aerial photography include the need to automate the images and to have an expert delimit the patches, which can introduce a subjective element into the delimitation of patches. In addition, photographs suffer from parallax error in that photographs are not to scale from their center to their edges, and there is no coordinate system inherent in photographs. Satellite images have the advantage of availability in automated form and an included coordinate system, but suffer from a coarser resolution than aerial photography. The six TM bands usually used from Landsat images limit the degree of subdivision unless it is possible to combine the reflectance with other kinds of information. For instance, in many studies forest types usually are divisible only into evergreen versus deciduous versus mixed; however, combining spectral data with other layers of information allows refined delimitation of ecosystem types.

Descriptions of patch and mosaic structure based on remote sensing can both have a large subjective component based on the need of the investigator to delimit patches in many cases, and those mosaics can differ markedly from the patchiness to which organisms respond. But the range of organismal responses to mosaics is so great that it would be unproductive to describe a patch hierarchy objectively from a quantification of the spatial responses of each of the constituent species. The more pragmatic option is to construct subjectively the mosaic from multiple criteria of structure so that different aggregations can be tested against the distribution, abundance, and flux or organisms, materials, and information. For example, determining the relative mix of hemlocks and hardwoods in different stands can be used to describe patches that then can be used to predict or explain spread or intensity of infestation by the hemlock wooly adelgid, whereas the susceptibility of trees to other sorts of disturbance-caused mortality such as wind, or more rarely fire, might require a different kind patch descriptor.

Standard landscape methods are introduced by Holland et al. (1991) and Turner and Gardner (1991). Once the structure of patchiness is quantified and mapped, the detailed ecological characteristics of the various patches in a mosaic must be determined. At relatively fine scales, appropriate to quantifying within-patch structure, the standard methods of population, community, ecosystem, and evolutionary ecology can be used to character-

FIGURE 4.2. The relationship of the patch hierarchy in the Sabie River, South Africa, to geomorphology, management, prediction, mosaic type, and biotic responses. The scales are given in orders of magnitudes of kilometers or meters. The Sabie River arises on the highveld and flows through the Kruger National Park (KNP) in the lowveld. It enters Mozambique at the eastern boundary of the park, and empties into the Indian Ocean.

ize key features of landscape mosaics and their components. In order to look at patch characteristics at higher hierarchical levels or coarser scales, some method of aggregation is required. Because the ecological characteristics of all patches or all areas in the mosaic are rarely available, models that make assumptions about the scaling of the processes or phenomena are used to aggregate the measurements made on finer scales to coarser scales (Pielke and Avissar 1990, Dale et al. 1993). The issue of scaling is one aspect of the problem of emergent versus collective properties (see Bissonette, this volume).

A critical insight of landscape ecology is that not only is spatial heterogeneity, per se, significant to the structure and function of the natural world, but the exact way that the components of natural systems are arranged is important. For example, an old-growth broadleaved forest adjacent to a mown meadow may experience different seed inputs from those embedded in a younger broadleaved forest or pine plantation. At a finer scale, the susceptibility of a population to a predator or herbivore may differ depending on the identity of its neighboring patches. Therefore, spatially explicit models and assessments of natural and managed ecosystems are a real advance in patch dynamics. Many patch dynamics approaches and models in the past have merely used or modeled the average patch creation, change, and extinction in the aggregate. In other words, the average values of the parameters of populations of patches were measured. However, because the actual arrangement of patches—which patches abut others, especially when strongly directional flows are of concern—helps to control the impact of the flows, averaging may not explain the function of land mosaics. For example, suitability of certain temperate forest remnants in an agricultural matrix depends on their orientation relative to the general direction of bird migration (Pearson 1993).

## 4.3  What Use Is Patch Dynamics to Managers?

Patch dynamics is an exciting development for natural resource management because it has important practical advantages. First, it is realistic. Patches can represent actual land-cover types, land uses, and parcels of land that managers have to deal with. Second, patchiness is pervasive. We live and work in a fragmented world in which patches are an obvious component at the scale of everyday landscapes (Harris 1984). People and their institutions have broken continuous natural landscapes into fragments or human-generated patches. Third, patchiness is increasing. In many parts of the world as agriculture, suburbia, exurban development, and industry spread, anthropogenic patchiness is increasing. Determining how these new levels, extents, and patterns of human-generated fragmentation affect the quality and sustainability of natural resources, ecosystem services, and wildlife is a crucial research need. Natural patch dynamics and anthropogenic patch dynamics may have very different effects on ecological systems.

Patch dynamics has immense utility for management of natural resources, including sustainable resource use, conservation, and restoration (National Research Council 1986, White 1994). Such utility emerges from the number of concrete handles that patch dynamics provides managers and policy makers. The key is that patchiness is spatially explicit, and because mosaics and patches can be decomposed, patchiness is highly relational. That is, elements of diversity, resources, and functions that managers are interested in can be linked to appropriate patch types or patch states across scales. There seem to be few, if any, other approaches to the structure and function of the natural world that are as flexible but clearly organized as patch dynamics. The handles that patch dynamics provides are of general application. As we noted in the introduction, all environments, especially on the scales that humans most often practice planning and management, are patchy. Therefore, patchiness is convenient and appropriate in conducting management.

It is easy to measure the features of a patch—the size and shape of the patch, the position and relationship of the patch relative to others of similar and different types, its species composition, the numbers of organisms of a species of interest that use or inhabit the patch (Turner and Gardner 1991) are easily documented. These are standard kinds of things that ecologists measure about environments, and the only difficulty is amassing enough expertise and hands to do the work. The spatial assessments are somewhat more difficult than assessment of individual patches since they require remote sensing or extensive surveys and relating the survey data to specific spots on the ground. But even this is becoming easy with portable global positioning systems and geographically registered electronic data acquisition and management (e.g., Geographic Information Systems). The principal advantage of such analyses is that the spatial context in which patches must be managed is made explicit. Even in cases where data are not immediately available on the spatial mosaic, the awareness of a framework for assessing patches and their changes within the context of the mosaic in which they appear, helps prevent scientists, managers, and policy makers from neglecting spatial context as an important aspect of the environment.

Taking a patch perspective in management helps make explicit the goals for management. In many cases measuring and assessing the patch mosaic will be the means to important management ends (Christensen 1997). For example, ecologists recognize that mosaics are the basis of much of the biodiversity in a region. The diversity of a region is not simply the result of adding together all the individual point diversities. Rather, the spatial structure of the landscape, including the variety of successional stages it contains, the number, type and position of disturbed patches, the permanent and mobile hot spots for resources, the sources and sinks for animals, and the divided structure of populations are all critical conditions that determine the diversity of the landscape and its ability to support certain species and to supply resources.

The way a patch mosaic operates to provide all these services is critical knowledge that managers must have, and which will improve their ability to provide, maintain, or restore those values in a landscape. In addition, patchiness provides a tool to optimize the values provided by a landscape. Because various species and various functions are concentrated in only one or a few of the patch types that exist in an area, maintaining all the values in a landscape requires maintaining all the appropriate patch types as well. Understanding the mosaic provides the means by which different values can be maintained in a landscape. In cases where there are conflicting values, a sufficient total area is required to permit the sustainable persistence and availability of all the patch types needed to support the whole spectrum of species and desired functions (National Research Council 1995). The principal lesson is that mosaics are the origin and maintainers of natural diversity and ecosystem services. Therefore, using mosaics and their dynamics is a smart tool for management.

Patch dynamic mosaics may be a management goal in themselves. The structure of the natural world is predominantly patchy. Even the oceans and large lakes, which look superficially uniform, are in fact patchy. Great currents, eddies, and layers exist in water bodies, just as mountain slopes, moisture gradients, disturbance patches, and resource patches determine the structure of the terrestrial world. Therefore, one major way to ensure that the natural component of populated and human-controlled landscapes persists, is to manage its heterogeneity. Without knowing the specific ecological and evolutionary functions or societal benefits of a particular environmental mosaic, it is certain that some benefits exist. Hence, in the absence of complete knowledge, management of mosaics is an environmentally wise strategy. Mosaic heterogeneity, and patch dynamics is a large component of how the natural world works.

In some cases, a landscape mosaic will have been structurally and functionally modified by human introduced exotic species or pollutants. Similarly, humans almost always modify the temporal and spatial patterns of disturbance, which ecologists call the disturbance regime, when they inhabit and use a landscape. In fact, modification or substitution of a natural disturbance regime is one of the most common human strategies for wildlife and resources management. In ecological terms, agriculture is a suite of specialized disturbance regimes combined with plants adapted or bred to exploit that disturbance regime and provide specialized outputs easily harvested by humans. Because of such modification of patch mosaics by humans, management will often have to compensate for missing phenomena (White and Bratton 1980, Baker 1992), such as dispersal, or control of some quickly reproducing animals by predators that no longer have access to parts of the mosaic. Alternatively, managers may be required to remove untoward introduced elements that persist in the modified mosaic, such as some harmful exotic species. This points out that the mere existence of the coarse-scale structure of a mosaic does not represent successful management. The details of the composition and processes in the mosaic must be known and perhaps specifically managed. Hence, a patch dynamic and mosaic approach permits assessment of both the ends and the means targeted for management. Ends involve trade-offs between different species, elements of the mosaic, and functions of those patch elements, and ends can be explicitly incorporated into the management plan. Means include activities that modify patch structure and resource flow. Patch structure can be altered by modifying physical form, geomorphology, or biotic composition. Biotic structure can involve species composition, population age and size structure, and genetic structure, for example, Management activities that alter these aspects of patch structure include bulldozers, planting, hunting, and the like.

One of the most important uses of the mosaic and patch dynamics perspectives in management is to indicate that the dynamic landscape extends beyond the boundaries of the target area. Thus, there must be no blank areas on a planning and management map, either in the parcel of concern or beyond it. The manager cannot afford to stop seeing the processes and structures at the edge of the property or parcel of interest. Ecological systems are most often open to outside influence, and the strength and identity of those influences are determined to a large extent by the patch mosaic. Neither boundaries on maps nor fences in the real world hold back those influences. What is going on in adjacent lands and waters always affects reserves and management units. Cooperation and partnerships among different land owners and uses of neighboring patches may be required for success. Plans and analyses that incorporate entire mosaics will be required to ensure that the greatest utility of the patch dynamics perspective is realized in wildlife management (Rogers 1997). In the case of the Kruger Park rivers, for example, the changes in the ability of the Sabie River upstream of the park to provide water for communities undergoing post-apartheid development, is an important input into management planned in the park. The context of the management unit thus becomes key in the patch dynamics perspective.

In assessing the current state of patch dynamics as an ecological specialty, we noted that refinements were necessary. Management concerns may help guide the refinements in an adaptive management sense. Patches need to be considered in the context of their inclusive mosaics. For example, even when the mangement is aimed at a particular patch or patch type, there is a need to know the nature of the surrounding matrix. Without improved knowledge of the mosaic, in fact, the patch dynamics perspective reduces to a simple successional trajectory within a patch. While accounting for succession in the patch(es) of interest is crucial information for management, it is not enough for successful management. Populations that participate in the succession may come from other patches. Species that play keystone roles as pollinators or predators may reside in other patches or the matrix (Gilbert 1980). Pollutants or resources may flow in from other patch types. Even species that perform important engineering roles can reside elsewhere. These cases suggest very practical reasons for refining and enhancing knowledge of mosaic patch dynamics.

An excellent example of the various nested scales of patchiness appears in the streams and watersheds in a globally important wildlife conservation and management area in South Africa, the Kruger National Park (Figure 4.2). Starting at the coarsest scale, there are five perennial rivers that flow through the Kruger National Park (KNP), resolvable on a scale of hundreds of km². A single river, for example the Sabie, appears on a linear scale of 100s of kilometers, and can be divided into four zones based on hydrology and sediment yield. The zone of the Sabie River that appears within the Kruger National Park occupies a scale of 10 km and accommodates several macro-reaches. Macro-reaches are defined by a specific geology, stream gradient, and broad type of vegetation on their banks. Within the river, a macro-reach does not repeat. A reach, in contrast, occupies the lower end of the 10 km to 100 m scale, and is the repeating unit within a river. Reaches contain one or more channel types defined by channel morphology and substratum type. It is the channel type that is the main kind of patch mosaic to repeat within the river. Within a channel type, morphological units, defined by position on the elevational cross section of the channel, are the elements of the mosaic. Such elements are subjected to different frequencies and duration of flooding and of stability over time. Within morphological units are microsites, which are characterized partly by their elevation and partly by the specific local features of the substratum. Each of these seven scales has particular geomorphological correlates, focus for managers and planners, utility and target of prediction for conservation and resource exploitation, type of mosaic, and biotic and ecological response. This richness of patch structure, hierarchy, and implications provides a valuable framework for integration of interdisciplinary research and demonstrates how important patch mosaics and patch dynamics can be for management.

Patch scaling and hierarchy (King, this volume) offers a useful intersection with the concerns of managers. For instance, a manager may be responsible for a wildlife species on the scale of a catchment. The ecologists working in the area may focus on processes at the scale of small patches within the catchment. Explaining how the larger and smaller scales affect one another is a step toward increasing understanding and communication between managers and researchers. Such a linkage is especially critical if there are functional links between the state of populations and resources at the fine scale with the resources and populations at the coarser scale (Bissonette, this volume). Both researchers and managers may be forced, for practical or biological reasons, to focus on a certain scale. Understanding that these scales can be linked through a hierarchy of patches, may allow the patch hierarchy to serve as a useful way to explore whether and how processes and populations at the contrasting scales of management and research are linked (Rogers 1997). Thus, patch dynamics is one of ecology's generalizable concepts that can engage specialists and application in any ecological topic, and at many scales.

Refinement of patch dynamics information is guided by practical concerns of management. Managers, due to limited budgets, time, and authority may have relatively few tools at their disposal. Use of fire in a prairie surrounded by a suburban area may be limited by the need to maintain visibility on adjacent highways by minimizing smoke. Use of pesticides to control exotic species may be limited by the food requirements of other species or by water quality concerns. Therefore, one goal of patch dynamics research should be to determine alternative tools that managers can use to manipulate patch condition, and to compensate for processes that may no longer be acting in a mosaic or in particular patches. For example, historic disturbance regimes often are missing from contemporary nature reserves. Questions that might guide refinements to knowledge on patch dynamics especially relevant to management include the following: How can the impact of native herbivores and browsers be replaced by domestic animals or mechanical effects? How can the variety of microhabitats for establishment of plant species structuring riparian vegetation be maintained under altered flood regimes? Can the structure or function of forest edges be manipulated to maintain the integrity of processes in the interior? What arrangement of planted patches in arid lands permits the installation of trees in small catchments without diminishing water yield downstream? Answers to such questions will also advance the management of patch structure and function.

## 4.4 What Are the Frontiers in Patch Dynamics?

This section will identify the outstanding issues surrounding patch dynamics in management and summarize the needs for future ecological research.

Ecological research must better document the structure and dynamics of entire mosaics of patches and do so in a spatially explicit fashion. It is not enough to know what patches are present and how many of each kind or age there are. How they are arranged in space can govern how each patch and the entire mosaic behaves in regulating biodiversity and ecosystem function (Harrison and Fahrig 1995). It is ensuring these overall functions that determines the ecological success of management, because sustainability must be judged in part on the maintenance of community function, the continuation of adaptive evolution, and the processing of ecosystem fluxes. Most attention so far has been devoted to describing the structure of patch, edges, and some few entire mosaics (Murcia 1995). Effort will have to be ratcheted up to this rank of completeness so that patchwise and structural descriptions are complemented by functional and dynamic understanding of patches and the entire mosaics of which they are a part. Lack of information on the function of patch mosaics is currently the largest limit to ecological knowledge needed to manage patch dynamics effectively. Three-dimensional structure, filtering, and pumping functions of patches and boundaries especially need ecological attention.

As the complete spatial structure of mosaics is investigated, their hierarchical structure must be exposed. Examining various scales also is valuable because it can show how, on a particular scale, patches are affected by the dynamics of smaller patches nested within them and by the context provided by the mosaic in which they are nested. Such hierarchical studies can show how influences that act first on one scale promulgate to another (Kolasa and Pickett 1989, Pickett et al. 1989). Such data can illuminate small-scale actions that managers can use to influence their larger targets. If managers and researchers share a hierarchical model of patchiness, they will have a powerful tool for comparing concerns, sharing data, and formulating questions across the range of scales affecting their areas.

A final need is the integration of the gradient approach with the patch perspective. Not all patch boundaries will be sharp; some may be fuzzy or gradual (Ims 1995). Examining patches and mosaics across several scales of resolution is one approach to unifying the gradient and patch unit approaches. For some structures and functions in landscapes, recognizing gradual patch boundaries may be helpful, while for other processes, sharp boundaries may be more productive. In management at a certain scale, some organisms or functions may respond better to sharp boundaries, while others will respond to gradual boundaries. Little knowledge of such contrasts or their scale relationships exists.

Management based on patch dynamics is limited not only by scientific knowledge but also by practical constraints. Practical limits may be driven by the fact that zoning and planning regulations do not recognize the need to manage in a mosaic and dynamic context and that the public and policy makers do not appreciate actions that will generate inconveniences or entail some economic cost, even if that would make patch dynamic management more effective. Thus there is need for education about the new approach. Another limitation is that the kind of knowledge needed by managers may not be the knowledge that researchers would automatically think of generating. In order for the priorities of managers and researchers to be jointly set so that the management knowledge base is advanced at the same time that the scientific knowledge is enhanced, better and more regular communication between the two groups is necessary. A successful example is the Kruger Park Rivers Program, in which the interaction of managers and scientists has focused the scientific effort first on those scales of patchiness that managers find useful and has built a hierarchical framework of patchiness that links the practical needs of managers with the more complete understanding of the systems the scientists desire (Figure 4.2). In addition, the hierarchical framework has permitted better communication and integration of the models and approaches of physical scientists and ecologists. Because patch hierarchies can help in understanding heterogeneity in any habitat, can help link management and scientific concerns, and can help unify different sciences that support managers, they can be used in many situations.

Using the basic knowledge of patch dynamics in the context of heterogenous mosaics can advance the practice of management (Wiens 1997). The metaphorical use of patch dynamics and mosaics can help the public and policy makers understand the need to manage within a broad, inclusive spatial context. Patch dynamics also requires that management plans not be static, but assess the changes in systems and relate them to the stated goals of management. The tactics or goals of management can be altered if the monitoring reveals shortcomings in the effectiveness of the management (Holling 1978). Hierarchical frameworks of patchiness provide a meeting ground for managers and scientists and different scientific specialties in which data and modeling needs can be prioritized and information shared. Patch dynamics thus meets the needs of managers in many important ways.

*4.5 Acknowledgments.* We thank M.L. Cadenasso for a helpful reading of the manuscript. S.T.A.P. thanks the National Science Foundation (DEB 9307252) for support of research on which this paper draws.

## 4.6 *References*

Abbott, I. 1980. Theories dealing with land birds on islands. Advances in Ecological Research 11:329–371.

Allen, M. 1991. The ecology of mycorrhizae. Cambridge University Press, New York, New York, USA.

Alverson, W.S., W. Kuhlmann, and D.M. Waller. 1994. Wild forests: conservation biology and public policy. Island Press, Washington, DC, USA.

Angelstam, P. 1992. Conservation of communities—the importance of edges, surroundings and landscape mosaic structure. Pp. 9–70 in L. Hanson, editor. Ecological principles of nature conservation: applications in temperate and boreal environments. Elsevier Applied Science, New York, New York, USA.

Baker, W.L. 1992. The landscape ecology of large disturbances in the design and management of nature reserves. Landscape Ecology 7:181-194.

Beatty, S.W. 1984. Influence of microtopography and canopy species on spatial patterns of forest understory plants. Ecology 65:1406–1419.

Bell, S.S., E.D. McCoy, and H.R. Mushinsky, editors. 1991. Habitat structure: the physical arrangements of objects in space. Chapman and Hall, New York, New York, USA.

Bormann, F.H., and G.E. Likens. 1979. Pattern and process in a forested ecosystem. Springer-Verlag, New York, New York, USA.

Breen, C.M., K.H. Rogers, and P.J. Ashton.ᵉ1988. Vegetation processes in swamps and flooded areas. Pp. 223–247 in J.J. Symoens, editor. Vegetation of inland waters. Kluwer, Dordrecht, Holland.

Brown, J.H. 1986. Two decades of interaction between the MacArthur–Wilson model and the complexities of mammalian distributions. Biological Journal of the Linnean Society 28:231–251.

Burgess, R.L., and D.M. Sharpe, editors. 1981. Forest island dynamics in man-dominated landscapes. Springer-Verlag, New York, New York, USA.

Caldwell, M.M., and R.W. Pearcy, editors. 1994. Exploitation of environmental heterogeneity by plants: ecophysiological processes above and below ground. Academic Press, New York, New York, USA.

Christensen, N.L. 1997. Managing for heterogeneity and complexity on dynamic landscapes. Pp. 167–186 in S.T.A. Pickett, R.S. Ostfeld, M. Shachak, and G.E. Likens, editors. The ecological basis of conservation: heterogeneity, ecosystems, and biodiversity. Chapman and Hall, New York, New York, USA.

Dale, V.H., F. Southworth, R.V. O'Neill, A. Rose, and R. Frohn. 1993. Simulating spatial patterns of land-use change in central Rondonia, Brazil. In R.H. Gardner, editor. Some mathematical questions in biology. American Mathematical Society, Providence, Rhode Island, USA.

Dunn, C.P., D.M. Sharpe, G.R. Guntenspergen, F. Stearns, and Z. Yang. 1991. Methods for analyzing temporal changes in landscape pattern. Pp. 173–198 in M.G. Turner and R.H. Gardner, editors. quantitative methods in landscape ecology: the analysis and interpretation of landscape heterogeneity. Springer-Verlag, New York, New York, USA.

Forman, R.T.T., and M. Godron. 1986. Landscape ecology. John Wiley and Sons, New York, New York, USA.

Forman, R.T.T., and P.N. Moore. 1992. Theoretical foundations for understanding boundaries in landscape mosaics. Pp. 236–258 in A.J. Hansen and F. di Castri, editors. Landscape boundaries, consequences for biotic diversity and ecological flows. Volume 92. Springer-Verlag, New York, New York, USA.

Foster, R.B. 1980. Heterogeneity and disturbance in tropical vegetation. Pp. 75–92 in M.E. Soulé and B.A. Wilcox, editors. Conservation biology: an evolutionary-ecological perspective. Sinauer Associates, Sunderland, Massachusetts, USA.

Franklin, J.F. 1993. Preserving biodiversity: species ecosystems, or landscapes. Ecological Application 3:202–205.

Gilbert, L.E. 1980. Food web organization and the conservation of neotropical diversity. Pp. 11–33 in M.E. Soulé and B.A. Wilcox, editors. Conservation biology: an evolutionary-ecological perspective. Sinauer Associates, Sunderland, Massachusetts, USA.

Groffman, P.M., D.R. Zak, S. Christensen, A.R. Mosier, and J.M. Tiedje. 1993. Early spring nitrogen dynamics in a temperate forest landscape. Ecology 74:1579–1585.

Grumbine, R.E. 1994. What is ecosystem management? Conservation Biology 8:27–38.

Harris, L.D. 1984. The fragmented forest: island biogeography theory and the preservation of biodiversity. University of Chicago Press, Chicago, USA.

Harris, L.D. 1988. Edge effects and conservation of biotic diversity. Conservation Biology 2:330–332.

Harrison, S. 1994. Metapopulations and conservation. Pp. 111–128 in P.J. Edwards, R.M. May and N.R. Webb, editors. Large-scale ecology and conservation biology. Blackwell Scientific Publications, Boston, Massachusetts, USA.

Harrison, S., and L. Fahrig. 1995. Landscape pattern and population conservation. Pp. 293–308 in L. Hansson, L. Fahrig, and G. Merriam, editors. Mosaic landscapes and ecological processes. Chapman and Hall, New York, New York, USA.

Heinselman, M.L. 1973. Fire in the virgin forests of the Boundary Waters Canoe Area, Minnesota. Journal of Quaternary Research 3:329–382.

Hibbs, D.E. 1982. Gap dynamics in a hemlock-hardwood forest. Canadian Journal of Forest Research 12:522–527.

Holland, M.M., P.G. Risser, and R.J. Naiman, editors. 1991. Ecotones: the role of landscape boundaries in the management and restoration of changing environments. Chapman and Hall, New York, New York, USA.

Holling, C.S. 1978. Adaptive environmental assessment and management. John Wiley and Sons, New York, New York, USA.

Huston, M.A. 1994. Biological diversity: the coexistence of species in changing landscapes. Cambridge University Press, New York, New York, USA.

Ims, R.A. 1995. Movement patterns related to spatial structures. Pp. 85–109 in L. Hansson, L. Fahrig, and G. Merriam, editors. Mosaic landscapes and ecological processes. Chapman and Hall, New York, New York, USA.

Jones, C.G., J.H. Lawton, and M. Shachak. 1994. Organisms as ecosystem engineers. Oikos 69:373–386.

Kingsland, S.E. 1985. Modeling nature: episodes in the history of population ecology. University of Chicago Press, Chicago, USA.

Kolasa, J., and S.T.A. Pickett. 1989. Ecological systems and the concept of organization. Proceedings of the National Academy of Science of the United States of America 86:8837–8841.

Kolasa, J., and S.T.A. Pickett, editors. 1991. Ecological heterogeneity. Springer-Verlag, New York, New York, USA.

Kotliar, N.B., and J.A. Wiens. 1990. Multiple scales of patchiness and patch structure—a hierarchical framework for the study of heterogeneity. Oikos 59:253–260.

Levins, R. 1970. Extinction. Lectures on Mathematics in the Life Sciences. American Mathematical Society, Providence, Rhode Island, USA. 2:75–101.

Little, S. 1979. Fire and plant succession in the New Jersey pine barrens. Pp. 297–314 in R.T.T. Forman, editor. Pine barrens: ecosystem and landscape. Academic Press, New York, New York, USA.

MacArthur, R.H., and E.O. Wilson. 1967. The theory of island biogeography. Princeton University Press, Princeton, New Jersey, USA.

McDonnell, M.J., and S.T.A. Pickett, editors. 1993. Humans as components of ecosystems: the ecology of subtle human effects and populated areas. Springer-Verlag, New York, New York, USA.

McIntosh, R.P. 1991. Concept and terminology of homogeneity and heterogeneity. Pp. 24–46 in J. Kolasa and S.T.A. Pickett, editors. Ecological heterogeneity. Springer-Verlag, New York, New York, USA.

Murcia, C. 1995. Edge effects in fragmented forests: implications for conservation. Trends in Ecology and Evolution 10:58–62.

National Research Council. 1986. Ecological knowledge and environmental problem-solving: concepts and case studies. National Academy Press, Washington, DC, USA.

National Research Council. 1995. Science and the endangered species act. National Academy of Sciences Press, Washington, DC, USA.

Niering, W.A. 1987. Vegetation dynamics (succession and climax) in relation to plant community management. Conservation Biology 1:287–295.

O'Neill, R.V., R.H. Gardner, B.T. Milne, M.G. Turner, and B. Jackson. 1991. Heterogeneity and spatial hierarchies. Pp. 85–96 in J. Kolasa and S.T.A. Pickett, editors. Ecological heterogeneity. Springer-Verlag, New York, New York, USA.

Pearson, S.M. 1993. The spatial extent and relative influence of landscape-level factors on wintering bird populations. Landscape Ecology 8:3–18.

Pickett, S.T.A. 1976. Succession: an evolutionary interpretation. American Naturalist 110:107–119.

Pickett, S.T.A. 1996. Natural processes. Pp. 1–22 in P.A. Opler, editor. National status an trends of our living resources. National Biological Service, Washington, DC, USA.

Pickett, S.T.A., and M.L. Cadenasso. 1995. Landscape ecology: spatial heterogeneity in ecological systems. Science 269:331–334.

Pickett, S.T.A., and P.S. White, editors. 1985. The ecology of natural disturbance and patch dynamics. Academic Press, Orlando, Florida, USA.

Pickett, S.T.A., J. Kolasa, J.J. Armesto, and S.L. Collins. 1989. The ecological concept of disturbance and its expression at various hierarchical levels. Oikos 54:129–136.

Pielke, R.A., and R. Avissar. 1990. Influence of landscape structure on local and regional climate. Landscape Ecology 4:133–155.

Pulliam, H.R. 1988. Sources, sinks, and population regulation. American Naturalist 132:652–661.

Rogers, K.H. 1997. Operationalizing ecology under a new paradigm: examples from African savannas and their rivers. Pp. 60–77 in S.T.A. Pickett, R.S. Ostfeld, M. Shachak, and G.E. Likens, editors. The ecological basis for conservation: heterogeneity, ecosystems, and biodiversity. Chapman and Hall, New York, New York, USA.

Shachak, M., and S. Brand. 1991. Relations among spatiotemporal heterogeneity, population abundance, and variability in a desert. Pp. 202–223 in J. Kolasa and S.T.A. Pickett, editors. Ecological heterogeneity. Springer-Verlag, New York, New York, USA.

Shafer, C.L. 1990. Nature reserves: island theory and conservation practice. Smithsonian Institution Press, Washington, DC, USA.

Turner, B.L., W.C. Clark, R.W. Kates, J.F. Richards, J.T. Matthews, and W.B. Meyer, editors. 1990. The Earth as transformed by human action: global and regional changes in the biosphere over the past 300 years. Cambridge University Press, New York, New York, USA.

Turner, M.G., and R.H. Gardner, editors. 1991. Quantitative methods in landscape ecology: the analysis and interpretation of landscape heterogeneity. Springer-Verlag, New York, New York, USA.

Turner, M.G., R.H. Gardner, V.H. Dale, and R.V. O'Neill. 1989. Predicting the spread of disturbance across heterogeneous landscapes. Oikos 55:121–129.

Turner, S.J., R.V. O'Neill, W. Conley, M.R. Conley, and H.C. Humphries. 1991. Pattern and scale: statistics for landscape ecology. Pp. 17–49 in M.G. Turner and R.H. Gardner, editors. Quantitative methods in landscape ecology: the analysis and interpretation of landscape heterogeneity. Springer-Verlag, New York, New York, USA.

Waring, R.H., and W.H. Schlesinger. 1985. Forest ecosystems: concepts and management. Academic Press, Orlando, Florida, USA.

Watt, A.S. 1947. Pattern and process in the plant community. Journal of Ecology 35:1–22.

White, P.S. 1979. Pattern, process, and natural disturbance in vegetation. Botanical Review 45:229–299.

White, P.S. 1994. Synthesis: vegetation pattern and process in the Everglades ecosystem. Pp. 445–458 in S.M. Davis and J.C. Ogden, editors. Everglades: the ecosystem and its restoration. St. Lucie Press, Delray Beach, Florida, USA.

White, P.S., and S.P. Bratton. 1980. After preservation: philosophical and practical problems of change. Biolgical Conservation 18:241–255.

Wiens, J.A. 1989. Spatial scaling in ecology. Functional Ecology 3:385–397.

Wiens, J.A. 1992. Ecological flows across landscape boundaries: a conceptual overview. Pp. 216–235 in F. di Castri and A.J. Hansen, editors. Landscape boundaries. Volume 92. Springer-Verlag, New York, New York, USA.

Wiens, J.A. 1995. Landscape mosaics and ecological theory. Pp. 1–26 in L. Hansson, L. Fahrig, and G. Merriam, editors. Mosaic landscapes and ecological processes. Chapman and Hall, New York, New York, USA.

Wiens, J.A. 1997. The emerging role of patchiness in conservation biology. Pp. 93–107 in S.T.A. Pickett, R.S. Ostfeld, M. Shachak, and G.E. Likens, editors. The ecological basis for conservation: heterogeneity, ecosystems, and biodiversity. Chapman and Hall, New York, New York, USA.

Wiens, J.A., C.S. Crawford, and J.R. Gosz. 1985. Boundary dynamics: a conceptual framework for studying landscape ecosystems. Oikos 45:421–427.

Wu, J., and S.A. Levin. 1994. A spatial patch dynamic modeling approach to pattern and process in an annual grassland. Ecological Monographs 64:447–464.

Wu, J., and J.L. Vankat. 1995. Island biogeography, theory and applications. Pp. 371–379 in W.A. Nierenberg, editor. Encyclopedia of environmental biology. Volume 2. Academic Press, orlando, Florida, USA.

# 5
# Disturbance and Diversity in a Landscape Context

PETER S. WHITE and JONATHAN HARROD

## 5.1 Introduction

Natural disturbances like windstorm, fire, and flooding are part of the ecological and evolutionary setting of all ecosystems and are sources of variation in ecosystem structure and composition (White 1979). It is not surprising, then, that many plant and animal species are dependent on or otherwise affected by the conditions created by disturbance. As a result, species richness within ecosystems and landscapes is, in part, a function of the history and characteristics of disturbances. Changes in the frequency and intensity of natural disturbances constitute one of the major ways that humans have altered ecosystems and thus the biological diversity that occurs in them.

While some disturbances are outside management control (e.g., hurricanes), others are among the natural processes most often affected by people (e.g., fire and flooding). Even for those disturbances that are outside the bounds of management, human influence can indirectly affect disturbance rate. We often have influenced the susceptibility of ecosystems to disturbances because we have changed the pattern of ecosystem structure

---

Peter S. White is a Professor in the Department of Biology and the director of the North Carolina Botanical Garden at the University of North Carolina at Chapel Hill. He has served as an NEA Postdoctoral Fellow at the Missouri Botanical Garden, as a research biologist for the National Park Service at Great Smoky Mountains National Park, on the faculty of the Graduate Program in Ecology at the University of Tennessee, Knoxville, and was the leader of the Cooperative Park Study Unit at the University of Tennessee. His research interests include vegetation dynamics, vegetation–landscape relations, species richness, and conservation biology.

Jonathan Harrod is a graduate student in the Curriculum in Ecology at the University of North Carolina at Chapel Hill. His research interests include the effects of land use and natural disturbances on vegetation patterns and processes in the southern Appalachians.

128

and composition across landscapes (Baker 1992a, Franklin and Forman 1987). Human-caused disturbances may superficially resemble natural disturbances but may differ in several important respects. It is critical that managers and scientists work to understand the influence of disturbance on biological diversity.

In this paper, we first define *disturbance*, describe the concept of the *disturbance regime*, and discuss the way disturbances affect resource availability in ecosystems. We then discuss the relation of biological diversity to disturbance, with special reference to species richness and variation in ecosystem composition and structure. A final section discusses the implications of disturbance for management at landscape scales. We focus on vegetation because it forms the physical structure of terrestrial habitats and provides resources for higher trophic levels, but we seek to describe general principles concerning the relationship of disturbance and biological diversity.

## 5.2 Disturbance and the Disturbance Regime

### 5.2.1 Definitions

*Disturbance* is any relatively discrete event in time that disrupts ecosystem, community, or population structure and changes resources, substrate availability, or the physical environment (White and Pickett 1985). We emphasize that disturbance produces abrupt and measurable changes and reject definitions that suggest that disturbance should be defined relative to the natural processes within an ecosystem (see next paragraph). Inevitably, what we recognize as disturbance depends on the temporal and spatial dimensions of our observations. For example, a disturbance to a streamside moss community will not necessarily be a disturbance to streamside trees. In addition, *abruptness* can be defined only with reference to the rates of change that characterize the ecosystem before and after disturbance.

An alternative to the absolute definition of *disturbance* is a relative one: disturbance is a departure from the normal domain (environmental, biological) of an ecosystem. The most frequent example of this is the suggestion that the absence of fire, rather than fire, is a disturbance in the prairie. The problem with a relative definition of disturbance stems from the difficulty of defining the "normal" range of conditions for an ecosystem. While data are often short-term, large and infrequent events may have an important influence on ecosystem and landscape structure. Furthermore, year-to-year climate variation, decade-scale climatic cycles, directional climate change on time scales of thousands of years, spatial shifts in the distribution of particular ecosystems, and functional differences between current and former ecosystems make the definition of *normal* problematic at best and dependent on the time span and resolution of the data.

The importance of dynamics such as those caused by disturbance have led some workers to suggest "natural range of variation" as a basis for ecosystem management (Landres 1992, Hunter 1993). Because of long-term climatic shifts, the generally short-term nature of the data, and difficulties in defining *natural*, some ecologists focus on the last several centuries and thus the historical rather than natural range of variation as a guide to management (Swetnam 1993, Morgan et al. 1994, Wright et al. 1995). Even when these efforts are successful, the mechanisms of disturbance and disturbance response are, to us, more important than whether the events are known to be within a normal, natural, or historical range of variation. Thus, we emphasize the physical effects of disturbance and the mechanisms of species and ecosystem response. Nonetheless, the concept of the historical range of variation (and other measures of variation in addition to range) can help us understand the bounds of ecosystem variation within time frames (past decades to centuries or longer) that are most relevant to management. For example, an investigation of historical changes in lodgepole pine forests in Yellowstone National Park revealed that large, infrequent crown fires are a natural feature of that system (Figure 5.1; Romme and Despain 1989). Managers will often want to know whether a given disturbance is unprecedented or within the bounds of recent or historic variation.

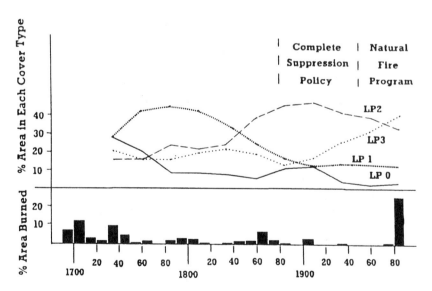

FIGURE 5.1. Disturbance and succession on a 129,600 ha study area in Yellowstone National Park (Romme and Despain 1989). Changes in the percent of the study area in each of four successional types (top) and the percent of the study area burned each decade (bottom) over the past three centuries. LP0 are lodgepole pine stands less than 40 years old; LP1 are even-aged stands 40–150 years old; LP2 are even-aged stands 150–300 years old; and LP3 are mixed-age pine-fir-spruce.

In sum, because of the difficulties of applying a relative definition of *disturbance*, because the absolute definition requires measures of real and absolute change (thus, stressing a mechanistic approach), and because even one kind of disturbance can exhibit a range of effects (e.g., fires can vary in intensity and season of occurrence, and considerable patchiness may exist within the boundary of a single fire), we use an absolute definition here.

## 5.2.2 The Patchiness and Heterogeneity of Ecosystems

Individual plants in an ecosystem take up space and use resources. Mortality is inevitable. Given these truisms, ecosystems will have an inherent patchiness at the spatial scale of individuals and a disturbance frequency that corresponds with the temporal scale of the life span. Patchiness occurs, of course, over a wider range of spatial scales: some disturbances affect only part of an individual (e.g., part of a tree crown), while others fell dominant plants over a wide area (e.g., many thousands of individuals).

Relatively small patches are "gaps" (e.g., gaps in a forest canopy). Colonization and succession in gaps is termed *gap dynamics* (Figure 5.2). Small

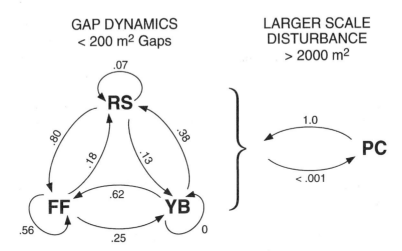

FIGURE 5.2. Species response to gap dynamics and larger magnitude wind disturbance in southern Appalachian spruce–fir forests (based on data in White et al. 1985). FF is Fraser fir (*Abies fraseri*), RS is red spruce (*Picea rubens*), YB is yellow birch (*Betula alleghaniensis*), and PC is pin cherry (*Prunus pensylvanica*). The values by the arrows indicates the probability of the particular transition taking place based on the observation of gap formation and capture rates. Thus, Fraser fir has a 0.56 chance of being replaced by itself in the canopy, a 0.25 chance of being succeeded by a yellow birch, and a 0.18 chance of being succeeded by a red spruce. Larger disturbances are rare. Pin cherry's dominance of these stands leads back to the mix of red spruce, Fraser fir, and yellow birch.

scale patchiness is sometimes considered a within-community phenomenon (e.g., Forcier 1975); the successions that result are sometimes said to be microsuccessions. Where these reestablish predisturbance structure and composition, they have been called cyclic successions (Churchill and Hanson 1958). Small-scale, within-community, patchiness is near one end of a continuum; at the other extreme, when patches are very large, patchiness is more likely to be described as a between-, rather than within-community phenomenon. *Successional change* can be described as a gradient in time, along which there may be great changes in structure and complete turnover of composition (Peet 1992). These changes may be arbitrarily described as a series of successional communities or seral stages (the temporal gradient is the sere). Oliver (1980) and Peet and Christensen (1987) describe four structural phases of recovery after disturbance in forests: stand initiation/establishment, stem exclusion/thinning, understory reinitiation/transition, and old growth/steady state.

The general phrase for the changes within individual patches and the distribution and interaction of patches in space, regardless of scale and regardless of whether these are seen as within or between communities, is *patch dynamics* (Thompson 1978, Pickett and White 1985). Patch dynamics is the subject of another chapter in this volume.

Disturbances may promote heterogeneity at multiple scales (Figure 5.3). Within a single patch, disturbances may produce a variety of conditions. For example, a treefall gap contains a range of light levels (gap edge versus center) and substrates (root mound, trunk, forest floor). In the larger landscape, disturbances create patches that differ in age, history, and successional state.

Heterogeneity in ecosystems also is caused by variation in the physical environment, including the effects of elevation, slope, aspect, soil texture, and soil chemistry. These environmental gradients influence the composition of plant communities. They affect disturbance rates directly (e.g., warmer, drier sites may be more flammable) and indirectly via the vegetation (e.g., one forest type may produce more fuel than another). Landscape heterogeneity thus reflects interactions between disturbance and the physical environment.

Disturbances also differ greatly in their effects on living and dead plant material. For example, fire in tallgrass prairie incinerates most aboveground biomass but causes little or no mortality to native grasses and forbs. In contrast, bark beetles in a pine forest may cause high rates of mortality but have little short-term impact on total biomass. While most disturbances produce patchiness and heterogeneity, at least at large scales of observation, individual disturbances may result in relatively homogeneous conditions, particularly at small scales of observation. For example, fires in tallgrass prairie may erase fine-scale patterns created by grazing (Hobbs et al. 1991). The great range of disturbance effects again suggests to us that the best definition of *disturbance* is one that stresses the absolute characteristics

a.

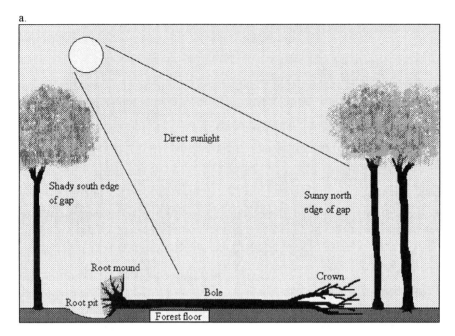

b.

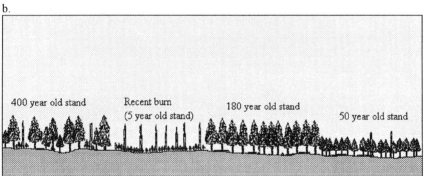

FIGURE 5.3. Disturbances produce heterogeneity at several scales. (a) Small scale heterogeneity in an eastern deciduous forest. Disturbance creates a range of light levels and microsites. (b) Large-scale heterogeneity in a lodgepole pine forest. Crown fires create a mosaic of even-age stands.

of disturbance effects and focuses on the mechanisms of ecosystem response.

## 5.2.3 The Concept of the Disturbance Regime

Disturbances are not equivalent. The kinds of disturbance that are important vary geographically (e.g., across climatic gradients) with topography and geological substrate and because of biological differences among systems (White 1979, Harmon et al. 1983, White and Pickett 1985). Hurricanes and typhoons characterize the eastern sides of the major North Temperate continents. Tornados are most frequent in midcontinental regions where warm and cold air masses meet. Fires are most important in areas in which production of organic fuels is relatively high and seasonal drought allows the drying of those fuels. Fires thus have a major role in grasslands, savannahs, and in all but the most mesic forests and sometimes control whether closed forest, open forest, or grassland dominate. Fire frequency and intensity vary across regions; they may also vary within a landscape between dry, exposed ridges and moist, sheltered valleys (Figure 5.4). The erosive power of rivers varies from region to region (as a function of rainfall and substrate) and from the headwaters to mouths of individual rivers. Steep slopes are prone to debris and snow avalanches. Freeze–thaw cycles disturb and churn soils in arctic and alpine areas. Some disturbances are caused or influenced by biological attributes of the system. Burrowing animals, grazers, and nest builders can be major creators of disturbance patches and mosaics (Coppock et al. 1983, Hobbs et al. 1991). Outbreaks of native insects and pathogens are a recurrent periodic disturbance in some ecosystems (Schowalter 1985). Perhaps our most enduring image of natural dynamic ecosystems comes from unstable coastal substrates: the shifting sands of time.

Differences among disturbances are described by the parameters of the disturbance regime (Table 5.1). The disturbance regime includes the type of

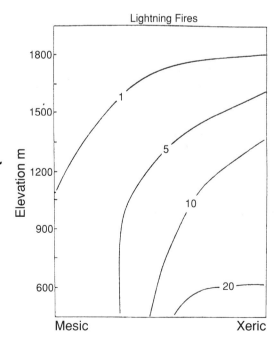

FIGURE 5.4. Distribution of lightning fires in Great Smoky Mountains National Park as a function of elevation and site moisture gradients (from Harmon et al. 1983).

TABLE 5.1. Parameters of disturbance regimes (expanded from White and Pickett 1985)

---

Kind
Spatial characteristics
  Size: Patch size, area per event, area per time period, area per event per time period,
    total area per disturbance per time period
  Distribution: Spatial distribution including relationship to geographic, topographic,
    environmental and community gradients
  Landscape context: Patch dispersion, contiguity, matrix
Temporal characteristics
  Frequency: Number of events per time period
  Rotation period: Time needed to disturb an area equivalent to the study area
  Return interval, cycle, or turnover time: Interval between disturbance events
  Predictability: A scaled inverse function of the variance in return interval
  Seasonality: Seasonal distribution
Specificity
  To species: Probability of disturbance by species
  To age or size classes: Probability of disturbance by age or size classes
  To successional time or community state: Probability of disturbance by successional stage
    and feedback between community state and disturbance rate
  To landforms: Probability of disturbance by landform element
Magnitude
  Intensity: Physical force of the event per area per time
  Severity: Impact on the organism, community, or ecosystem
Ecosystem effects
  Internal heterogeneity: Degree of internal patchiness within disturbed areas
  Ecosystem legacies: Structures, dead and living biomass remaining after disturbance
Synergisms
  Interactions between disturbances, including influence on subsequent disturbance rate

---

disturbance, the sizes of the patches disturbed, the frequency (and related concepts like *return interval* and *rotation period*), and the magnitude of disturbance (separated into measures of disturbance force or intensity and the effect on the ecosystem or severity). Human influence alters each of these aspects of disturbance regime.

Disturbance effects are not just the conversion of living to dead biomass. A critically important factor in disturbance recovery is the nature of the ecological elements that survive the disturbance itself and are present in the early stages of recovery (Franklin 1989, Swanson and Franklin 1992). These "biological legacies" include the residual plants not killed or removed by the disturbance, woody debris and other forms of organic matter, seeds stored in the soil, soil organisms, and animals that find refuges in one way or another from the disturbance itself. Disturbances differ in the ecosystem legacies they leave; when evaluating the effect of human disturbances, we must also examine these legacies. For example, both logging and windstorms open the forest canopy and increase light availability in the forest understory. However, they differ in their effects on woody debris, soils, and forest herbs and seedlings. Even if logging and windstorm produce patches of similar size, rates of recovery and impacts on plant and animal populations may differ greatly.

There is often a feedback between patch state and the probability of subsequent disturbance. In some cases, susceptibility decreases immediately after a disturbance and gradually increases through successional time. For example, a hot fire in chaparral consumes most available fuel. Flammability is low in the young stand and increases as the stand matures (Minnich 1983). One disturbance can also increase the likelihood of subsequent disturbances, as when fire-scarred trees are more vulnerable to insect attack (Geiszler et al. 1980) and wind breakage (Matlack et al. 1993). The tallgrass prairie provides a particularly intriguing example of interactions between fire and grazing disturbance. Patches of closely grazed prairie do not burn well. However, the vigorous new growth on recently burned patches attracts grazers. Patches that have been both burned and grazed are more diverse than patches with only a single disturbance or no disturbance at all (Collins 1987). The way patches respond to disturbance may depend, in large part, on their disturbance history. The presence of multiple types of disturbance may promote landscape and patch-scale diversity.

While natural processes like disturbances influence landscape pattern, pattern can also affect process (Turner 1989). Some disturbances are contagious in space: they start at one point and spread outward (e.g., fire and insect outbreaks). The occurrence of such disturbances at a point in the landscape may be a function not only of the conditions at that point but also of the conditions in the surrounding area (Turner et al. 1989). For example, the occurrence of fire in one habitat may depend on the vulnerability to fire of the surrounding matrix. A relatively nonflammable patch may burn in a year when the surrounding matrix supports high intensity fires (e.g., the Yellowstone fires of 1988; Romme and Despain 1989). Likewise, a relatively flammable patch may escape fire if it is surrounded by fire-resistant patches. When humans homogenize ecosystems (e.g., producing one or a few age classes of stands over large areas), a large area may become vulnerable, all at once, to a single disturbance (either because vulnerability is a direct function of age or because some successional stages are so competitive that many trees are under stress). Thus, some blame large outbreaks of insect pests on artificially monotonous landscapes with large contiguous areas of forest in susceptible age classes.

In sum, patches vary in vulnerability to disturbance. The landscape matrix can either increase or decrease the probability that a disturbance will spread to a particular point. Process not only creates pattern, but pattern influences process. The mutual interaction of pattern and process is the fundamental reason that landscape ecology has coalesced as a distinct subdiscipline in ecology.

Disturbance rate is a function of the overall environment (e.g., climate), topography, landscape pattern, and ecosystem history. Interannual and longer term variation in climate may result in variation in disturbance

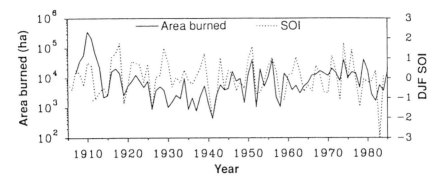

FIGURE 5.5. The association between area burned in Arizona and New Mexico between 1905 and 1985 and an index of the Southern Oscillation (SOI) for the December to February (from Swetnam and Betancourt 1990). High values of SOI are associated with droughts and higher fire incidence across large areas.

frequency (Figure 5.5; Swetnam and Betancourt 1990). The regional scale of climate variation may cause synchronization of disturbance dynamics over very large areas such that a bad fire year in one management unit is a bad fire year in all management units. This has consequences for management response, emergency planning, and ecosystem resilience. Synchronization of disturbances would tend to lead to fluctuations in landscape composition and would tend to make rare species more vulnerable to extirpation. Ecosystems would be less resilient because all patches of a given ecosystem type would be of similar age and structure and would have parallel rather than independent dynamics through time.

Humans have also affected historical disturbance rates and ecosystem structure (Baker 1992a, Harmon et al. 1983). In some areas, Native Americans set fire to vegetation to improve game habitat, increase berry production, clear brush, and drive game. They also cleared land for cultivation. The magnitude of Native American impacts is poorly understood; it was probably greatest in coastal areas and major river valleys, where populations were highest. European Americans also have altered disturbance regimes through logging, agriculture, setting and suppressing fires, and fragmenting natural areas. Understanding the role of humans in the disturbance regime is important for two reasons. First, both historic and modern ecosystems have been shaped by human activity. Second, by changing ecosystem structure and composition, humans have changed the way that ecosystems will respond to future natural disturbance.

## 5.2.4 Disturbance and Resource Availability

The effects of disturbances on environmental conditions and resource availability will vary with the kind, size, and intensity of disturbance. These disturbance effects are superimposed on spatial variation in resources due to climate, topography and soils. On relatively productive sites like most deciduous forests and prairies, plant growth is dense and competition for limiting resources (e.g. light, nitrogen) is severe. In these systems, disturbances that remove community dominants may produce a dramatic increase in the growth of smaller plants. In less productive systems like deserts, where plant growth is limited more by harsh environmental conditions than by competition, disturbances that remove one element may have little effect on the growth of others.

Perhaps the most interesting cases are those natural disturbances that increase resource availability in ecosystems in which intense competition leads to low resource levels (Vitousek 1984). For example, older patches of forest have shady forest interiors and soils that are exploited by the root systems of many individuals. Over time, soil nutrients may have accumulated in living biomass. A windstorm that opens a gap in the canopy results in higher light on the forest floor and may also produce changes in other resource levels. Tree death may result in lower rates of evapotranspiration and thus increased soil moisture (although increased insolation and temperature may also cause evaporative drying). Tree death also results in lower rates of nutrient uptake, and increased temperature hastens decomposition rates. The result of these changes may be increased nutrient availability in soils. If trees are uprooted rather than snapped, the disturbance also exposes mineral soil and creates open, plant-free patches. The resulting pit and mound topography characterizes forests disturbed in this way.

The effects of disturbance on site environment will vary with patch size, shape, and location. For example, light and temperature levels in windfall gaps vary with latitude, gap size, slope, and aspect, and height of the surrounding forest (Canham et al. 1990). In the tropics, the sun is more directly overhead. As a result, a small tropical gap allows more light to reach the forest floor than an equivalent gap at higher latitudes. In the north temperate zone, the northern edge of a gap receives more light than the southern edge, which is in the shadow of the intact forest canopy. Similarly, a gap on a south-facing slope receives more light than a gap on a north-facing slope. A given-sized gap size admits less light in a tall forest than in a short one. A gap in a sparse, open forest contrasts less with the surrounding matrix than a gap in a dense forest. Thus, environmental changes produced by disturbances depend on the structure and location of the ecosystem.

Small, within-community disturbances (e.g., small gaps in a forest) may not create sharp contrasts in resource abundance. In small gaps that do not include soil disturbance, light may be the only resource that increases within the patch, and the primary response may be the release of understory saplings. The same species that dominate the canopy may dominate the gap; as a result, we may not recognize any species as disturbance specialists. At the other extreme, very large disturbance patches may result in such different environmental conditions and may put such a premium on rapid dispersal that the species that colonize and dominate are ones not found in older patches. In such a situation, we recognize some species as disturbance specialists.

Species have evolved that thrive in, and in some cases require, the high-resource conditions produced by disturbance. Many of these species follow a "fugitive" life-history strategy that corresponds to the patchy and ephemeral nature of early-successional habitats. They grow quickly, mature early, and produce large numbers of seeds. They typically senesce rapidly and do not tolerate intense competition; thus, they disappear in the course of succession. The persistence of these species depends on their ability to disperse their seeds to newly disturbed patches ("dispersal in space") or to remain in the soil seed bank until a site is disturbed again ("dispersal in time"). Species with rapid growth rates during periods of high resource availability provide an important ecosystem function: they assimilate nutrients that might otherwise leach from the system (Marks 1974).

Cases in which a relatively slow-growing, competitive, and resource-limited system becomes a place of high resource availability (and at larger spatial scales there is a resulting patchwork of resource abundance) are one of the most interesting situations in the study of natural disturbance. However, disturbances do not always increase resource levels. For example, in northern Appalachian spruce–fir forests, a debris avalanche may expose bedrock, creating an erosional zone in which the environment is harsh and succession is extremely slow (Figure 5.6; Lang 1985). The same avalanche

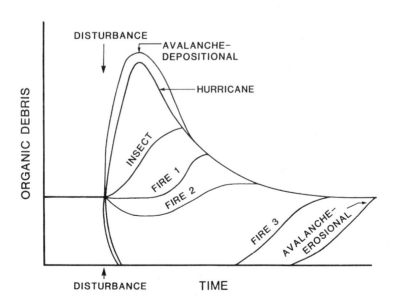

FIGURE 5.6. The effects of different kinds of disturbance on organic debris in one ecosystem, northern Appalachian spruce–fir forests (redrawn from Lang 1985).

creates a depositional zone in which soil, organic matter, and nutrients are enriched. Fires vary substantially in how much fuel they consume and how much they create by causing tree death without consuming the debris. This is a further argument for the absolute definition of *disturbance* and for a quantitative and mechanistic approach to measuring disturbance effects.

In sum, understanding the role of disturbance requires an understanding of how specific resources are affected and an assessment of how the biota respond. Structural recovery (e.g., leaf area, the density and biomass of living individuals, and the amount and distribution of coarse woody debris) is a predictable and straightforward response from disturbance effects on resource availability and site environment. We take up the subject of composition and species richness in the next section.

## 5.3 Disturbance and Diversity

Noss (1990) includes three axes (composition, structure, and process) and four levels (genes within populations, species within communities, ecosystems, and landscapes) in his definition of *biological diversity*. Disturbance can affect all these dimensions. In this discussion we focus on the species level (and in terms of the assemblage of species, the community level) as a fundamental element of biological diversity. The species level has several advantages: it is reasonably straightforward (versus the nature of communities or the definition of natural process), it is often central to management goals, and it is fairly easy to monitor. Managers and conservationists often seek the persistence of communities of native species, on the assumption that ecosystems and landscapes, though dynamic, are characterized by bounded variation within which native species have a low risk of extinction despite changes in abundance and spatial shifts. We have already argued that disturbance effects usually produce heterogeneity at several scales. In this section we will explore the consequences of this heterogeneity for species and communities.

### 5.3.1 Scale Dependence and Species Richness

Before we consider the response of species richness to disturbance, we must discuss the issue of scale dependence. The number of species we observe always depends on our scale of observation. As we will see below, disturbances may have different effects on local, stand, and landscape-scale species richness. For this reason, statements about the relation of diversity to disturbance can be made only with explicit reference to scale.

We divide scale into two components, grain and extent (Allen and Hoekstra 1992, Palmer and White 1994). *Grain* is the size of a contiguous observation or unit of resolution (e.g., the size of a plot). *Extent* is the area over which observations (regardless of grain size) are made. At a given level of sampling effort, we can vary the grain and extent of our observations. For example, we can inventory $16\,m^2$ of the forest floor by sampling sixteen $1 \times 1\,m$ plots or one $4 \times 4\,m$ plot (Figure 5.7). In the former case, the quadrats are noncontiguous and thus cover a larger spatial extent. The latter sample will be at a single place within the overall study area; it will probably encompass a narrow range of environmental conditions and only a single successional age and will receive seed input only from nearby populations. For these reasons, the total number of species inventoried is likely to be lower with one $4 \times 4\,m$ than with 16 scattered $1 \times 1\,m$ plots. A comparable effect is likely when sample effort is measured in time: 100 hours of observation of bird populations is likely to produce different results if one hour is sampled every day for 100 days versus a sample of 100 hours of continuous observation over 4.16 days. The former sample will constitute a narrower range of daily activities for birds, but will sample seasonal variation of bird populations over one quarter of the year.

The species–area relation describes the increase in species richness with area examined. For species with large daily and seasonal changes, we could define a comparable species-time relation. In some cases, the number of species found is better expressed as a function of the number of individuals observed than of the area or duration of the observations. Species richness data may be described as a curve (i.e., the accumulation of species with area, time, or numbers of individuals); more often, it is reported as a single value at a certain summed sample effort (i.e., the number of species observed in a fixed sample area, time of observation, or number of individuals). Regardless of how species richness is reported, values are contingent on both the grain and extent of the observations. We will use the species-area curve for illustration.

As we examine larger areas we encounter more individuals, a greater range of environmental conditions, and a greater range of patch ages. For all these reasons, species richness increases with area sampled. The actual rate of species accumulation with area, however, is a function of the grain and extent of the observations (Figure 5.8). Let us take two extremes. If we sample a series of nested quadrats, species number will increase more slowly than if we sample a series of quadrats, all of the same grain size, across a sharp environmental gradient. The nested quadrats start at one point in space and, thus, are likely to sample a narrow range of environmental conditions, disturbance histories, and dispersing individuals. Each successive quadrat includes an area that has already been sampled. By contrast, a series of quadrats distributed along an environmental gradient is likely to sample a wider range of environmental conditions, disturbance histories, and dispersing individuals.

As a consequence of scale dependence, any observation of species richness is an arbitrary value that depends not only on explanatory variables such as environment and disturbance, but also on the grain and extent of

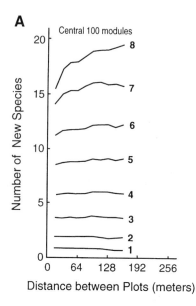

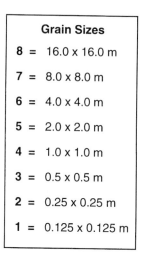

**A**

Central 100 modules

Number of New Species

| Grain Sizes |
|---|
| **8** = 16.0 x 16.0 m |
| **7** = 8.0 x 8.0 m |
| **6** = 4.0 x 4.0 m |
| **5** = 2.0 x 2.0 m |
| **4** = 1.0 x 1.0 m |
| **3** = 0.5 x 0.5 m |
| **2** = 0.25 x 0.25 m |
| **1** = 0.125 x 0.125 m |

Distance between Plots (meters)

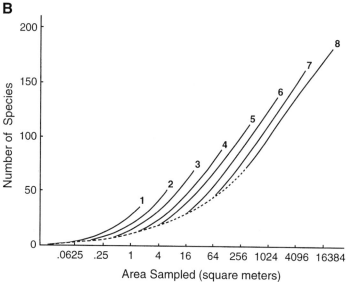

**B**

Number of Species

Area Sampled (square meters)

FIGURE 5.7. The number of species observed depends on the grain and extent of the observations in a 2.56 ha plot of forest in North Carolina (from Palmer and White 1994). The numbers 1–8 indicate the grain size of the quadrats used to construct the curves. (A) The distance between two quadrats affects how many species they collectively sample, but only at larger grain sizes (quadrats were a minimum of 16 m and a maximum of 160 m apart). (B) At any total area sampled, the number of species observed increases from large to small grain sizes. For example, the number of species observed on a total area of 16 m$^2$ more than doubles as one moves vertically on the graph from one 4 × 4 m quadrat to 64 0.5 × 0.5 m quadrats.

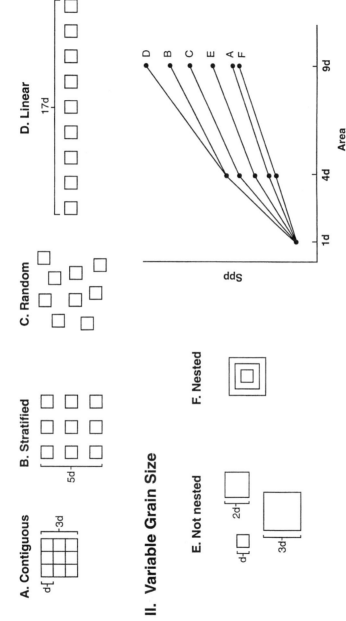

FIGURE 5.8.  A family of species-area curves can be produced depending on sampling design (and the grain and extent inherent in the design). The number of species that accumulates with area is slow when quadrats are clustered in space and therefore cover a low spatial extent (A and F) and fast when quadrats are arrayed to sample environmental gradients and therefore cover a large spatial extent (B and D).

the sampling window. Even when there is some justification for a single scale of observation, there is an important consequence of scale dependence: if we seek to detect change in time, we must compare two samples of the same grain and extent (i.e., the earlier methods must be repeated). More often, however, there is no justification for a single scale of observation. The underlying scale dependence may vary among samples because of the very effect (e.g., disturbance) we seek to investigate. This suggests that we should explicitly analyze the scale dependence itself. That is, we should not compare the number of species between two places or two times at a single scale of observation, but we should compare the curves that describe the rates of accumulation of species with grain and extent of observation.

Understanding scale dependence is important because disturbance will affect the way species richness is distributed. Species richness may change in different directions and different ways at different scales (Figure 5.9). For example, small-scale species richness may be higher in a canopy gap than in intact forest because light-demanding species colonize after the disturbance and share the gap with the shade-tolerant saplings not affected by the windstorm. At a larger scale, those same light-demanding species may have

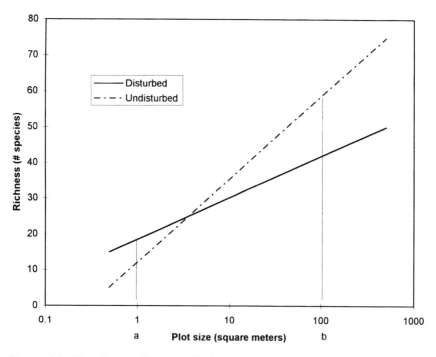

FIGURE 5.9. Disturbance effects are likely to be scale dependent. In this hypothetical example, studies conducted at different spatial scales would conclude that disturbance (a) increases or (b) decreases species richness.

already been present in older tree-fall gaps or in areas of permanently high sunlight (edges of cliffs and rivers). Thus, disturbance may increase diversity locally, but not at a landscape scale. In other cases, local diversity may drop with disturbance. For example, a shady, mixed-species forest may be replaced over large areas by an early successional forest dominated by a single species. The early successional stand may pass through a dense self-thinning phase in which few resources are available at the forest floor. This is often a time of very low species richness within the stand. This loss of diversity, however, may not be expressed at a larger scale if the landscape includes patches of all ages.

Scale dependence has implications for understanding the edge effect. Where communities of two growth forms (usually this is open field or meadow bordered by forest) join, local species richness is usually high. In addition, when an open area (permanent or successional) is produced adjacent to a forest stand, light increases within the forest edge, productivity of the understory increases, stem densities increase, and species richness usually increases. Wildlife species may respond to the increase of productivity, the greater availability of food resources, or the increase in structural diversity. Creation of edge can be a management goal for these reasons. However, the increase in species richness may be purely a local phenomenon, with no increase at larger scales (Figure 5.9). Further, some of the increase may be contributed by exotic species that have the potential to replace native species, and the fragmentation of undisturbed habitats may cause loss of forest interior species over time. In the latter case, a gain in species richness at the local scale may actually be accompanied by a loss of species richness at large scales (if the amount of edge relative to interior habitat changes greatly).

For all of these reasons, it is critical that we address the issue of species richness and disturbance at a series of scales. Ideally, we would address the underlying issue of scale dependence itself. The scale of disturbance (patch size, dispersion of patches in space) is likely to vary from one disturbance to another and from one landscape to another. Hence, we should approach the issue of species richness and disturbance carefully. In addition, other aspects of ecosystem structure exhibit scale dependence (Figure 5.10; Busing and White 1993).

## 5.3.2 Species Responses to Disturbance

No two species are precisely alike. They may differ in the breadth of their tolerance to environmental conditions, the gradient positions at which they reach optimum growth and reproduction, seasonality, competitive ability, colonizing ability, and life history. As a result, even species that occur together are unlikely to respond in precisely the same way to a given disturbance. Furthermore, individuals within a population may respond differently to a given disturbance depending on life stage. We could make

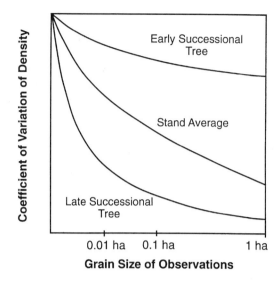

FIGURE 5.10. A hypothesis about scale-dependent variation in the structure of an old growth forest characterized by gap dynamics (after Busing and White 1993). Shade-intolerant trees reproduce only in scattered gaps, whereas shade-tolerant trees reproduce throughout the forest. As a result, quadrats of small grain size are likely to be less variable for shade-tolerant trees and more variable for shade-intolerant trees.

similar statements about competition and succession during intervals between disturbances: species niche differences will result in different performances over time. Ecologists have attempted to organize the variation of species responses to disturbance and succession in two ways: first, by examining life history strategies and, second, by examining the responses themselves.

Traits related to response to disturbance, competition, and the sudden availability of unexploited resources have always played a role in the classification of life-history strategies. The logistic equation of population growth was used to suggest an $r$–$K$ continuum in life-history traits (MacArthur and Wilson 1967). The $r$ end of this continuum is named from the population growth rate parameter, $r$, whose value dominates the rate of population growth when resources are not limiting. Traits at this end of the continuum include fast growth rates, early reproductive maturity, high reproductive output, high dispersal distances, and short life span. Species with these traits are good colonizers and require high resource levels for rapid growth and high reproductive output. These conditions are often found in recently disturbed patches. The $K$ end of the spectrum is named for the carrying capacity, $K$, the parameter that suggests limits to population density. At the $K$ end of the continuum, traits include slow growth rates, delayed reproductive maturity, low reproductive output, slow dispersal rates, and long life spans. Such species can compete successfully when resource levels are low; thus, they dominate undisturbed patches.

Grime (1979) proposed an alternative to the $r$–$K$ scheme that used two axes to classify plant strategies: disturbance and the severity of environmental conditions. He defined three extreme strategies, while also suggesting that intermediates may exist. The ruderal strategy would prevail where resources are high and disturbance frequent; the competitive strategy would prevail where resources are high and disturbance infrequent; and the tolerant strategy would prevail where resources were low and disturbance infrequent. A fourth combination of high disturbance and low resources was considered too unfavorable to support a strategy in this scheme. The first two strategies suggest the $r$–$K$ division. Grime's third strategy and his resources axis emphasize that environmental quality differs among sites, independent of disturbance regime.

In productive habitats, the $r$–$K$ continuum and Grime's scheme both predict successional turnover with time since disturbance. At a very general level, they are useful in predicting the life-history strategies that should dominate the early and late stages of succession. But these models are less useful in predicting the way that a particular species will respond to disturbance or the exact course of successional change. Species with intermediate life-history strategies are difficult to classify. In addition, species respond individualistically to disturbance; two long-lived, shade-tolerant trees may differ considerably in their susceptibility to fire or windthrow. Life-history strategies involve large numbers of trade-offs, including maximum growth rate versus shade tolerance, early versus delayed maturity, annual versus perennial life cycle, allocation of resources to growth versus reproduction, and many small versus a few large seeds. Variation occurs in several dimensions, and species may not fit neatly into classifications based on one or two axes. For these reasons, both the $r$–$K$ continuum and Grime's scheme are often difficult to apply.

A classification based on species' performance may be more useful than one based on life-history strategies. The labels "early," "mid-," and "late" successional are derived from the sequence of a species' appearance or its association with patches of a particular age. Vogl (1974) adopted a different approach, classifying species based on their response to disturbance. In Vogl's scheme, "increasers" are present before a disturbance and increase afterwards; "decreasers" are also present before the disturbance but decrease. "Invaders" are species that are initially absent but appear following disturbance, while "retreaters" are present before a disturbance but disappear. The abundance of "neutrals" is unaffected by disturbance. To these we add "integrators", species that range over patches of different successional age, like many grazers and omnivores. Younger patches often have the highest productivity and quality of food, while old patches have the best cover and den sites. One important aspect of Vogl's classification is that it is specific to particular disturbances; one species may be an increaser after windthrow but a decreaser or retreater after fire. Another is that the classification is sensitive to patch state prior to disturbance; whether a species is classified as an increaser or invader depends on its presence in the undisturbed patch.

The problem of scale dependence is implicit within Vogl's classification and would have to be addressed in any application of the scheme. First, the landscape must be able to supply colonists for those species that are invaders; this implies a landscape heterogeneity of patch states. Similarly, if retreaters are to be available for later invasion during succession (during disturbance-free times), they must persist somewhere else in the landscape. In a sense, then, Vogl's scheme makes assumptions about the patch dynamics at larger scales. Species that are invaders or retreaters at small scales would be classified as increasers or decreasers at large scales.

Within each of Vogl's categories, species may respond to disturbance by various mechanisms. For example, among species that are increasers in an eastern deciduous forest after wind disturbance, some plants may respond to increased resource levels by increasing growth rates, whereas other species may allocate the products of increased photosynthesis into increased flowering and seed set. Among invaders, some species may colonize from seeds produced during the current year, while other species may colonize from seeds long dormant in the soil.

In a competitive and resource-limited ecosystem, all species, including late successional species, may respond positively to disturbance. For example, if they survive disturbance, relatively slow-growing late successional trees show growth releases. While both early- and late-successional trees may be increasers in Vogl's scheme, there is variation in the absolute magnitude of response. Shade-tolerants generally do not respond as dramatically to increased resource levels as early successional species (White et al. 1985, Brokaw 1985).

Which species benefit most from disturbance may depend on the size and structure of the gap. In tropical rain forests, growth rates of late-successional trees increase with gap size along a gradient from small gaps to full sunlight. Early-successional trees cannot survive in intact forest or small gaps. However, their growth rates in larger openings greatly surpass the growth rates of late-successional trees (Figure 5.11). The smallest gaps are often filled by lateral expansion of crowns of adjacent canopy trees (Runkle 1985). Shade-tolerant trees form a seedling or sapling bank that is in place to exploit any gaps that form. Shade-tolerants may need several small-scale gaps events to reach the canopy. Early successional species are dependent on dispersing to and growing quickly in larger disturbance patches. When resources are abundant, their growth rates exceed those of late successional species, although both increase in response to increased resource abundance (Figure 5.11).

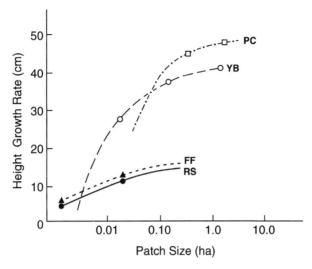

FIGURE 5.11. Height growth rates as a function of disturbance patch size for four species of varying shade tolerance in southern Appalachian spruce–fir forests (data from White et al. 1985). Height growth rates for the two shade-tolerant species, Fraser fir (FF; *Abies fraseri*) and red spruce (RS; *Picea rubens*) show the effects of the added light levels of gaps. however, yellow birch (YB; *Betula alleghaniensis*), an intolerant species, though virtually absent in the shade (in-the-shade life span average four years), reaches over twice the height growth rates of these species in forest gaps. Pin cherry (PC; *Prunus pensylvanica*) is absent from both the shady understory and forest gaps but has the highest height growth rates of all in large blow downs. Presumably, physiological trade-offs underlie these curves: species that excel when resources are abundant are not able to sustain growth when resources are limiting, and species that tolerate low resource levels are less able to exploit abundant resources.

    Species response to a given disturbance will be a function of the actual conditions created. In cool temperate forest gaps, soil moisture may be favorable for seedling establishment even in large openings. In these forests, both shade-tolerant and intolerant trees may respond positively to increased sunlight levels following disturbance. In tropical forests, soil moisture may be low in intact forest because of competition, may increase in small gaps with reduced competition, but may decrease again in large openings because of increased insolation, temperature, and air movement. Under these conditions shade-tolerant trees may increase in growth and density in gaps, but be unable to tolerate the doughtiness in large openings. Some early successional trees that are intolerant of low soil moisture may be found only in small gaps where they compete with established shade-tolerant species. Only those species tolerant of low soil moisture may survive in large openings.

## 5.3.3  Disturbance and Species Richness

Species vary individualistically in life-history traits and response to disturbance. Changing the disturbance regime may thus affect the presence and abundance of species in an ecosystem or landscape. In an ecosystem with an evolutionary history of disturbance, and in which some species are dependent on the conditions created by disturbance, eliminating disturbance will result in the decline and eventual extirpation of species. At the other extreme, increasing the rate of disturbance will eventually result in the loss of disturbance-sensitive species, including those dependent on older successional patches. The maximum species richness will occur between the extremes of no disturbance and very high frequency disturbance. These results arise, in part, because the scale of the landscape or study area is much larger than an individual patch. The landscape contains patches of many ages. While species appear and disappear from individual patches, they persist in the larger landscape because it always contains some patches of suitable habitat.

The "intermediate disturbance hypothesis" that species diversity is highest at some intermediate frequency and intensity of disturbance was first suggested by Connell (1978). Huston (1979) produced results similar to Connell's and proposed that species richness is highest at intermediate disturbance frequencies and population growth rates (Figure 5.12).

The intermediate disturbance hypothesis is difficult to apply because *intermediate* is difficult to define. The only clear expectation is that diversity drops at the end points of the axis. It is less clear how to test the hypothesis over the broad range of disturbance frequencies that are possible. If diversity is highest at either end (which would contradict the prediction), one can always assume that there exists a yet more extreme disturbance case. There is no reason to assume that a natural disturbance regime is the one that produces maximum species richness at a particular scale.

The intermediate disturbance hypothesis makes no explicit statement about landscape size relative to the size of disturbance. In a landscape context the issue of landscape size becomes critical. For all species to persist, the landscape must be large enough to ensure that all patch ages are present. For the landscape to retain species capable of responding to new disturbances and to retain disturbance-sensitive species, refuges for all species must always exist within dispersal distances. For plants, an ability to store dormant seeds in the soil lessens the need for immediate dispersal after the disturbance.

Finally, the intermediate disturbance hypothesis makes no mention of evolutionary constraint: presumably disturbance is needed in ecosystems in which prior evolutionary history of disturbance has resulted in disturbance-dependent species but in which some patches escape long enough to support disturbance-sensitive species. Disturbance-sensitive species would not exist in an ecosystem if the disturbance rate was so high that they could not

FIGURE 5.12. Two
theoretical models that
predict highest species
richness at intermediate
rates of disturbance.
(A) Connell's
intermediate
disturbance hypothesis
(Connell 1978). (B)
Huston's model of
species richness as a
function of disturbance
rate (rate of reduction)
and rate of
displacement (Huston
1979).

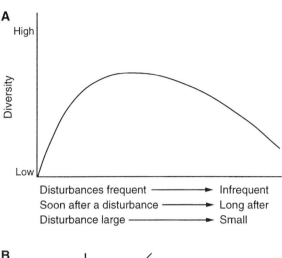

**A**

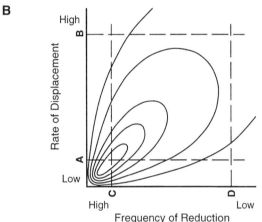

**B**

evolve or could not persist if they invaded from elsewhere. In this sense, disturbance maintains diversity that evolved because of a long history of disturbance. This argument suggests a natural, historic, and evolutionary basis for the relationship of diversity and disturbance.

Regardless of these criticisms, the basic point is that to maintain species with different adaptations, diverse environmental conditions are needed, including those created by disturbance. It should also be clear that late successional ecosystems (which may be stable or only very slowly changing in the intervals between disturbances) support only one component of the diversity of any region. Managing for diversity requires that there be a diversity of patch states.

While disturbance is essential for the maintenance of diversity, not all species benefit from a given disturbance; some may be driven to extirpation by too high a frequency or intensity of disturbance. These disturbance-sensitive species have persisted through time because landscapes were large relative to patch size, because individual disturbances had heterogeneous effects (e.g., a single fire has variable intensity in time and space), and because some parts of the landscape had very low disturbance frequencies (e.g., moist valleys may burn less often than drier sites). While disturbances occurred over much of the landscape on a regular basis, undisturbed "refuges" allowed the persistence of disturbance-sensitive species.

Because species have different adaptations to the conditions created by disturbance and succession, human alteration of the natural disturbance regime is a threat to biological diversity. Perhaps the two most manipulated natural processes in this regard are fire and hydrology. Reductions in fire frequency over the last 100 years have caused widespread changes in forest and landscape structure (Covington and Moore 1994, Kilgore and Taylor 1979). Changes to water levels and the natural dynamics of river channels through channelizing, damming, and draining have affected many major river systems. Restoration of fire and hydrologic fluctuation is a frequent focus of ecosystem management (Baker 1994, Dahm et al. 1995). Whenever humans monotonize the natural world (through suppression of disturbance or through creating large patches of uniform age), the result is low diversity. Only some components of the regional biota are adapted to such conditions; other species decline or are extirpated.

Another major human influence that affects diversity is the alteration of landscape scale through habitat loss and fragmentation. The reduction in size of natural areas may mean that not all successional ages are present within easy dispersal distance of newly disturbed sites (Pickett and Thompson 1978). The alteration of spatial context may thus impair response to disturbance itself. Further, human alteration of the landscape matrix may thus affect the propagation of disturbances across landscapes. The altered size and spatial context of natural areas represents an important challenge for the management of biological diversity in the future.

## 5.3.4 Deterministic and Stochastic Responses to Disturbance

Plant community response to disturbance can be deterministic or stochastic. Deterministic responses result from the adaptations of species to the conditions created by disturbance. These responses are predictable from a knowledge of the available species and the effects of the disturbance. When several alternative outcomes are possible, each described by a probability, the response is stochastic. If probabilities are relatively even among the possible responses, the response in a particular place may be unpredictable. If stochastic responses dominate, disturbance will tend to increase the role of chance and, hence, the amount of unexplained variation in composition.

For example, a spring freeze that damages early-blooming flowers may influence the species available to colonize patches disturbed the following fall.

The deterministic model derives from the observation that species have different adaptations to the physical conditions created by disturbance and have different dispersal abilities. Species therefore vary in abundance among disturbed patches and during succession. If we know enough about the site, the disturbance event, and species characteristics, we can predict the response to disturbance. Furthermore, the results can be repeated: a given disturbance will produce the same responses at different locations in different years. The deterministic model suggests niche partitioning with regard to the range of conditions created by disturbance. Species should be found in predictable locations along gradients within disturbed areas and at predictable times during succession.

The stochastic model proposes that after disturbance, probabilities come into play. Opportunistic species adapted to high resource levels engage in scramble competition. The first species to arrive may be able to establish and hold that site against further invasion. The outcome may be sensitive to the seasonal timing of the disturbance, weather, or year-to-year variation in seed production by a particular species. Because we cannot know these variables, we cannot predict the response to disturbance.

The simplest case of the stochastic model is one in which the species are all equivalent in their response to environmental factors. In this case, there is no niche partitioning. Species can occur anywhere; actual occurrences are governed by chance events. In the more complex case, species differ in their optimum environmental conditions but overlap broadly in their tolerances. Niche differences are not sufficient to explain species distributions, and chance events control which species dominate a particular site early in succession.

Increased species richness in gaps compared to the forest interior may not reflect niche partitioning per se. When disturbance results in increased stem density (e.g., the density of saplings in gaps in many times higher than that in the shady forest), the increased number of individuals alone may result in an increase in richness. Gaps and shady forest may have the same curves of species number as a function of the number of individuals observed. On the other hand, if species in gaps are specially adapted to gap conditions or partition the environment within the gap (e.g., some species in the centers of the gaps, others exploiting the intermediate environment of gap edges), then species richness curves should be different in gaps than in the shady forest. In an old growth forest with which we are familiar (Busing and White, 1997), species richness is explained mostly by the increase in stem density (there was one gap-dependent specialist), but this study covered only relatively small-scale disturbances. Niche partitioning may be more evident when larger gaps with their higher resource levels and greater environmental extremes are included.

It has also been argued that competitive sorting of initially random species will occur over successional time as the best competitors dominate each environmental setting (Peet 1992). The slow and deterministic process of competition during succession and the potential importance of chance events immediately after disturbance combine to suggest that predictability of composition may increase during succession, though this increase may not be monotonic (Christensen and Peet 1984).

## 5.4 Summary: Disturbance and Diversity in a Landscape Context

The landscape perspective is fundamental to understanding the role of disturbance. Disturbance regimes vary across landscapes as a function of topography and substrate. Even in ecosystems with relatively homogeneous topographic settings, disturbance can produce heterogeneity by creating a range of patch ages and states. The spatial scale of this heterogeneity is important to understanding the role of disturbance in maintaining diversity.

The "shifting mosaic" (Heinselman 1973) was an early description of the distribution and dynamics of disturbance patches across a landscape. The particular patches disturbed each year shift in space, but all patch ages are always present. Under ideal conditions (a very large landscape and a constant disturbance regime), the percentage of the landscape in each successional state would be expected to remain constant through time. While individual patches would undergo disturbance and successional change, the landscape as a whole would remain in compositional equilibrium. In real-world systems, year-to-year climate variation (e.g., hurricanes, droughts) and the ability of contagious disturbances (e.g., fires, insect outbreaks) to spread over large areas may cause systems to deviate from equilibrium. Even very large landscapes may show considerable fluctuations in composition (Romme and Despain 1989, Baker 1989).

Patch dynamics is the subject of another chapter in this book and we will not discuss this issue in detail here. However, the nature of the patch dynamics in a landscape is important to diversity (Pickett and Thompson 1978, White 1987, Baker 1992b). If all patches and patch states persist in a landscape, then the persistence of species dependent on different patch states is also likely. We hypothesize that the landscape need not, strictly speaking, be in equilibrium; species and communities may persist in landscapes characterized by bounded variation rather than equilibrium, as long as sufficient areas of all patch types are present and distances between refuge populations and appropriate sites do not limit dispersal (cf. Turner et al. 1993, Baker 1989).

Disturbances may produce heterogeneity within a single patch (e.g., the hot and cool spots within a fire or the pit and mound topography that creates islands of mineral soil in forests) and create a variety of ecological opportunities for differently adapted species. Even when disturbances act to homogenize small patches, they may increase heterogeneity in the larger landscape and provide opportunities for species with different traits.

By contrast, spatial homogeneity and temporal monotony (even if these result form ongoing natural disturbances) may reduce diversity because only a component of the regional biota will be able to persist. Homogeneity also affects the size and intensity of subsequent disturbances, as when a uniform forest of one age class supports large pest outbreaks or widespread and heavy fuel loads support large and intense fires. If some patch types are absent or become rare in the landscape, the species found in those patch types may no longer be able to colonize patches that become suitable. This would further destabilize landscapes.

We define *resilience* as the persistence of native species and communities in the face of local fluctuation. Such persistence requires a variety of environmental conditions to allow spatial shifts in populations and create gradients of disturbance frequency and intensity. It also requires enough patches and patch states to provide opportunities during favorable times and refuges during unfavorable ones. For both these reasons, we expect larger landscapes and natural areas to be more resilient.

Humans have not only changed disturbance regimes, but also have imposed a new scale on ecosystem processes through habitat loss and fragmentation. Natural ecosystems usually have become smaller and their surroundings often have been altered. The reduced range of patch types in the diminished system may lead to the disappearance of some species. How do we manage for disturbances when we cannot allow the events of natural magnitude and intensity to occur? In such cases, we may need to manage individual species populations for persistence.

Disturbances are not equivalent in their effects. Ecosystems differ in their predisturbance resource levels and in the effect that a given disturbance has on those resource levels. Any general understanding of disturbance effects must include variables such as the amount and percentage of biomass removed, the amount and percentage change in resource levels, and the amounts and kinds of legacies that persist from the predisturbance ecosystem.

Since we must start with the assumption that species responses are individualistic and may involve stochasticities, we must be prepared to investigate the mechanisms of response. We may find that we can group species into response categories and we may find that species traits (e.g., life-history characteristics) help us to establish such groups. We should always begin by considering such classifications as hypotheses to be tested, rather than established facts.

Historical disturbance regimes and dynamics of species and communities should be addressed in management plans. No one management action will benefit all species. Species richness is likely to be a function of the mosaic of conditions that are produced. Disturbance is a prerequisite for diversity, but we must be careful about the scale at which we analyze this effect. The scale-dependent nature of species richness and the complex effects of disturbance require that we address questions at multiple spatial and temporal scales.

*5.5 Acknowledgements.* We would like to thank the following colleagues for stimulating discussions of these topics: J. Betancourt, P. Landres, R. Peet, W. Romme, N. Stephenson, T. Swetnam, and J. Walker.

## 5.6 *References*

Allen, T.F.H., and T.W. Hoekstra. 1992. Toward a unified ecology. Columbia University Press, New York, New York, USA.

Baker, W.L. 1989. Landscape ecology and nature reserve design in the Boundary Waters Canoe Area, Minnesota. Ecology 70:23–35.

Baker, W.L. 1992a. Effects of settlement and fire suppression on landscape structure. Ecology 73:1879–1887.

Baker, W.L. 1992b. The landscape ecology of large disturbances in the design and management of nature reserves. Landscape Ecology 7:181–194.

Baker, W.L. 1994. Restoration of landscape structure altered by fire suppression. Conservation Biology 8:763–769.

Brokaw, N.V.L. 1985. Treefalls, regrowth, and community structure in tropical forests. Pp. 53–71 in S.T.A. Pickett and P.S. White, editors. The ecology of natural disturbance and patch dynamics. Academic Press, New York, New York, USA.

Busing, R.T., and P.S. White. 1993. Effects of area on old-growth forest attributes: implications for the equilibrium landscape concept. Landscape Ecology 8:119–126.

Busing, R.T., and P.S. White. 1997. Species diversity and small-scale disturbance in an old-growth temperate forest: a consideration of gap partitioning concepts. Oikos 78:562–568.

Canham, C.D., J.S. Denslow, W.J. Platt, J.R. Runkle, T.A. Spies, and P.S. White. 1990. Light regimes beneath closed canopies and treefall gaps in temperate nd tropical forests. Canandian Journal of Forest Research 20:620–631.

Christensen, N.L., and R.K. Peet. 1984. Convergence during secondary succession. Journal of Ecology 72:25–36.

Churchill, E.D., and H.C. Hanson. 1958. The concept of climax in arctic and alpine vegetation. Botanical Reviews 24:127–191.

Collins, S.L. 1987. Interaction of disturbances in tallgrass prairie: A field experiment. Ecology 68:1243–1250.

Connell, J.C. 1978. Diversity in tropical rain forests and coral reefs. Science 199:1302–1310.

Coppock, D.L., J.E. Ellis, J.K. Detling, and M.I. Dyer. 1983. Plant–herbivore interactions in a North American mixed-grass prairie II: Responses of bison to modification of vegetation by prairie dogs. Oecologia 56:10–15.

Covington, W.W., and M.M. Moore, 1994. Southwestern ponderosa pine forest structure: changes since Euro-American settlement. Journal of Forestry 92:39–47.

Dahm, C.N., K.W. Cummins, H.M. Valett, and R.L. Coleman. 1995. An ecosystem view of the restoration of the Kissimmee River. Restoration Ecology 3:225–238.

Forcier, L.K. 1975. Reproductive strategies and the co-occurrence of climax tree species. Science 189:808–810.

Franklin, J. 1989. Towards a new forestry. American Forests, November/December:37–44.

Franklin, J.F., and R.T.T. Forman. 1987. Creating landscape patterns by forest cutting: ecological consequences and principles. Landscape Ecology 1:5–18.

Geiszler, D.R., R.I. Gara, C.H. Driver, V.F. Gallucci, and R.E. Martin. 1980. Fire, fungi, and beetle influences on a lodgepole pine ecosystem of south-central Oregon. Oecologia 46:239–243.

Grime, J.P. 1979. Plant strategies and vegetation processes. John Wiley and Sons, New York, New York, USA.

Harmon, M.E., S.P. Bratton, and P.S. White, 1983. Disturbance and vegetation response in relation to environmental gradients in the Great Smoky Mountains. Vegetatio 55:129–139.

Heinselman, M.L. 1973. Fire in the virgin forests of the Boundary Waters Canoe Area, Minnesota. Quaternary Research 3:329–382.

Hobbs, N.T., D.S. Schimel, C.O. Owensby, and D.S. Ojima. 1991. Fire and grazing in the tallgrass prairie: contingent effects on nitrogen budgets. Ecology 72:1374–1382.

Hunter, M.L. 1993. Natural fire regimes as spatial models for managed boreal forests. Biological Conservation 65:115–120.

Huston, M. 1979. A general hypothesis of species diversity. The American Naturalist 113:81–101.

Kilgore, B.M., and D. Taylor. 1979. Fire history of a seqoia–mixed conifer forest. Ecology 60:129–142.

Landres, P.B. 1992. Temporal scale perspectives in managing biological diversity. Transactions of the North American Wildlife and Natural Resources Conference 57:292–307.

Lang, G.E. 1985. Forest turnover and the dynamics of bole wood litter in subalpine balsam fir forest. Canadian Journal of Forest Research 15:262–268.

MacArthur, R.H., and E.O. Wilson. 1967. The theory of island biogeography. Princeton University Press, Princeton, New Jersey, USA.

Marks, P.L. 1974. The role of pin cherry (*Prunus pensylvanica* L.) in the maintenance of stability in northern hardwood ecosystems. Ecological Monographs 44:73–88.

Matlack, G.R., S.K. Gleeson, and R.E. Good. 1993. Treefall in a mixed oak–pine coastal plain forest: Immediate and historical causation. Ecology 74:1559–1566.

Minnich, R.A. 1983. Fire mosaics in southern California and northern Baja California. Science 219:1287–1294.

Morgan, P., G.H. Aplet, J.B. Haufler, H.C. Humfries, M.M. Moore, and W.D. Wilson. 1994. Historical range of variability: A useful tool for evaluating ecosystem change. Pp. 87–111 in R.N. Sampson and D.L. Adams, editors. Assessing forest ecosystem health in the inland West. The Haworth Press, Bimington, New York, USA.

Noss, R.F. 1990. Indicators for monitoring biodiversity: a hierarchical approach. Conservation Biology 4:355–364.

Oliver, C.D. 1980. Forest development in North America following major disturbances. Forest Ecology and Management 3:157–168.

Palmer, M.W., and P.S. White. 1994. Scale dependence and the species–area relationship. The American Naturalist 144:717–740.

Peet, R.K. 1992. Community structure and ecosystem function. Pp. 103–151 in D.C. Glenn-Lewin, R.K. Peet, and T.T. Veblen, editors. Plant succession: theory and prediction. Chapman and Hall, London, UK.

Peet, R.K., and N.L. Christensen. 1987. Competition and tree death. Bioscience 37:586–595.

Pickett, S.T.A., and J.N. Thompson. 1978. Patch dynamics and the design of nature reserves. Biological Conservationist 13:27–37.

Pickett, S.T.A., and P.S. White, editors. 1985. The ecology of natural disturbance and patch dynamics. Academic Press. New York, New York, USA.

Romme, W.H., and D.G. Despain. 1989. Historical perspective on the Yellowstone fires of 1988. Bioscience 39:695–699.

Runkle, J.R. 1985. Disturbance regimes in temperate forests. Pp. 17–34 in S.T.A. Pickett and P.S. White, editors. The ecology of natural disturbance and patch dynamics. Academic Press. New York, New York, USA.

Schowalter, T.D. 1985. Adaptations of insects to disturbance. Pp. 235–252 in S.T.A. Pickett and P.S. White, editors. The ecology of natural disturbance and patch dynamics. Academic Press, New York, New York, USA.

Swanson, F.J., and J.F. Franklin. 1992. New Forestry principles from ecosystem analysis of Pacific Northwest forests. Ecological Applications 2:262–274.

Swetnam, T.W. 1993. Fire history and climate change in giant sequioa groves. Science 262:885–889.

Swetnam, T.W., and J.L. Betancourt. 1990. Fire–southern oscillation relations in the southwestern United States. Science 249:1017–1020.

Thomspon, J.N. 1978. Within-patch structure and dynamics in *Pastinaca sativa* and resource availability to a specialized herbivore. Ecology 59:443–448.

Turner, M.G. 1989. Landscape ecolgoy: the effect of pattern on process. Annual Review of Ecology and Systematics 20:171–197.

Turner, M.G., R.H. Gardner, V.H. Dale, and R.V. O'Neill. 1989. Predicting the spread of disturbance across heterogeneous landscapes. Oikos 55:121–129.

Turner, M.G., W.H. Romme, R.H. Gardner, R.V. O'Neill, and T.K. Kratz. 1993. A revised concept of landscape equilibrium: disturbance and stability on scaled landscapes. Landscape Ecology 8:213–227.

Vitousek, P.M. 1984. A general theory of forest nutrient dynamics. Pp. 121–135 in G.I. Agren, editor. State and change of forest ecosystems—indicators in current research. Swedish University of Agricultural Science, Report Number 13, Uppsala, Sweden.

Vogl, R.J. 1974. Effects of fire on grasslands. Pp. 139–194 in T.T. Kozlowski and C.E. Ahlgren, editors. Fire and ecosystems. Academic Press. New York, New York, USA.

White, P.S. 1987. Natural disturbance, patch dynamics, and landscape pattern in natural areas. Natural Areas Journal 7:14–22.

White, P.S. 1979. Pattern, process, and natural disturbance in vegetation. Botanical Reviews 45:229–299.

White, P.S., M.D. MacKenzie, and R.T. Busing. 1985. A critique of overstory/understory comparisons based on transition probability analysis of an old growth spruce–fir stand in the Appalachians. Vegetatio 64:37–45.

White, P.S., and S.T.A. Pickett. 1985. Natural disturbance and patch dynamics, an introduction. Pp. 3–13 in S.T.A. Pickett and P.S. White, editors. The ecology of natural disturbance and patch dynamics. Academic Press. New York, New York, USA.

Wright, K.A., L.M. Chapman, and T.M. Jimerson. 1995. Using historic range of vegetation variability to develop desired conditions and model forest plant alternatives. Pp. 258–266 in Analysis in support of ecosystem management: analysis workshop III. U.S. Department of Agriculture Forest Service, Ecosystem Management Analysis Center, Washington, DC, USA.

# 6
# Populations in a Landscape Context: Sources, Sinks, and Metapopulations

MARK E. RITCHIE

## 6.1 Introduction

Traditionally, a population is considered a single, well-mixed collection of individuals in an arbitrarily defined space (Elton 1927). From this view, ecologists have sought detailed understanding of dynamics by measuring the survival and fecundity of different-sized or aged individuals. However, this approach ignores the explicit spatial distribution of individuals and the resources to which they respond. Many recent studies demonstrate that populations can exhibit well-defined distributions, or "structure," in space, and this spatial structure may have considerable influence on population dynamics (Pulliam 1988; Kareiva 1990, Pulliam and Danielson 1991, Kareiva and Wennergren 1995). Landscape features, i.e., spatial distribution of habitats, barriers to movement, etc., are likely to affect strongly the spatial structure of populations. In this chapter, I develop a simple mathematical theory to generate predictions of how landscape pattern can have large consequences for species' population dynamics and hence the management of those species. These predictions suggest an expanded set of research questions that encourage the measurement of different population characteristics and the use of different criteria for management.

Mark E. Ritchie is an Assistant Professor in the Department of Fisheries and Wildlife and Ecology Center at Utah State University. His specialty is population ecology and wildlife management. He teaches undergraduate courses in general ecology, population ecology, and field techniques, as well as·a graduate course in foraging and habitat selection. Ritchie's research focuses on plant–herbivore interactions and the application of chaos and fractal theory to behavior, population dynamics, and community ecology. He recently has applied ecological theory to help develop a new recovery plan for the Utah prairie dog and multiple herbivore production models for holistic grazing systems in western rangelands.

## 6.2  Spatial Structure in Populations

Ecologists typically treat a population as a single unit, defined as a collection of individuals in an arbitrarily defined space. For example, we can define a population of pronghorn antelope (*Antilocaprus americana*) as the number of antelope in the Rocky Mountains, in Wyoming, or in a particular mountain valley. This arbitrary definition assumes that the population is "panmictic," i.e., represented by the *average* density or total number of individuals divided by the size of the arbitrary area. If the population is spatially structured, i.e., individuals are clustered in some locations and absent from others, the assumption of a mixed population still applies if differences in "local" densities are responses to spatial differences in habitat. For example, a total population of size $N$ might be distributed unequally across the landscape. If individuals distribute themselves freely according to the relative quality of different habitats (i.e., an ideal free distribution, Fretwell 1972), a proportional reduction in $N$ will yield the same proportional reduction in each habitat. Thus population dynamics across the entire landscape can be understood from the average dynamics at a particular location.

Spatial complications in population dynamics were first recognized in the 1950s and 1960s via "island biogeography theory" (MacArthur and Wilson 1967) and "metapopulation dynamics" (Andrewartha and Birch 1954, Levins 1969, 1970). These seminal papers fostered the idea that population dynamics depend on spatial variability in habitat, the explicit spatial location of some habitats relative to others, and the relative size of habitat patches. Although these ideas have been applied mostly to explain patterns in species diversity across landscapes, they form the foundation for understanding population dynamics in a landscape context.

Many ecologists have used simulation models to explore the consequences of local population size, local dynamics, and explicit spatial locations for population dynamics (Wiens et al. 1993, Day and Possingham 1995, Kareiva and Wennergren 1995). These simulations suggest that population dynamics in whole landscapes cannot be understood by averaging dynamics at local scales; "nonaveraging" population dynamics arise when population changes at one locality (e.g., outbreaks, extinction) are at least somewhat independent of changes at another. In this case, population dynamics may not respond to average density, but instead to densities and conditions at particular localities. Local populations may be small enough that they experience significant risks of stochastic extinction (Goodman 1987, Lande 1993). If so, the same number of individuals aggregated into a single locality may show very different dynamics than if dispersed into many localities. In addition, the spatial arrangement of local populations can influence dynamics if separation or isolation of localities influences dispersal and colonization rates. Landscape features define the distribution of habitats and individuals (Wiens et al. 1993), so population dynamics viewed in a landscape context must consider the explicit distribution of individuals in space and interactions among populations from different localities.

Reviews of field studies (Harrison 1991, 1994, Pulliam et al. 1992) suggest two major types of spatially explicit population dynamics: source-sink and metapopulation. Source-sink dynamics (Pulliam 1988, Lomnicki 1988, Pulliam and Danielson 1991) refer to landscapes with two or more habitats, where birth rates exceed death rates in one or more favorable "source" habitats, but death rates exceed birth rates in one or more remaining "sink" habitats. Sink habitats are occupied because individuals in source habitats control access to resources and thus force less dominant individuals into sink habitats. Metapopulation dynamics, on the other hand, occur when local populations are isolated sufficiently to produce low colonization rates. Dynamics result from changes within local populations, including periodic extinction, coupled with intermittent colonization from other localities (Levins 1969, Quinn and Hastings 1987, Harrison 1991, Kareiva and Wennergren 1995). Metapopulations grade into source-sink populations if certain local populations consistently sustain net positive growth rates, have little risk of extinction, and produce colonists to other localities with net negative growth rates and/or high risks of random extinction (Pulliam 1988, Schoener and Spiller 1987, Harrison 1991).

Landscape features such as habitat isolation and heterogeneity are likely to be key elements in spatially structuring populations and in determining whether populations exhibit source-sink vs. metapopulation dynamics. The density, degree of aggregation, and connectivity of suitable habitat are likely to control colonization rates, and the size of remnant habitat patches will likely control the number of dispersers and the risk of local extinction (Bascompte and Solé 1996, Hess 1996). Ecologists are just beginning to explore these relationships, and there are few accepted general hypotheses. However, Kareiva and Wennergren (1995) recently proposed two principles that emerge from simulation models which explicitly account for the location of organisms in space (e.g., Hassell et al. 1991, Dytham 1995): (1) spatially structured populations may live with a threshold requirement for habitat, below which they face extinction even though some habitat remains; (2) the details of how habitats are arranged and how organisms move between them control whether species persist or go extinct, or whether populations fluctuate wildly or remain stable.

These principles can be viewed as the beginnings of a predictive theory for connecting landscape pattern to population dynamics. However, they clearly suggest room for more specific predictions, particularly with regard to the effects of details in landscape patterns. In this chapter, I promote these principles of landscape effects as ideas to be explored. I develop some simple models that consider the separate effects of density and configuration of habitat on population sizes and their likely persistence. I explore these models' predictions as they apply to population management for conservation (Wennergren et al. 1995) and/or harvesting (McCullough 1996). Field data for testing these ideas is still scarce, and has been reviewed in detail elsewhere (Harrison 1991, 1994, Pulliam et al. 1992). However, I provide selective evidence for patterns predicted by these simple models to show that they may be useful in thinking about the consequences of spatial structure for population dynamics.

## 6.3 Population Dynamics in Complex Landscapes: A Simple Approach

To begin to explore the effects of landscape pattern, one must precisely define *pattern*. For population dynamics, a simple definition is the amount and spatial configuration of "suitable" habitat for a given species. This definition dichotomizes arrays of different vegetation cover types, ponds, streams, etc. on a landscape into "habitat" and "nonhabitat," and simplifies the analysis of landscape effects. With this classification, the density and pattern of habitat on a landscape can very often be described through fractal geometry (Milne 1991, Hastings and Sugihara 1994). The amount of habitat ($H$) in a landscape can be related to the extent, or size, of the landscape by the simple power relation

$$H = vx^{F}, \tag{1}$$

where $x$ is the extent, $v$ is the proportion of landscape occupied by habitat (habitat density), and $F$ is the fractal dimension (configuration) of habitat on the landscape (see Milne, this volume). $F$ can vary between 0 (a single point) and 3 (habitat dispersed throughout a cube) to represent different landscape types (Figure 6.1). For simplicity, I will consider $F$ or represent a

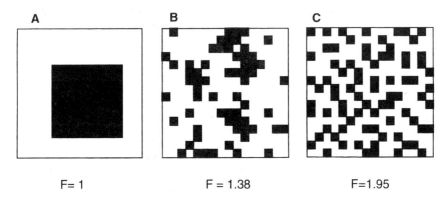

A          B          C

F= 1          F = 1.38          F=1.95

FIGURE 6.1. Hypothetical hand-drawn landscapes with different fractal dimensions of habitat (black areas) within a two-dimensional landscape. For a given habitat density of 35% of the landscape, habitat distributions range from a single patch (A: $F = 1$) to fragmented (B: $F = 1.38$), to fully dispersed (C: $F = 1.95$). Thus, the fractal dimension of the habitat describes a full range of habitat configurations.

single habitat patch ($F = 1$) to a random scattering of mostly small patches in a two-dimensional landscape ($F = 2$). The parameters $v$ and $F$ can describe a wide variety of landscapes, with $F$ determining the degree of aggregation of habitat patches (Milne 1992, Palmer 1992, Ritchie, in press). The landscape extent, $x$, represents the length of one side of the map that represents the landscape and is measured in units of the scale of resolution, such that $x = 1$ is the scale of resolution of information about the landscape (i.e., the length of one landscape unit). For example, a square map composed of pixels would have an extent equal to the number of pixels along one side.

Fractal geometry allows the central features of a landscape pattern to be incorporated into familiar population models. It also incorporates decisions of scale explicitly into such models, and yields solutions in which the separate effects of habitat density, $v$, and habitat configuration, $F$, can be considered. Although landscapes may not be perfectly fractal (i.e., self-similar across scales), many appear to be approximately so (Hastings and Sugihara 1994, Milne et al. 1996, Milne, this volume).

I use fractal geometry to incorporate landscape pattern explicitly into basic models of source-sink (Pulliam 1988) and metapopulation dynamics (Levins 1969). The effects of landscape pattern can be predicted relative to the already well-known behavior of these models. With the incorporation of landscape extent, the fractal approach also explores the effect of scale of management, and as it turns out, the scale at which a species responds to the environment. The effects of scale-dependent ecosystem patterns and processes are only beginning to be explicitly considered (Holling 1992), and have not been addressed in studies of population dynamics. This approach also explicitly incorporates size structure; in fact, the power relation describing landscape pattern implicitly defines the frequencies of different patch sizes (see Boxes 6.1 and 6.2) and thus local population sizes. Such size structure may characterize the majority of spatially structured populations (Harrison 1991, 1994, Hanski 1994) and is important in obtaining realistic estimates of colonization and extinction rates.

## 6.3.1 Source-Sink Dynamics and Landscape Pattern

In a source habitat, Pulliam (1988) envisioned that $V$ resources exist and each individual occupying the habitat controls $c$ resources. Thus, a fixed number $\frac{V}{c}$ of individuals each produce offspring at a net rate $\beta j$, where $\beta$ is the production rate of offspring and $j$ is juvenile survivorship in the sink habitat, where offspring are driven by the lack of breeding territories in source habitat. If adults in either habitat survive at a rate $a$, then the total number of individuals in the following generation (from both habitats) is

$$N_{t+1} = aN_t + \beta j \frac{V}{c}. \tag{2}$$

## Box 6.1.  Patch Size and Local Extinction in Source—Sink Dynamics.

If patches of habitat retain the pattern of the landscape, then the size of a patch with linear length $w$ is $bw^F$, where increasing fractal dimension $F$ indicates increasing dispersion of habitat into more smaller patches and $b$ is a proportionality constant. If population size within the patch is proportional to patch size, then the expected extinction rate for that patch is

$$e(w) = \frac{q}{bw^F} \qquad (B1.1)$$

where $q$ is the probability of extinction due to demographic stochasticity in the smallest size patches ($w = 1$). The average extinction rate for all patches is

$$e(F) = \frac{q}{\int_z^m bw^F w^{-F} dw} \qquad (B1.2)$$

where $w^{-F}$ is the frequency of each size patch (Barnsley 1988, Milne 1991, 1992, Ritchie, in press), $m$ is the length of the maximum patch size in the landscape. Evaluating the integral yields

$$e(F) = \frac{q}{b(m-1)} \qquad (B1.3)$$

Assuming that the largest patch must have a frequency equal to the ratio of its size to landscape area,

$$m = x^{\frac{F}{3F-1}} \qquad (B1.4)$$

(Ritchie in press), then

$$e(F) = \frac{q}{b\left(x^{\frac{2F}{3F-1}} - 1\right)}. \qquad (B1.5)$$

Suppose a source-sink population exists where patches of source habitat may go extinct at a rate $e(F)$ (eq. 3 in the text). Substituting eq. B1.5 for $e(F)$ and solving for population size at equilibrium ($N^*$) yields

$$N* = \frac{j\beta V}{(1-a)c}\left(1 - \frac{q}{b\left(x^{\frac{F}{3F-1}} - 1\right)}\right) \qquad \text{(B1.6)}$$

Because the ratio $F/(3F-1)$ decreases as $F$ gets larger, $N*$ will decline as $F$ increases. However, for reasonable landscape extent values ($x > 10$) and $q < 1$, the reduction in $N*$ from dispersion of habitat into smaller patches (i.e., as $F$ increases) will be small (see Figure 6.2).

## Box 6.2. Metapopulation Dynamics as a Function of Habitat Fragmentation

To assess metapopulation dynamics in response to both habitat density and configuration, the functions $c(F)$ and $e(F)$ must be specified in more detail. To specify colonization rate, suppose a colonist disperses from its patch of origin and searches for a new habitat patch. Since the habitat in which it searches is presumably inhospitable, assume that it has some expected amount of time before it succumbs to starvation, predation, etc. Suppose a species has a minimum grain or scale of resolution, $z$, at which it detects new habitat patches. If $z$ is measured in the same units as landscape features, then the minimum suitable size patch selected by colonists is size $az^F$ (Ritchie, in press), where $a$ is a proportionality constant. Furthermore, if the colonist can move a distance $kz$ in this fixed time, then it will encounter new habitat patches of at least size $za^F$ at a rate $(\frac{k}{a})vz^{1-F}$ (Ritchie, in press). This assumes it will encounter different-sized patches according to their relative frequency in the environment. For a patch of a given size, the probability it will be colonized is

$$c(F) = \frac{kvz^{1-F}}{a} \int_z^m w^{-F} dw, \qquad \text{(B2.1)}$$

where $w^{-F} dw$ is the frequency of patches of exactly size $aw^F$ (Ritchie, in press). Evaluating the integral yields

$$c(F) = \frac{kvz^{1-F}}{a(F-1)}\left(z^{1-F} - m^{1-F}\right). \qquad \text{(B2.2)}$$

In a recent paper, Ritchie (in press) shows that a forager maximizing its resource intake rate should select $z = m(\frac{3-F}{4-F})$. Using the solution for $m$ from Box 6.1 and substituting for $z$ and $m$ in equation B2.2 yields a complicated-looking but rather simple nonlinear function for colonization rate that declines with increasing habitat dispersion ($F$) (see Figure 6.3)

$$c(F) = \frac{kv}{a}\left(\frac{3-F}{4-F}\right)^{1-F} x^{\frac{2F(1-F)}{3F-1}} \frac{1}{F-1}\left[\left(\frac{3-F}{4-F}\right)^{1-F} - 1\right].$$  (B2.3)

To find extinction rate, the approach used in Box 6.1 is useful, except that the population is now assumed to use only patches above the size threshold $az^F$. Consequently,

$$e(F) = \frac{q}{b(m-z)}.$$  (B2.4)

and substituting for $m$ and $z$ yields

$$e(F) = \frac{q(4-F)}{bx^{\frac{F}{3F-1}}}.$$  (B2.5)

The solutions for $c(F)$ and $e(F)$ equation imply that both rates are functions of the speed or dispersal ability of a species (as reflected by the parameter $k$), habitat configuration (as reflected by fractal dimension $F$), and landscape extent $x$. Extinction rate is also a function of the inherent annual extinction risks faced by a local population in a small patch $q$, and colonization rate is a function of habitat density, $v$.

The equilibrium proportion of "suitable" habitat patches occupied (patches smaller than $az^F$ will never be occupied) $p*$ is then determined from eq. 6 in the text. As seen in Figure 6.2, $p*$ declines with increasing $F$ and decreasing $v$, such that for a species with poor dispersal ability (small $k$), the metapopulation will go extinct even when some habitat remains (Figures 6.3 and 6.4).

In this basic model, individuals are assumed to disperse freely among habitats, and demography in all source habitat patches is assumed to be the same. Landscape pattern therefore matters little, except to the extent that it determines the amount of source habitat and thus the total amount of resources $V$.

Suppose, however, that source habitat is fragmented into patches, and that some patches are sufficiently small that demographic stochasticity (MacArthur and Wilson 1967, Goodman 1987, Lande 1993) within these patches produces local extinction. Given that local extinction rates are likely to be an inverse function of local population (and thus patch) size (Goodman 1987, Lande 1993, Hanski 1994), and patch size is a function of the landscape's fractal dimension, $F$, the dynamics of the population can be rewritten as

$$N_{t+1} = aN_t + \beta j\left(1 - e(F)\right)\frac{V}{c}. \tag{3}$$

This model assumes that only the number of reproductive individuals are affected by local extinction, because local extinctions occur only in source habitat. Local extinction already characterizes sink habitat. The function $e(F)$ might take on any number of forms (Hess 1996), but the derivation in Box 6.1 suggests that average extinction rate should increase with $F$. Consequently, as habitat becomes increasingly dispersed into small patches ($F$ gets larger), equilibrium population size should decrease because some source habitat patches fail to produce offspring and go locally extinct.

If extinction rate is related to patch size, habitat configuration turns out not to have large effects on population size (Figure 6.2). This prediction arises mainly because extinct patches are immediately recolonized. Thus, even with local extinction incorporated, the configuration of a landscape in which a species moves rapidly and easily among habitats is unlikely to affect population dynamics. Rather, the density of source habitat is most important.

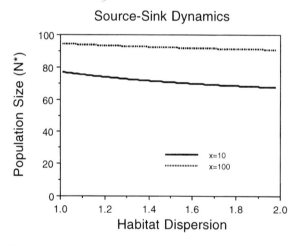

### Source-Sink Dynamics

FIGURE 6.2. Simulation example of the effects of landscape extent on the influence of habitat configuration on population size for a source-sink population. Fractal dimension quantifies the degree to which source habitat is fragmented into many dispersed patches ($F$ nearly 2) or aggregated into a few large patches ($F$ nearly 1). I simulated equation B1.6 in Box 1 for a small ($x = 10$) and large ($x = 100$) landscape, where population size in the absence of extinctions was 100 and the extinction probability of very small patches ($q$) was 0.5. Local extinction of patches and habitat dispersion has relatively little impact on population size, especially if landscape extent is large.

## 6.3.2 *Metapopulation Dynamics and Landscape Pattern*

When habitat becomes sufficiently fragmented or species are poor dispersers, colonization rates may decline sufficiently for extinct habitat patches to remain unoccupied for one or more generations. If so, the relevant population metric becomes the proportion of suitable habitat patches occupied ($p$) rather than population size ($N$). The proportion $p$ is determined by a balance between colonization and extinction rates, and its rate of change is simply described by Levins's (1969) original meta-population model

$$\frac{dp}{dt} = cp(1 - p) - ep, \tag{4}$$

where $c$ is colonization rate of unoccupied patches $(1 - p)$ and $e$ is extinction rate of occupied patches. Variations of this simple model have been explored by Hanski (1982) and Gotelli (1991) to produce some well-known alternative hypotheses about how metapopulations function.

The core-satellite hypothesis (Hanski 1982) assumes that extinction rates depend on the proportion of occupied patches, since patch extinctions followed immediately by recolonization will not be counted as extinctions. This model yields an unstable equilibrium and therefore predicts alternative stable states: large "core" areas of habitat with little local extinction and "satellite" areas of small, high-turnover patches. While such "core-satellite" situations appear to exist often in nature (Harrison 1991), they may result from the underlying spatial pattern of habitat as often as from regional recolonization. If habitat distribution is fractal, then the population distribution will resemble that predicted by Hanski (1982). Specifically, large, extinction-resistant patches interspersed with many small extinction-prone patches may serve as a source of colonists for small patches.

The propagule-rain hypothesis (Gotelli 1991) assumes that colonization is not limiting due to a "rain" of colonists throughout the landscape. Thus, only extinction affects the proportion of patches occupied. This model is conceptually the same as Pulliam's (1988) source-sink model with local extinctions. Unless local extinction rates are very high due to environmental variability or strong species interactions (e.g., predation, disease), the derivation in Box 6.1 suggests that the proportion of patches occupied will remain relatively high regardless of habitat configuration (Figure 6.2).

The original Levins (1969) model and its various modifications have led to useful conceptual predictions about population patterns in space. However, they ignore the separate contribution of the density and configuration of habitat to extinction and colonization rates. More recent models (Andren 1994, Kareiva and Wennergren 1995, Bascompte and Sole' 1996) explore the consequences of habitat destruction for metapopulations, i.e., the effect of the *density* of habitat on metapopulations. These models predict thresholds in the proportion of landscape covered by suitable habitat, below which the population goes extinct. These thresholds are interesting because the proportion of patches occupied goes to zero even though some suitable habitat remains (see Bissonette et al., this volume). However, these models do not explicitly explore the independent effects of habitat configuration on patch occupancy.

Using an assumption of fractal geometry (eq. 1), the colonization and extinction rates of patches can be expressed as functions of habitat density ($v$) and configuration (fractal dimension $F$). Applied to the Levins (1969) model:

$$\frac{dp}{dt} = c(v, F)p(1-p) - e(F)p \tag{5}$$

Since colonization or extinction per patch are not functions of $p$, the equilibrium proportion of patches occupied is given by

$$p^* = 1 - \frac{e(F)}{c(v, F)}. \tag{6}$$

To determine the shapes of these functions, one must imagine how potential colonists search for suitable patches. From a colonist's perspective, not all habitat patches are suitable, and colonists with different spatial scales of perception will perceive different minimum patch sizes as suitable (Ritchie, in press). This selectivity also will be affected by the relative abundance and density of large and small patches, which is reflected by habitat configuration, i.e., the fractal dimension of habitat, for a given habitat density. These notions can be applied to derive reasonable specific functions for colonization rate and $p^*$ (Box 6.2).

The analysis in Box 6.2 suggests that extinction rate per patch should not change with habitat density but should increase at a decreasing rate as habitat becomes more dispersed, i.e., as $F$ approaches 2 (Figure 6.3). Colonization rate per patch should increase linearly with habitat density but decline exponentially as habitat becomes more dispersed (Figure 6.3). In many cases, there will exist an extinction threshold for habitat configuration (Figure 6.3) in addition to one for habitat density. Although "percolation theory" (Gardner et al. 1989) predicts dramatic thresholds in movement rates across a narrow range of habitat densities, this model predicts a gradual decline in $p^*$ as habitat becomes more dispersed. Nevertheless, for a variety of dispersal capabilities and landscape extents, the fractal dimension at which extinction occurs (1.4–1.6) corresponds to habitat densities of 0.5 to 0.6 in random landscapes (Gustafson and Parker 1992). These habitat densities are those at which individuals can no longer "percolate" through the landscape entirely through suitable habitat (Gardner et al. 1989). Thus, a small amount of fragmentation of low density, aggregated habitat (i.e., nonrandom landscapes) may be enough to prevent percolation of colonists.

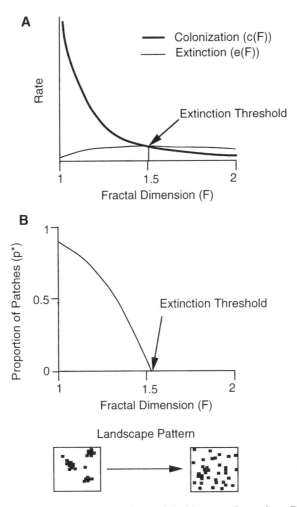

FIGURE 6.3. Metapopulation dynamics and habitat configuration: Part A shows basic relationships of extinction rate and colonization rate versus habitat dispersion (as measured by fractal dimension, $F$, and picture at the bottom). The two curves cross at an extinction threshold for habitat configuration, i.e., if habitat is dispersed more than the threshold amount, the metapopulation occupies zero patches and goes extinct, as shown in B. Note the rapid decline in $p^*$, the proportion of patches occupied as the fractal dimension shifts from 1.4 to 1.5; this indicates the potential sensitivity of population persistence to habitat configuration independent of habitat density. Interestingly, this range of fractal dimension corresponds to habitat densities of 0.5–0.6 in random landscapes (Gustafson and Parker 1992), which are in the range at which potential colonists begin to fail to percolate across the landscape through suitable habitat (Gardner et al. 1989).

For poorly dispersing species, an extinction threshold "isoline" exists that represents combinations of habitat density and habitat configuration below which the metapopulation goes extinct (Figure 6.4). Thus, as habitat destruction reduces the amount of habitat, a population becomes much more sensitive to dispersion of remaining habitat. Typical patterns of habitat loss such as through clear-cutting, agriculture, or wildfire are likely to increase $F$ at the same time as they decrease density. Thus, the decline of a metapopulation in response to habitat fragmentation may be quite sharp, and the spatial structure of habitat may be crucial in determining population persistence.

Most metapopulation models have focused on patch occupancy. However, managers need to know the expected persistence of metapopulations, which may depend more on the absolute number of patches occupied. More specifically, metapopulations that occupy many small patches at a low incidence might still have greater persistence than those that occupy a few larger patches because the risk of all patches going extinct simultaneously is spread across more patches. Expected persistence time, $T$, can be thought to be the inverse of the joint probabilities of extinction for each patch:

$$T = \frac{1}{e(F)^{n(v,F)}} \tag{7}$$

where $n$ is the equilibrium number of occupied patches and is a function of both colonization and extinction rates, and thus landscape pattern. The derivation of $T$ in Box 6.3 suggests that, for poor dispersers, persistence

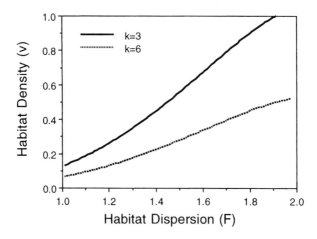

FIGURE 6.4. Relationship between the extinction threshold for habitat dispersion ($F$) and habitat density ($v$) for species with different dispersal abilities ($k = 3$ and $k = 6$). I set $q = 0.5$, $x = 100$, and $b = 1$ for the simulations. The relationship suggests that rarer habitats must be more aggregated (less dispersed) for metapopulations (particularly of poor dispersers) to persist.

## Box 6.3. Persistence Time of a Metapopulation in Fragmented Habitat

A simple relationship for expected persistence time $(T)$ of a metapopulation is the inverse of the joint probability of extinction of all occupied habitat patches (Harrison and Quinn 1989). This is simply

$$T = \frac{1}{e(F)^{n(F)}}, \qquad (B3.1)$$

where the function $n(F)$ is the number of occupied habitat patches and is a function of the total amount of habitat, $vx^F$, the proportion of "suitable" habitat patches occupied, $p^*$, and the degree to which the habitat is fragmented into many or few patches, which can be accounted for by the sum of the frequencies of different-sized patches. Thus

$$n(F) = p^* vx^F \int_z^m w^{-F} dw \qquad (B3.2)$$

Evaluating the integral and simplifying yields

$$n(F) = \frac{ax^F}{kz^{1-F}} \left[ c(v,F) - e(F) \right] \qquad (B3.3)$$

With this relationship for $n(F)$, persistence time becomes

$$T = \frac{1}{e(F)\left[ \dfrac{ax^F}{kz^{1-F}} \left[ c(v,\ F) - e(F) \right] \right]}. \qquad (B3.4)$$

A fully expanded version of this equation has little conceptual value, but simulations in Figure 6.5 reveal some of its predictions. Principally, species with poor colonization ability (low $k$) have persistence times that closely parallel their patch occupancy $p^*$. Species with good colonization ability (high $k$) may actually benefit from patch dispersion when habitat is relatively dense ($v > 0.5$), even though the average extinction rate per patch is higher. This interaction between habitat density, habitat configuration, and species' dispersal ability illustrates the need to consider habitat configuration separately from habitat density in assessing the effects of landscape pattern on populations. It also points out that the persistence of metapopulations may not be accurately reflected by patch occupancy. Although I do not explore the detailed consequences here, these predictions become more pessimistic if extinction events are spatially correlated, effectively reducing $n(F)$ (Harrison and Quinn 1989).

time closely mirrors equilibrium patch occupancy, as distinct thresholds for $T < 1$ (no persistence) exist for different combinations of habitat density, $v$, and configuration, $F$ (Figure 6.5A,B). For relatively good dispersers, however, persistence increases with $F$ at higher habitat densities (Figure 6.5B). This suggests that if colonization rates are high enough, more dispersed habitat may actually increase persistence. Such a pattern might explain the evolution of habitat "generalists" or "edge" species that thrive in frag-

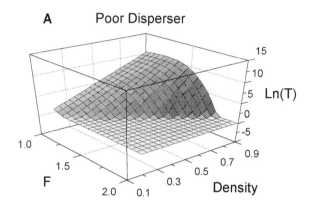

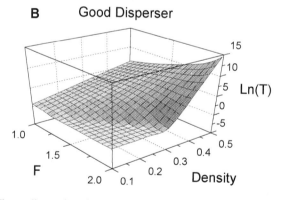

FIGURE 6.5. Three-dimensional representation of the natural logarithm of expected persistence times for different levels of habitat dispersion (fractal dimension, $F$) and habitat density. Simulations ($q = 0.5$, $x = 100$) were run for a poor disperser, $k = 10$, and a good disperser, $k = 3$, with eq. B3.4 in Box 6.3. For the good disperser, habitat dispersion ($F$) had relatively weak effects on persistence relative to habitat density, and at high habitat densities, persistence time actually increased with habitat dispersion. For the poor disperser, habitat dispersion had stronger effects than habitat density, and the two acting together produce an extinction "escarpment" of landscape pattern. The differences in landscape effects with dispersal ability points out the difficulty in using simple generalizations about effects of habitat fragmentation.

mented habitats. Considering spatial structure may be as important for understanding some species' success as it is for others' decline.

Theoretical effects of landscape pattern on patch occupancy and population persistence suggest that different species will respond very differently to the same habitat fragmentation pattern. Dispersal ability is a crucial parameter, because good dispersers may not have extinction thresholds for habitat density or configuration. A species' scale of perception is also important; the derivation in Box 6.2 suggests that larger scale species will be more sensitive to habitat configuration for a given habitat density because they may selectively avoid smaller patches and thus have less habitat available to them. However, their dispersal abilities might be better because they can move faster, and this advantage might compensate for greater habitat selectivity. The reproductive potential of a species may also play a role, because species with high reproductive capacities (e.g., annual weeds, insects, small mammals) should produce more potential colonists per patch and increase colonization rates. If larger scale species are also larger in body size, then they may have lower reproductive capacities and thus greater sensitivity to habitat fragmentation.

A final interesting prediction is that greater extent of a landscape can improve a species' persistence in addition to the obvious effect of having absolutely more habitat for a given habitat density. A fundamental property of fractal distributions is that landscapes of larger extent contain larger maximum patch sizes (Milne 1991, 1992). Thus, colonists in larger landscapes have a greater chance of encountering large patches and so should selectively ignore smaller patches (Ritchie, in press). Such selectivity would increase the mean size of "suitable" patches, lower expected local extinction rates, and increase persistence. This effect is akin to that observed in adding space to predator–prey or competitive interactions (Hassell et al. 1991, Tilman 1994); more space increases the stability (persistence) of the system. Thus, there may be good reasons for enlarging the extent of management efforts, even if overall habitat density declines.

## 6.4 Spatial Structure in the Real World

Theory has far outstripped field data in the effort to understand spatially structured populations (Wiens et al. 1993, Wennergren et al. 1995). The relative ease in performing computer simulations compared to gathering long-term census data certainly contributes to this discrepancy, but theoretical studies generally have not generated clear, interesting hypotheses (but see Kareiva and Wennergren 1995). Another difficulty is that spatially explicit data on vital rates, such as extinction and dispersal, are much more difficult to gather than demographic survival and fecundity data for panmictic, spatially unstructured populations. Dispersal and colonization have proven particularly hard to measure (Wennergren et al. 1995). A final problem is that population census efforts are usually focused locally to increase precision, when knowing the spatial variation among dispersed, less precise censuses might provide greater insight. Consequently, few studies (Fritz 1979, Sjogren Gulve 1994, Peltonen and Hanski 1991, Harrison 1991, Hanski 1994) have measured dynamics of local populations and also considered patch size and interpatch distance of habitat. Even these have seldom tested specific predictions from metapopulation models about patterns in patch occupancy (cf. Fritz 1979, Hanski 1994).

In spite of these difficulties, I believe that field data do support the basic assumptions of the simple source-sink and metapopulation models that ecologists commonly use. These data support two principal assumptions: (1) extinction rates are higher in smaller habitat patches or for smaller local populations (Hanski 1994); (2) local populations are strongly size-structured, i.e., habitat is typically available in a few large, extinction-resistant patches and in many small, extinction-prone patches (Schoener and Spiller 1987, Harrison 1991).

Support for these assumptions is crucially important to validate the current metapopulation modeling approach. In addition, data support three predictions of the models: (1) patch occupancy declines with increasing interpatch distance; (2) minimum patch sizes used by species increase with a species' body size; and (3) where colonization does not seem to be limiting, spatial dynamics appear to be driven entirely by local extinction, as predicted for source-sink dynamics with local extinctions.

However, these patterns are not sufficient to provide strong confidence in the predictions of popular, general models (e.g., Levins 1969, Hanski 1982, Lomnicki 1988, Pulliam 1988, Gotelli 1991) or complex, specific simulation models (e.g., Hassell et al. 1991, Dytham 1995, With and Crist 1995, Gustafson and Gardner 1996).

The concept of habitat *sources* and *sinks* for populations is popular, but rigorous tests are still limited. Source-sink dynamics require more than just the existence of one good and one poor habitat, because sink habitats should be occupied only if individuals control access to resources in source habitat through territoriality. For example, red squirrel (*Tamiasciurus* spp.) territories vary in their abundance of pine cones and epiphytic fungi to the extent that some territories support squirrel survival and reproduction while others may not even support survival (Smith 1968). Thus, individuals are forced into "sink" territories where they may survive but not reproduce. This would seem to be a textbook example for source-sink dynamics. On the other hand, juveniles of white-footed mice (*Peromyscus leucopus*) may occupy unsuitable habitat while dispersing to suitable habitat (Krohne 1989). It is not clear whether these mice are residents that may reproduce but at below-replacement rate or whether they are merely short-term transients searching for suitable habitat. As Pulliam et al. (1992) note, demographic data from multiple habitats need to be collected for many more species before the prevalence of territoriality-imposed occupancy of sink habitats is firmly established.

In virtually all cases where source-sink dynamics are invoked, it is assumed that individuals can readily disperse among source and sink habitats. No explicit criteria exist for differentiating a source-sink population (in the Pulliam [1988] sense) from a "true" metapopulation (in the Harrison [1991] sense). Colonization rate seems crucial in this distinction because source-sink populations are assumed to have full occupancy of all source habitat. If geographic or habitat barriers impede dispersal, then local source populations that go extinct may not be replaced (Howe et al. 1991). If so, then patch occupancy becomes a relevant population metric, and the familiar metapopulation models (Gilpin and Hanski 1991) apply.

If habitat patches are size-structured, i.e., patches of different size occur on a landscape, and within-patch population size is proportional to patch size, then local extinction rate should decline with increasing patch size. Data to test this hypothesis come from 20 years' unpublished population censuses for 35 colonies of Utah prairie dogs (*Cynomys parvidens*). This species has a restricted distribution in southern Utah that represents a typical metapopulation structure (Figure 6.6A). The number of extinctions per year (number of annual counts of zero per year) declines significantly as an inverse function of increasing colony size (as determined from the maximum annual census in 20 years at each colony, Figure 6.6B). This matches the shape predicted in Box 6.1. These data provide a particularly powerful test because they incorporate long-term (admittedly imprecise) censuses dispersed in space.

The models in Boxes 6.1 to 6.3 assume that species will select habitat patches during colonization on the basis of patch size. Specifically, species with larger scales of perception (e.g., body size) should be more selective, i.e., have a larger minimum threshold of patch size that they will occupy. Many descriptive data support this prediction, but censuses of passerine birds on grassland-barren remnants in Maine (Figure 6.6C), (Vickery et al. 1994) demonstrate the pattern nicely. The patch size at which a bird species attained 50% of its maximum incidence (percentage of patches of a given size class occupied) correlated strongly with body length for six bird species. This suggests that larger species may find a smaller proportion patches of available habitat as "suitable," resulting in a reduction in the total number of habitat patches occupied and population persistence. These data therefore support the predictions of the fractal model that patch selection will be scale dependent.

The ultimate test of the effect of landscape pattern on metapopulation structure is to compare patch occupancy or *incidence* (Hanski 1994) versus both habitat density and habitat configuration. Many studies suggest the negative relationship between patch incidence and habitat configuration predicted by the model, except that configuration is inferred most often from either patch size or interpatch distance (e.g., Gill 1978, Sjogren Gulve

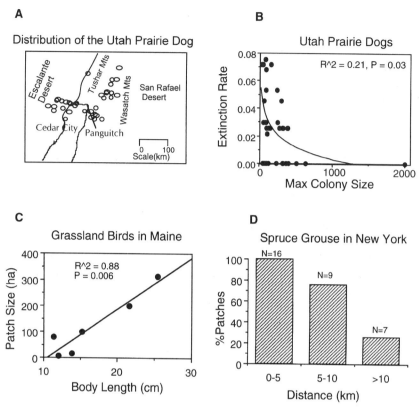

FIGURE 6.6. A smorgasbord of results from field studies that support assumptions and predictions of spatially structured population models. In A, the distribution of colonies of the Utah prairie dog in southern Utah are shown, demonstrating the fractal nature of the spatial distribution of habitat. Using the standard box-counting method (Barnsley 1988, Milne 1991), the fractal dimension of colonies is 1.22. In B, extinction rates over 20 years for 35 Utah prairie dog colonies are shown to decline significantly with maximum colony size (as measured by maximum annual census over 20 years). This pattern suggests that extinction rates decline as a nonlinear inverse function of habitat patch size (assumed to correlate with maximum annual census). In C, the patch size corresponding to 50% of maximum incidence correlates highly significantly with body length for six passerine bird species in grassland-barrens in Maine (Vickery et al. 1994). This pattern suggests that larger-bodied (and larger-scaled) species have higher patch-size thresholds for patch occupancy, as predicted by the model of fractal habitat dispersion in Box 6.2. Finally, in D, patch occupancy declines with increasing average distance from the nearest habitat patch, as predicted by many metapopulation models, for spruce grouse inhabiting forest patches in upstate New York (Fritz 1979).

1994, Hanski 1994). Either of these patch characteristics may reflect habitat density and/or configuration. Thus, the effects of specific types of landscape pattern on populations remain ambiguous. Nevertheless, these data at least qualititatively support model predictions. A nice demonstration for wildlife species is the study of patch occupancy by spruce grouse in remnant forest patches in upstate New York (Fritz 1979, Figure 6.6D). Not only does patch occupancy decline with interpatch distance, but the decline is nonlinear and hints at a "threshold" at which occupancy quickly drops to zero. Such a threshold is exactly that predicted by the models in Boxes 6.1 to 6.3 and by other models (Kareiva and Wennergren 1995, Bascompte and Solé 1996).

Although field studies have not directly measured colonization and extinction rates in direct relation to landscape pattern, results for extinction rates and patch occupancy in relation to patch size and isolation are in the direction predicted by theory. This correspondence suggests two major avenues of future research. First, there is a need to link processes affecting local dynamics, such as density depenence, competition, predation, disease, etc., with landscape pattern. New models that do this have already provided important insights (Nee and May 1992, Tilman 1994, Dytham 1995, Kareiva and Wennergren 1995). However, more general models are needed to develop broad-based hypotheses. Second, field researchers need to design studies that measure dispersal, colonization, extinction, and population dynamics in relation to overall landscape pattern rather than individual habitat patch characteristics (Wiens et al. 1993). This will allow direct tests of hypotheses about the effects of landscape pattern and more effectively identify successful models and ideas.

## 6.5 Management Implications

The analysis presented here suggests that management of spatially structured populations should view overall landscape pattern as a variable driving population dynamics, rather than the characteristics of individual patches (e.g., size or isolation). Management of source-sink population systems should be focused on the density of source habitat, as has always been suggested (Howe et al. 1991, Pulliam et al. 1992). This applies equally to threatened or harvested species (Day and Possingham 1995). However, source and sink habitat should be managed as a collective "landscape," since sink habitat can increase population size and persistence (Pulliam and Danielson 1991, Howe et al. 1991). Harvesting should not be focused necessarily on sink habitats under the rubric of taking "surplus" individuals. Instead, sink habitat should be seen as an important component of spatial structure, just as one particular age class may comprise an important component of age structure in maintaining a population subject to harvesting (McCullough 1996).

For metapopulations, theoretical models and reviews of field data in the past three years suggest strongly that the dynamics and persistence of metapopulations depend on landscape pattern. This means that focusing on the fates of populations in individual patches provides only a part of the story; the response of populations to entire landscapes needs to be studied more explicitly (Wiens 1995). Landscape-based approaches such as those employed in designing recovery goals for the spotted owl (*Strix occidentalis*, Lande 1988) serve as standards for management of other species. Long-term data sets like that for Utah prairie dogs may exist; if so, they need to be exploited. More reviews (e.g., Kareiva and Wennergren 1995) of the results of specific simulation models need to be made to synthesize a set of clear predictions that structure field work. Finally, models presented here and elsewhere (McCullough 1996) suggest that harvesting metapopulations remains a dicey business; additional mortality from harvesting is likely to act in much the same way as reducing habitat density, so, in the same way, metapopulations may be very sensitive to harvesting rates. Moreover, harvesting efforts focused in unsuitable habitat may indirectly limit colonization rates. For large or poorly dispersing species whose habitats are already fragmented (e.g., sage grouse *Centrarchus urophasianus*), harvesting may pitch populations over extinction escarpments (e.g., Figure 6.5).

To this point, I have ignored the effects of spatial autocorrelation in the dynamics of local populations, largely because few data exist. However, these effects need to be measured in the field and their consequences explored in models (e.g., Harrison and Quinn 1989). Calculations of expected persistence times of metapopulations are sensitive to spatial autocorrelation, because autocorrelation effectively reduces the number of independent "patches" that must go extinct simultaneously for the entire metapopulation to go extinct. The exact nature of this sensitivity awaits future modeling efforts. I am not aware of any studies that have measured spatial autocorrelation in extinction or colonization rates, so I propose this as a major task for future field studies. Until we know the magnitude of spatial autocorrelation, we will be overly optimistic in our assessment of metapopulation persistence.

## 6.6 Conclusions

The simple models presented here and a review of previous work suggest that both the amount and configuration of suitable habitat determine the spatial structure, density, and persistence of populations. For source-sink populations, density, but not configuration, of source habitat is most important in determining population size. For landscapes of reasonable size, sufficient numbers of large habitat patches will be present to minimize local extinction, and any unoccupied habitat patches should be rapidly recolonized. For metapopulations, habitat amount and configuration are predicted to have approximately equal effects, with configuration having the strongest effect on species that are poor dispersers or large in body size, and habitat density having the strongest effect on species with the opposite traits. Moreover, there may be "extinction thresholds" of minimum dispersion for habitats of a given density, especially for poor dispersers (Figures 6.4 and 6.5). On the other hand, increased habitat dispersion may actually increase the expected persistence time of good dispersers (Figure 6.5). These results make good intuitive sense, but the simple models presented here predict quantitative relationships that might be tested with field data. To date, such data suggest that assumed relationships between patch size and extinction and between patch configuration and colonization are reasonable and that their predictions have some chance of accuracy.

The modeling results suggest that information on landscape pattern (e.g., through GIS databases) may be crucial for assessing the persistence of spatially structured populations. The amount of source habitat for source-sink populations might be directly assessed from such maps, and spatial configuration of different habitat types can probably be assessed only with such data. The fate of threatened and endangered species that occupy fragmented habitats may be particularly tied to landscape pattern, and recovery of these species may involve landscape-level habitat manipulation or strategic planning. The existence of extinction thresholds in which a small increase in fragmentation may have drastic effects on population sizes is disturbing because it suggests that some populations may be highly sensitive to management. The models imply that species with large body size, relatively poor dispersal ability, or preference for relatively rare habitat may be most sensitive. Thus, managers need to be aware of these potential risks, especially when good information is lacking.

A final point is that the scale of habitat management must be matched to the species of interest. Since the simple models presented here rely on fractal geometry, they assume that landscape pattern is scale independent but that species' responses are not. Indeed, field data suggest that species have their own spatial scale at which they respond to the environment. If so, management tactics that operate at too fine a spatial scale may be entirely ignored by the target species. Larger species will require that larger blocks or patches of habitat be conserved. When only a fixed percentage of the landscape is (or can be) represented by desired habitat, the most impressive method of improving a species' persistence may be to conserve a few large habitat patches, rather than a higher density of smaller patches. Even so, the fractal dimension of a given landscape applies only for a restricted range of scales; the scale at which the fractal dimension changes may suggest the scale at which managers should structure their effort (Milne 1991, Wiens et al. 1993). Thus, landscape pattern and species' characteristics may suggest appropriate administrative scales for population management.

*6.7 Acknowledgments.* I thank W.C. Pitt, M. McClure, and J.A. Bissonette for helpful comments on the manuscript. B.T. Milne also influenced my ideas in applying fractal geometry.

## 6.8 *References*

Andren, H. 1994. Effects of habitat fragmentation on birds and mammals in landscapes with different proportions of suitable habitat: a review. Oikos 71:355–366.

Andrewartha, H.G., and L.C. Birch. 1954. The distribution and abundance of animals. University of Chicago Press, Chicago, Illinois, USA.

Barnsley, M. 1988. Fractals everywhere. Academic Press, New York, USA.

Bascompte, J., and R.V. Solé. 1996. Habitat fragmentation and extinction thresholds in spatially explicit models. Journal of Animal Ecology 65:465–473.

Day, J., and H.P. Possingham. 1996. A stochastic metapopulation model with variability in patch size and position. Theoretical Population Biology 48:333–360.

Dytham, C. 1995. Competitive coexistence and empty patches in spatially explicit metapopulation models. Journal of Animal Ecology 64:145–146.

Elton, C. 1927. Animal ecology. Sidgwick and Jackson, London.

Fretwell, S.D. 1972. Populations in a seasonal environment. Princeton University Press, Princeton, New Jersey, USA.

Fritz, R.S. 1979. Consequeces of insular population structure: distribution and extinction of spruce grouse populations. Oecologia 42:57–65.

Gardner, R.H., R.V. O'Neill, M.G. Turner, and V.H. Dale. 1989. Quantifying scale-dependent effects of animal movement with simple percolation models. Landscape Ecology 3:217–227.

Gill, D.E. 1978. The metapopulation ecology of the red-spotted newt, *Notophthalamus viridescens* (Rafinesque). Ecological Monographs 48:145–166.

Gilpin, M., and I. Hanski, editors. 1991. Metapopulation dynamics: empirical and theoretical investigations. Academic Press, San Diego, California, USA.

Goodman, D. 1987. The demography of chance extinction. Pp. 11–34 in M.E. Soulé, editor. Viable populations for conservation. Cambridge University Press, Cambridge, UK.

Gotelli, N. 1991. Metapopulation models: the rescue effect, the propagule rain, and the core-satellite hypothesis. American Naturalist 138:768–776.

Gustafson, E.J., and R.H. Gardner. 1996. The effect of landscape heterogeneity on the probability of patch colonization. Ecology 77:94–107.

Gustafson, E.J., and G.R. Parker. 1992. Relationships between landcover proportion and indices of landscape spatial pattern. Landscape Ecology 7:101–110.

Hanski, I. 1982. Dynamics of regional distribution: the core and satellite species hypothesis. Oikos 38:210–221.

Hanski, I. 1994. A practical model of metapopulation dynamics. Journal of Animal Ecology 63:151–162.

Harrison, S. 1991. Local extinction in a metapopulation context: an empirical evaluation. Biological Journal of the Linnaean Society 42:73–88.

Harrison, S. 1994. Metapopulations and conservation. Pp. 111–128 in P.J. Edwards, R.M. May, and N.R. Webb, editors. Large-scale ecology and conservation biology. Blackwell Scientific, Oxford, UK.

Harrison, S., and J.F. Quinn. 1989. Correlated environments and the persistence of metapopulations. Oikos 56:293–298.

Hassell, M.P., H.N. Comins, and R.M. May. 1991. Spatial structure and chaos in insect population dynamics. Nature 353:255–258.

Hastings, H.M., and G. Sugihara. 1994. Fractals: a user's guide for the natural sciences. Oxford University Press, New York, USA.

Hess, G. 1996. Linking extinction to connectivity and habitat destruction in metapopulation models. American Naturalist 148:226–236.

Holling, C.S. 1992. Cross-scale morphology, geometry and dynamics of ecosystems. Ecological Monographs 62:447–502.

Howe, R.W., G.J. Davis, and V. Mosca. 1991. The demographic significance of "sink" populations. Biological Conservation 6:239–255.

Kareiva, P. 1990. Population dynamics in spatially complex environments: theory and data. Philosophical Transactions of the Royal Society, London B 330:175–190.

Kareiva, P., and U. Wennergren. 1995. Connecting landscape patterns to ecosystem and population processes. Nature 373:299–302.

Krohne, D.T. 1989. Demographic characteristics of *Peromyscus leucopus* inhabiting a natural dispersal sink. Canadian Journal of Zoology 67:2321–2325.

Lande, R. 1988. Demographic models of the northern spotted owl (*Strix occidentalis caurina*). Oecologia 75:601–607.

Lande, R. 1993. Risks of population extinction from demographic and environmental stochasticity and random catastrophes. American Naturalist 142:911–927.

Levins, R. 1969. Some demographic and genetic consequences of environmental heterogeneity for biological control. Bulletin of the Entomological Society of America 15:237–240.

Levins, R. 1970. Extinction. Pp. 77–107 in M. Gerstenhaber, editor. Lectures in mathematics for biology, vol. 2; American Mathematical Society, Providence Rhode Island, USA.

Lomnicki, A. 1988. Population ecology of individuals. Princeton University Press, Princeton, New Jersey, USA.

MacArthur, R.H., and E.O. Wilson. 1967. The theory of island biogeography. Princeton University Press, Princeton, New Jersey, USA.

McCullough, D.R. 1996. Spatially structured populations and harvest theory. Journal of Wildlife Management 60:1–9.

Milne, B.T. 1991. Lessons from applying fractal models to landscape patterns. Pp. 199–235 in M.G. Turner, and R.H. Gardner, editors. Quantitative methods in landscape ecology. Springer-Verlag, New York, USA.

Milne, B.T. 1992. Spatial aggregation and neutral models in fractal landscapes. American Naturalist 139:32–57.

Milne, B.T., A.R. Johnson, T.H. Keitt, C. Hatfield, J. David, and P. Hraber. 1996. Detection of critical densities associated with piñon-juniper woodland ecotones. Ecology 77:805–821.

Nee, S., and R.M. May. 1992. Dynamics of metapopulations: habitat destruction and competitive coexistence. Journal of Animal Ecology 61:37–40.

Palmer, M.W. 1992. The coexistence of species in fractal landscapes. American Naturalist 139:375–397.

Peltonen, A., and I. Hanski, I. 1991. Patterns of island occupancy explained by colonization and extinction rates in shrews. Ecology 72:1698–1708.

Pulliam, H.R. 1988. Sources, sinks, and population regulation. American Naturalist 132:652–661.

Pulliam, H.R., and B.J. Danielson. 1991. Sources, sinks, and habitat selection: a landscape perspective on population dynamics. American Naturalist 137:S50–S66.

Pulliam, H.R., J.B. Dunning, and J. Liu. 1992. Population dynamics in complex landscapes: a case study. Ecological Applications 2:165–177.

Quinn, J.F., and A. Hastings. 1987. Extinction in subdivided habitats. Conservation Biology 1:198–208.

Ritchie, M.E. In press. Scale-dependent foraging and patch choice in fractal environments. Evolutionary Ecology.

Schoener, T.W., and D.A. Spiller. 1987. High population persistence in a system with high turnover. Nature 330:474–477,

Sjogren Gulve, P. 1994. Distribution and exinction patterns within a northern metapopulation of the pool frog, *Rana lessonae*. Ecology 75:1357–1367.

Smith, C.C. 1968. The adaptive nature of social organization in the genus tree squirrel, *Tamiasciurus*. Ecological Monographs 38:31–63.

Tilman, D. 1994. Competition and biodiversity in spatially structured habitats. Ecology 75:2–16.

Vickery, P.D., M.L. Hunter, and S. Melvin. 1994. Effects of habitat area on the distribution of grassland birds in Maine. Conservation Biology 8:1087–1097.

Wennergren, U., M. Ruckelshaus, and P. Kareiva. 1996. The promise and limitations of spatial models in conservation biology. Oikos 74:349–356.

Wiens, J.A. 1995. Habitat fragmentation: island vs. landscape perspectives on bird conservation. Ibis 137:S97–S104.

Wiens, J.A., N.C. Stenseth, B. Van Horne, and R.A. Ims. 1993. Ecological mechanisms and landscape ecology. Oikos 66:369–380.

With, K.A., and T.O. Crist. 1995. Critical thresholds in species' responses to landscape structure. Ecology 76:2446–2459.

# 7
# Hierarchy Theory: A Guide to System Structure for Wildlife Biologists

ANTHONY W. KING

## 7.1 Introduction

Hierarchy theory is part of the conceptual foundation of landscape ecology. Developments in landscape ecology over the past decade have cooccurred with developments of hierarchy theory as applied to ecological systems, and many individual authors have contributed to both. The seminal monographs of Forman and Godron, *Landscape Ecology*, and O'Neill et al., *A Hierarchical Concept of Ecosystems*, were both published in 1986 (Forman and Godron 1986, O'Neill et al. 1986). Robert (Bob) O'Neill in particular has contributed a great deal to both fields (e.g., O'Neill 1988, 1989, O'Neill et al. 1986, 1988, 1989). By 1987, Urban et al. (1987) had proposed a hierarchal perspective for landscape ecology. In 1989, *Landscape Ecology* (Volume 3, nos. $^3/_4$) published the proceedings of a workshop on scale. In this volume O'Neill et al. (1989) argued that hierarchy theory provides a framework for the analyses of scale in landscapes and other ecological systems.

The synergy of these coincident developments has contributed to the rapid growth of landscape ecology. Both fields have benefitted. However, the rapid expansion of ideas has contributed to some confusion, or at least lack of precision, in the use of some key concepts and terms. Allen (in press) has written elsewhere of this confusion. Here I present some fundamentals of hierarchy theory and their relationship to ecological landscapes. I wish to document the contribution of hierarchy theory to the conceptual underpinnings of landscape ecology and perhaps clear some of the concep-

Anthony W. King is a research staff member in the Environmental Sciences Division of Oak Ridge National Laboratory, Oak Ridge, Tennessee. His research interests broadly include ecological systems at large spatial scales, and more specifically spatially structured population dynamics and the role of terrestrial ecosystems in the Earth's climate and biogeochemistry. To address these research questions, Dr. King has explored hierarchy theory and scale theory for clues, if not answers, to the problem of translating fine-scale information and models to larger scales and higher levels of system organization.

tual clutter that has accumulated over a decade or more of vigorous expansion. Allen and Starr (1982) and O'Neill et al. (1986) provide excellent introductions to hierarchy theory and its application to ecological systems, especially community and ecosystem ecology. Here I will simply try to bring those same concepts to bear on the spatially distributed systems that are the subject of landscape ecology. My presentation shares much with Urban et al.'s (1987) discussion of landscape ecology. The reader should refer to this earlier work for a complementary perspective on hierarchy theory and landscape ecology.

## 7.2  Hierarchy Theory

### 7.2.1  Middle-Number Systems

Hierarchy theory is a theory of system organization (Box 7.1). Emerging from general systems theory (von Bertalanffy 1968), hierarchy theory is an attempt to deal with complex medium- or middle-number systems (Weinberg 1975, Allen and Starr 1982) (Box 7.2). Middle-number systems are those with too many components to describe each with its own equation and too few to obtain reliable mean properties by averaging or other statistical treatments. Small-number systems, on the other hand, are those that can be explicitly described by a set of equations with at least one equation for each component. The equations describe the states of all components and their relationships with one another. Classical Newtonian dynamics address small-number systems. The familiar systems of differential equations for predator–prey or competitive dynamics in community ecology, or compartment models in ecosystem ecology, are examples of the small-number approach. Large number systems are those in which individual properties of many components are replaced by averages. Laws describing the macroscopic variables of gases (Serway 1986) (e.g., temperature and pressure in the Ideal Gas Law [Rosen 1989]) are possible because

---

**Box 7.1.  System and System State**

SYSTEM:
A collection of objects joined in a constitutive relationship of interactions that forms a whole.

STATE VARIABLES:
The component parts or observable attributes of a system. The mathematical variables describing the state or condition of a system.

## Box 7.2.    A System Classification

SMALL-NUMBER SYSTEM:
A system with a small number of component parts (state variables). The state of a small-number system and the relationships among components can be completely described by a system of $2^n$ equations where $n$ is the number of system components.

MIDDLE-NUMBER SYSTEM:
A system with a medium number of component parts, too many to be described as a small-number system and to few to be described by the stable statistical properties of a large-number system.

LARGE-NUMBER SYSTEM:
A system with a large number of component parts. Large-number systems are described by the statistical properties of the collective set of components. These properties are stable and predictable because of the large number of components.

of the large-number properties of these systems (Weinberg 1975). Kerner (1957) has explicitly derived macroscopic ecosystem properties using a statistical mechanics approach, and ecologists frequently assume macroscopic properties (and implicitly large-number systems) for populations and other ecological systems. The attribution of a constant mean birth rate to a population is an example. These assumptions may be incorrect if the system is actually a middle-number system.

Middle-number systems are intractable because idiosyncratic properties of individual components are expressed such that statistical properties of the ensemble are unreliable or unstable. Yet, there are too many components and interactions to address them adequately as a small-number system—with a set of coupled differential equations, for example. How many is "too many" is ambiguous. Three particles interacting according to the laws of Newtonian dynamics can show complex unpredictable behavior (the three-body problem) (Weinberg 1975, Prigogine 1980), and computers are increasingly capable of describing systems of tens to hundreds of individual organisms (more in physical systems) with what are essentially small-number approaches (Huston et al. 1988). There are computational limits, however, for even the most advanced computers. A complete general description of the interactions between $n$ bodies requires $2^n$ equations (Weinberg 1975) (e.g., $2^{100}$, or approximately $10^{30}$, equations for 100 interacting individuals); simplifying assumptions normally are used to reduce the equations to a manageable number (Caswell et al. 1972, Weinberg

1975). In any case, there comes a point when, even if the solution of a massively large system of equations is technically feasible, the behavior of the solution is so complex that it provides minimal understanding, i.e., the small-number approach does not yield an explanation for the middle-number behaviors.

Allen and Starr (1982) and O'Neill et al. (1986) argue that ecological systems are middle-number systems. This presumably includes system descriptions of landscapes. Allen and Hoekstra (1992, pp. 65–66) explicitly cite evidence for middle-number behavior for at least some kinds of landscapes in the work of Dean Urban. In arguing for individual-based models in ecology, Huston et al. (1988) believe that state-variable models describing ecological systems with aggregated variables such as population size violate a biological principle of the unique individuality of organisms. This individuality can influence system behavior such that the averaging inherent in the state-variable models is inadequate. In other words, they are arguing for ecological systems as middle-number systems. Or, as Weinberg (1975, p. 20) would have it, "forests have medium numbers of trees, or flowers, or birds."

Individual-based, spatially explicit population models (Dunning et al. 1995) increasingly are being developed as tools to study and manage animal populations distributed across heterogeneous landscapes (Pulliam et al. 1992, Turner et al. 1993, 1995, Liu et al. 1995). The number of individuals in these models (e.g., 75–150 in BACHMAP [Pulliam et al. 1992]) and the number frequently involved in landscape management units (e.g., approximately 1100 breeding pairs of Bachman's sparrow, *Aimophila aestivalis*, for the U.S. Department of Energy's Savanna River Site [Liu et al. 1995]) easily fall within the realm of middle-number systems.

Landscapes often are conceptualized as a grid of interconnected cells. If these cells are viewed as interacting elements in a landscape system, the number of individual cells (tens to hundreds of thousands) again argues for landscapes as middle-number systems. As a theory of complex middle-number systems, hierarchy theory is a profitable approach to the middle-number problems of landscape ecology.

## 7.2.2 System Dynamics and Levels of Organization

Hierarchy theory asserts that the interactions among components of some middle-number systems can be ordered by differences in interaction strength and frequency (Simon 1962, 1973). This ordering of system dynamics partitions the system into levels of organization (Box 7.3). Higher frequency dynamics and stronger interactions characterize lower levels. Lower frequencies and weaker interaction strengths characterize higher levels.

Hierarchically organized systems are actually only partially ordered by frequencies of dynamic behavior. They are "nearly-decomposable" or "nearly completely decomposable," but not "completely decomposable"

## Box 7.3.    Hierarchical System Organization

HIERARCHICAL SYSTEM:
A system organized as a system of systems within systems. The dynamics of a hierarchical system can be ordered along a spectrum of frequencies. Higher frequencies belong to lower level subsystems; lower frequencies belong to higher levels of organization. Subsystems of similar frequency belong to the same level of organization within the hierarchy.

TRIADIC STRUCTURE:
Characterization of a hierarchical system as a system with three levels: the focal level L, the next lower level L–1, and the next higher level L+1. Mechanistic explanation of the focal level is found at level L–1; the L+1 level constrains and provides context for focal-level behavior.

(Simon 1973). The levels are not completely isolated or strictly independent. Rather, the members of the system at one level are themselves systems of elements of the next lower level and are at the same time components of the systems that occupy the next higher level (von Bertalanffy 1968). A hierarchically ordered system is a system of systems within systems. Subsystems within the system are organized into levels according to differences in rate structure: frequency of behavior, frequency of interaction, and interaction strength. The subsystems are Koestler's holons (Koestler 1967). The subsystem/holon has a dual nature. It is simultaneously a whole and a part of another whole. In a hierarchically organized system, this other whole exists at the next higher level of organization.

In a hierarchically organized system, the elements at one level emerge as a consequence of the interactions and relationships among elements of the next lower level. This emergent behavior is a fundamental property of hierarchically organized systems (see Bissonette, this volume). Members at one level are in a constitutive relationship (von Bertalanffy 1968) or determinate association (Grobstein 1973) that determines and creates the systems of the next higher level. The emergence of properties that accompany the three-dimensional configuration (secondary structure) of proteins is an excellent biological example. The behavior of the protein at the level of the secondary structure emerges from the relationships, i.e., linkages, among amino acids at the lower-level organization of the polypeptide chain (Grobstein 1973). Its properties are a consequence of the system organization and the relationships among components and not the simple composition of the lower level (i.e., the amino acid composition of the polypeptide chain). The behavior of an individual animal is a similarly emergent property. The individual is a consequence of relationships among its component systems. Substantially alter those relationships, and the individual's behavior will be altered or the individual will die, even though the individual components still exist in the same approximate spatial relationship.

The superposition of systems within systems characteristic of hierarchical systems is a necessary but not sufficient condition. Superpositioning is shared with nested Chinese boxes where a box contains a smaller box that itself contains a smaller box and so on (Simon 1973). But because these boxes are not interacting as part of a system to generate the next box in the ordered set, the boxes do not represent a hierarchical system. The relationship can be described as a hierarchical ordering, but it does not represent a hierarchically organized dynamical system in the sense that I use the term here. Likewise, hierarchical ordering of patches within patches in a landscape (e.g., Kotliar and Wiens 1990) is not sufficient evidence of hierarchical system organization for the landscape. It must be shown further that higher order patches emerge as a consequence of a constitutive relationship among lower order patches or at least that the patches are the impressions left on the landscape by a system with those relationships. Efforts to define the community as a level of organization made up of populations suffer the same burden of proof. A constitutive relationship among the populations from which the community emerges must be demonstrated. That the community is a collection or set of populations is insufficient. This is, of course, an old and long debate in ecology (e.g., the Clementsian-Gleasonian debate, see Allen and Starr 1982, O'Neill et al. 1986).

## 7.2.3 Constraint and Mechanism

Membership in higher level systems constrains the dynamics of subsystems at lower levels (Box 7.4). Immersed in complex relationships and system topology (Caswell et al. 1972), elements are not free to behave as they would in isolation or in a different system with different relationships among elements. Thus it can be said that higher levels constrain lower levels or that they impose boundary conditions (Allen and Starr 1982, Salthe 1985, O'Neill 1989). Similarly, because a higher level system is made up of lower level subsystems, lower levels also impose constraints, albeit of a different kind. A holon at one level can only be and do that which is feasible as a consequence of its component parts and their interactions (Salthe 1985, O'Neill 1989). In other words, there are no behaviors open to a subsystem at one level that are not open to its constituents at lower levels (Caswell et al. 1972) or that do not emerge from the interactions among lower level constituents. Constraints from lower levels are sufficiently different from the constraints of higher levels that Salthe (1985) has termed them *initiating conditions* to distinguish them from the boundary conditions of higher levels.

## Box 7.4.    System Constraints

BOUNDARY CONDITIONS:
External conditions that constrain the behavior and dynamics of a system. In hierarchy theory, higher levels of organization impose boundary conditions on lower levels. Boundary conditions originating at higher levels are sometimes refered to as constraints, or constraints from above. In modeling, boundary conditions are constants or variables that are not functions of the system state; they do not change with changes in the state of the system.

INITIATING CONDITIONS:
Limitations or constraints on the focal level of a hierarchically organized system that are imposed by lower levels of organization.

EXTRA-HIERARCHICAL CONSTRAINTS:
Boundary conditions that originate from outside a hierarchically organized system.

There are also what I consider extrasystem or extrahierarchical boundary conditions that should not be confused with the boundary conditions imposed by higher level system organization. These are frequently abiotic environmental conditions like soils, geology, or temperature that are external to the hierarchical system under consideration. They do not emerge as a consequence of interactions among system components, or at least they emerge at levels of organization so far removed from the levels under consideration that only the broadest conceptualization of the system and the most omnipresent observer would perceive them as such. I am comfortable calling these higher order constraints, but I think it is important to distinguish them from those that emerge from within the dynamic structure of system interactions. To borrow an example from O'Neill (1989), if a flock of birds represents a level of organization, the flight speed of that flock is limited from below by the speed of the individual birds. This is a lower level constraint, or initiating condition. At the same time the flight (speed and direction) of individual birds is constrained by their membership in the flock; this is a higher level constraint, a boundary condition. There are also boundary conditions imposed on the flock by gravity and the fluid dynamics of the atmosphere; I would consider these extrahierarchical. Or, to borrow a landscape example from Urban et al. (1987), the compositional dynamics of a forested watershed are constrained from below by the species composition and dynamics of the stands in that watershed, which are in turn constrained by the dynamics of the gaps in the stands. The context of membership in the watershed constrains the stand and gap dynamics from above. Soil properties can be viewed as a higher level constraint, a slowly changing boundary condition that is a consequence of slowly changing vegetation. I would, however, consider the topographic constraints to be extra-hierarchical.

## 7.2.4 Rate Structure and the Description of Hierarchical System

The rate structure of hierarchical systems has important consequences for observation of these systems (Simon 1973). If the system is observed over a period $T$ at intervals of $\tau$, system behaviors of frequency much less than $\frac{1}{T}$ will not be observed as dynamics at all. Rather, they will appear as constants. As the frequencies approach $\frac{1}{T}$, those properties of the system may be observed as slowly changing boundary conditions. Neither will high-frequency behaviors, those greater than $\frac{1}{\tau}$, be observed as dynamics. These high-frequency behaviors belong to lower level internal dynamics of system components, and they do not determine the frequencies of the interactions among the subsystems to which they belong. Only the middle-frequency behaviors will be seen as the dynamics of interaction of major subsystems. Consider that if an oldfield system is observed once a month ($\tau = 1$ month) for one year ($T = 1$ yr), long-term successional dynamics of species replacement from oldfield to forest cannot be observed; the community type or seral stage will appear as a constant. Neither will diel patterns of flowering nor birth and death of short-lived individuals be observed. Only the middle frequencies of seasonal changes in species dominance will be observed as system dynamics.

Simon (1973) argues that success in building a theory (or model) of a hierarchically organized system depends upon how well the high-frequency, middle-frequency, and low-frequency components of the system can be separated. The sharper the separation, the better the approximation of the level of observable dynamics with a theory or model that ignores both the details of the next level down and the very slow behaviors of the next level up. This separation is determined by $T$ and $\tau$, a point to which I will return below. It is simply worth noting here that choices of $T$ and $\tau$ will determine the observable dynamics.

The middle-frequency observable dynamics of a hierarchical system can be described by a three-level triadic structure (Box 7.3): a focal level L, the next lower level L−1, and the next higher level L+1 (Figure 7.1). The focal level L is the level of observable dynamics, chosen by the investigator/observer as the level of interest (Allen et al. 1984, O'Neill 1989). The dynamics of the focal level, the middle-frequency behaviors of the system, are the dynamics of the interactions among major subsystems at the next lower level L−1. Thus it is that the dynamics of the focal level are "ex-

**level L+1** (**not depicted here**)

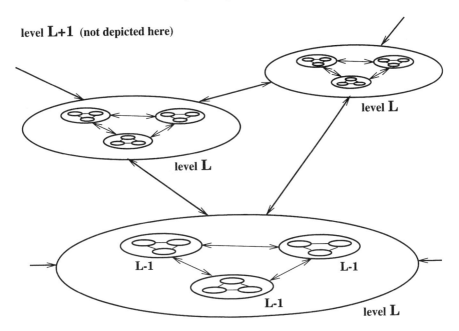

FIGURE 7.1. A hierarchical system is a system of systems within systems organized into levels. The ellipses represent systems or holons of a particular level of organization. In this figure, the 3 large ellipses represent holons at level L, the focal level from which all other levels are referenced. Lines between both large and small ellipses indicate dynamic interactions. The level L holons interact as components of the next higher level L+1 not depicted here. The next smaller ellipses represent the next lower level L–1. Interactions among level L–1 systems combined with the constraints of membership in level L+1 form the level L systems. The smallest ellipses are the L–2 components of the L–1 systems. Their components are not shown for clarity.

plained by looking to the next lower level" (O'Neill 1989, p. 144) or that mechanistic explanation can be found at the next lower level (O'Neill et al. 1986, Urban et al. 1987). The dynamics of an age-structured population (the focal level), for example, are explained by interactions (e.g., transition of individuals) among the age classes at the next lower level, L–1.

Focal-level dynamics occur within the context of a higher level system at level L+1. As noted above, this context constrains the behavior of level L members, and constraints on focal-level dynamics are thus found by reference to the next higher level. Moreover, because interactions among focal-level members determine the dynamics of level L+1 (i.e., they are components, parts, of the next level system), the consequences or significance of focal-level behavior also are found by reference to the next higher level (O'Neill et al. 1986, Urban et al. 1987). The focal-level population, for example, operates within the context of an assemblage of interacting populations, which is the next higher level of organization at level L+1, and the population dynamics are constrained by this context. Changes in the size of the population contribute to changes in the diversity/dominance structure of the L+1 assemblage and reflect the significance of changes at level L, the focal level. In some cases, reference to the higher level context explains the function or relevance of focal-level behavior (O'Neill 1989). For example, sexual behavior at the focal level of the individual is explained by its significance for persistence of the population at level L+1 (O'Neill 1989). What is the significance of individual sexual behavior, its functional or adaptive relevance (O'Neill 1989)? Answer: population persistence. In other cases, this sense of explanation from the higher level is less apparent. Change in "community" dominance/diversity at level L+1 does not carry the same sense of explanation for changes in population size at level L as population persistence does for sexual behavior.

The dynamics of interaction among the components of the focal level are independent of their internal structure and the dynamics of their own internal components. They appear and interact as essentially rigid bodies or indivisible particles. Their internal behaviors, the high-frequency behaviors of the system, will be observed at or near equilibrium and will appear as integrated mean properties of level L−1. They are not involved in the interactions that determine the dynamics of the focal level L. Indeed, the near decomposability of the hierarchically organized system isolates the focal level from dynamics below the L−1 level. Lower level dynamics are filtered or buffered through level L−1 by integration and can thus be ignored except for their determination of the stable L−1 properties. Consider the earlier example of the flock of birds. The dynamics of that flock as it flies across a field are explained by the interactions among individual birds. The internal dynamics of those individuals are high-frequency behaviors that do not determine the dynamics of interaction among individuals. The birds can be treated as rigid interacting particles. Similarly, the detailed lower level behavior of individuals within age classes in an age-structured population need not be considered. Variations in individual attributes or high frequency interactions among members of the same age class are not important to the dynamics of the focal-level population.

If a middle-number system can be described as a hierarchical system of relatively distinct and nearly decomposable levels, it can be dealt with as a small-number system. Hierarchy theory provides a set of rules or principles for simplification to reduce the number of equations needed to describe the dynamics of the system (Caswell et al. 1972). Accordingly, the focal level is decomposed into a set of interacting objects of level L−1 organization. These objects are systems in their own right (i.e., holons), but for the purposes of describing the dynamics of the focal level, they are not further subdivided but treated as whole objects with only aggregate properties. These L−1 objects and their interactions might be described by a system of coupled state cquations (e.g., a set of differential or difference equations). The state variables are the L–1 objects; the system of equations describes the focal level. The state of an object will change as a consequence of interactions with other objects, and their state may influence their interactions. However, the interactions are mediated only through this holistic state description. The interactions are not influenced by internal details of the object or the interactions among their constituent parts. Lower level properties are expressed as parameters in the system of equations. They may be constants, rapidly changing (high frequency) functions of time, or be described by frequency distributions or statistical moments, but they are not functions of the system state.

Consider, as a simple example, an age-structured population as the focal level. The $s$ age classes are the L–1 holons. The entire system of equations (Figure 7.2) describes the focal level. The state variables, $n_1, n_2, \ldots, n_s$ represent the L–1 components of the level L population. The individuals in each age class belong to a lower level ($< L–1$), and they are not represented as state variables in the level L system. They are represented only by the holistic state variable of number of individuals ($n_i$). Interactions between individuals in an age class are not expressed as changes in the state of the age class, the L–1 state, and individual interactions do not influence the other age classes. They are isolated from the focal level dynamics by the near decomposability of the hierarchical system. The fertility coefficients, $f_i$ and the survivorship probabilities, $p_i$ are system parameters. In this case, they are constants representing stable mean properties of individuals in the class, but they could be functions of time. They are not functions of $n_i$ or of other $n$. They are not functions of the system state. If they are, for example, if survivorship is density dependent, i.e., some function of $n_i$ such that $p_i = g(n_i)$, then survivorship is no longer a parameter or a constant lower-level property of individuals. It is, instead, a state variable and a property of the age-class and the population.

The level L system also exists as a subsystem, an object, in the L+1 level of organization. Interactions with other subsystems in the L+1 level will appear as constant or slowly changing boundary conditions on the system. The difference depends on the strength of the loose horizontal coupling between subsystems of the L+1 level. But, importantly, they are not functions of the system's state variables. In the example of the age-structured population (Figure 7.2), the population is shown as part of a higher level assemblage of interacting populations. The interactions in this case are assumed to be weak and lead to constant boundary conditions that can be ignored as an explicit part of the model. However, they could be represented by a loss or mortality term that affects all age classes equally. Note that the interaction with other populations is mediated entirely through the aggregate property of the population size $N$. Changes in other populations are not functions of the age-classes in population $i$. If they were, for example, if a predator selectively fed on one age class, then the near

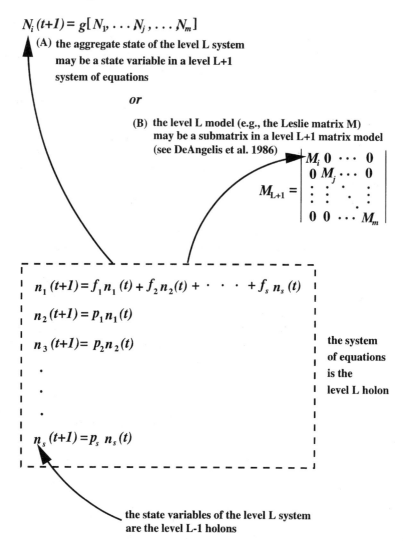

$$N_i(t+1) = g[N_1, \cdots N_j, \cdots N_m]$$

(A) the aggregate state of the level L system may be a state variable in a level L+1 system of equations

*or*

(B) the level L model (e.g., the Leslie matrix M) may be a submatrix in a level L+1 matrix model (see DeAngelis et al. 1986)

$$M_{L+1} = \begin{vmatrix} M_i & 0 & \cdots & 0 \\ 0 & M_j & \cdots & 0 \\ \vdots & \vdots & \ddots & \vdots \\ 0 & 0 & \cdots & M_m \end{vmatrix}$$

$$n_1(t+1) = f_1 n_1(t) + f_2 n_2(t) + \cdots + f_s n_s(t)$$

$$n_2(t+1) = p_1 n_1(t)$$

$$n_3(t+1) = p_2 n_2(t)$$

.

.

.

$$n_s(t+1) = p_s n_s(t)$$

the system of equations is the level L holon

the state variables of the level L system are the level L-1 holons

FIGURE 7.2. An example of hierarchical organization in an ecological model. The focal level is an age-structured population. The system of equations describing that population represents the level L organization. The age-class state variables are the level L–1 holons. The level L+1 may be represented by (A) a system of equations in which the total population is one of $m$ population state variables, or (B) a matrix model in which the population's Leslie matrix is a submatrix; other populations are represented by similar submatrices (DeAngelis et al. 1986). Terms are defined in the text.

decomposability of the hierarchical structure would be violated. Either the system is not hierarchically structured, a legitimate possibility, or we have failed to describe the system organization adequately. The population might not, in this circumstance would not, be the level of organization above age classes. That level, instead, is a system of predator–prey holons that cut across more traditional population boundaries.

This description resembles the familiar system of equations of a small-number approach. However, there are "constraints" that emerge from the explicit hierarchical approach. For example, all of the state variables are limited to the same level of organization. Interactions are of approximately the same frequency and strength. In addition, the assumption of hierarchical organization means that higher level boundary conditions are applied almost equally to all state variables. Higher level boundary conditions that affect state variables too strongly or frequently violate the assumptions of the nearly decomposable hierarchical structure. Too much variability in the influence that a boundary condition has on different state variables also violates that assumption. Careful adherence to these rules of hierarchical organization can lead to nontraditional, even radical, representations of ecological systems.

The principles and implications of a truly hierarchical approach to modeling ecological systems have been discussed elsewhere (Caswell et al. 1972, Overton 1977, O'Neill et al. 1986, 1989). They merit further attention. My presentation here is intended simply as an introduction and to encourage explicit consideration of hierarchical organization in the construction of ecological models.

## 7.3  Scale

### 7.3.1  Scale and Hierarchical Organization

Considerations of scale arise naturally in the consideration of ecological hierarchies. Given the definition of a *hierarchically organized system* as one ordered by differences in process rates, dynamics, and frequency of behavior, consideration of hierarchical organization is inseparable from a consideration of temporal scale. Higher levels of organization exhibit lower frequency behaviors; they are "slower." The flock's response to a disturbance is slower than the response of the individual birds. Changes in the size of the entire population are of lower frequency (slower) than the birth and death of individuals or the turnover of age classes. Higher levels of organization are of larger temporal scale.

Similarly, considerations of spatial scale are inseparable from consideration of nested hierarchical systems, a special but common class of hierarchical systems. Nested hierarchical systems are those in which higher levels physically contain the lower levels (Allen and Starr 1982, Allen and

Hoekstra 1984). The hierarchical system from cells to organism is an example of a nested hierarchy. So are the hierarchical organization of individuals within a spatially distributed population and the hierarchical ordering of trees within gaps in forested stands of a landscape (Urban et al. 1987). To be physically nested or contained within a higher level, the lower levels must be smaller than, and occupy less space than, the higher levels. Thus, lower levels of organization in a nested hierarchy are of smaller or finer spatial scale than higher levels. Because the nested hierarchy is also a rate-structured hierarchy, lower levels of a nested hierarchy are also of smaller temporal scale. Ecological hierarchies are commonly nested, and unless otherwise noted, subsequent references to hierarchy or hierarchical organization below should be taken to mean nested hierarchies.

## 7.3.2 Scale, Level, and Scale of Observation

Unfortunately, the close relationship between scale and hierarchy has led to some confusion or lack of precision in the use of the terms *level* and *scale* (Allen and Hoekstra 1990, 1992, Allen in press, O'Neill and King in press). Some definitions are in order (also see Turner et al. 1989b and Box 7.5).

First, as I have described above, *level* refers to level of organization in a hierarchically organized system. The level designation is a rank ordering relative to other levels in the same hierarchically organized system. For example, the components of the focal level L are level L–1, their components are level L–2, and so on. Similarly, levels above the focal level are designated L+1, L+2, and so on. The designation has meaning only within that particular hierarchical system and relative to some arbitrary level chosen as a starting point. I could just as easily elect to designate the focal level as level 0; the lower levels would then be designated –1, –2, etc. Despite its ordinal character, the level designation is almost a qualitative characterization (i.e., higher than, lower than).

*Scale*, on the other hand, refers to physical dimensions. Scale is recorded as a quantity and involves measurement and measurement units. Things, objects, processes, and events are characterized and distinguished from others by their scale. Reference to the size of an object, for example, is a characterization by scale. So, too, is the duration or frequency of a process. A bird can be characterized by weight (a mass scale with Système Internationale, SI, units of kilogram) or its length (a spatial dimension, a length scale, with SI units of meters). A population can be characterized by the number of individuals (a count scale), biomass (a mass scale), or density (a derived scale combining count and space, i.e., area). A patch of vegetation or a landscape can be characterized by its area (a length [area] scale with SI units of square meters). The life span of a bird is a characterization on a temporal scale (perhaps in units of years), as is the turnover time of the population.

*Scale* also refers to the scale of observation, the temporal and spatial dimensions at which and over which observations are made and measure-

---

# Box 7.5.   Level and Scale

LEVEL:
Level of organization in a hierarchically organized system.

SCALE:
The physical dimensions of a thing or an event.

SCALE OF OBSERVATION:
The spatial and temporal scales at which observations of natural systems are made.

EXTENT:
The area over which observations are made and the duration of those observations.

GRAIN:
The spatial and temporal resolution of observations.

---

ments are taken. Scale of observation has two parts: grain and extent. Grain is the temporal or spatial resolution of an observation set. Extent is the areal expanse or the length of time over which observations are made. The $\tau$ and $T$ in Simon's (1973) separation of low-, middle-, and high-frequency behavior described above are grain and extent, respectively. Sampling frequency in both space and time contributes to observation grain. Instrument and observer response time also contribute to the grain of observation, as does the time and space a single measurement integrates. For example, the size of a sampling frame or the duration of a flux measurement is part of the grain of the observation set. Sampling a transect at 100m with a sampling frame of $0.1\,m^2$ generates a finer grained observation set than sampling with a $1\,m^2$ frame. But with either frame, sampling the transect at 10-m intervals instead of 100m also generates a finer grained observation set. Mapping vegetation cover or habitat with a grid of 30m × 30m cells produces a finer-grained observation set than mapping with a grid of 1km × 1km cells. Observation for a year generates an observation set of larger extent than observation for one week. Mapping an area of $1000\,km^2$ produces an observation set of larger extent than mapping an area of 1000ha. Grain and extent are generally correlated; observation sets of large extent will generally be coarser grained than those of smaller extent. For practical reasons, sampling a large spatial extent for a long time will generally require sampling at a coarse spatial resolution and infrequently. A smaller extent can be sampled more frequently with finer spatial resolution. The relative grain (the ratio of grain to extent) might be similar, but the absolute grain will generally be finer in an observation set of smaller extent.

*Scale* and *level* (level of organization) are not the same things. Scale, whether in reference to the size of something or to scale of observation, involves physical dimensions, quantification, measurement, and units. Level is a relative ordering of system organization. There are no units for level designation. It is inappropriate to use the terms *scale* and *level* interchangeably as if they were synonymous. They are not.

It is true that the scale of observation determines the observable, and it therefore influences both the perception of hierarchical organization and its description (Simon 1973, Allen and Starr 1982, Allen et al. 1984). Choices of $\tau$ and $T$ (temporal grain and extent) will determine the system behaviors observed as middle-frequency dynamics (Simon 1973); different dynamics will be observed with different choices of $T$ and $\tau$. Similarly, the choice of spatial grain and extent will determine the observed spatial patterns. Too coarse a grain will obscure fine-scale patterns; too small an extent will miss larger scale patterns. The sharpness of the separation of middle-frequency behaviors from low and high frequencies is determined by the scale of observation, choice of $T$ and $\tau$. The sharper the separation, the more distinctive the separation between layers and the better the approximation of the observable dynamics with a theory or model of hierarchical organization that ignores both very slow and very fast (low- and high-frequency) behaviors, treating them as constants. If the choice of $T$ and $\tau$ does not lead to sharp separation, the frequency range of observable dynamics will be broad, and explanation will be hard won. The middle-number characteristics of the system will be most apparent.

Similarly, while the choice of focal level is in some sense arbitrary, the choice will often dictate the scale of observation. Observation of the chosen level will require observations at appropriate scales. You cannot see the forest if you are looking only at the trees. You cannot see the landscape if you are looking only at the patches. Furthermore, linking observations at the focal level with the next higher and lower levels places additional requirements on the grain and extent of the observation set. An increase in extent is required to observe higher levels of organization; finer grained observation is required to observe lower levels of organization. Manipulation of the scale of observation (both grain and extent) is an essential tool for extracting and linking levels of organization (Allen et al. 1984).

It is also true that a subsystem or holon of level L organization can be characterized by reference to scale. A level can, of course, be characterized by the temporal scale of the frequencies associated with that level because the frequencies define that level. The system of oldfield components with dynamics revealed by observation of extent $T$ and grain $\tau$ can be characterized as a level with time scales between $\frac{1}{T}$ and $\frac{1}{\tau}$. In a nested hierarchical organization a level can also be characterized by the spatial extent over which behaviors of certain frequencies operate or by the spatial scales of observation set used to discern that level of organization. The more tangible

FIGURE 7.3. The conventional biological/ecological hierarchy. This conventional view of hierarchical organization is misleading and should be abandoned.

the holons (sensu Allen and Starr 1982), the easier (and more tempting) it is to characterize them by scale. For example, an individual organism, a tangible level of organization in the hierarchically organized system of cells, tissues, and organs, is readily characterized by its physical dimensions in space or its mass. With more difficulty, the population level of organization that emerges from the interactions among individuals can be characterized by its scale, the spatial extent it occupies, or the time it takes for all members of the population to be replaced.

Because the tangible objects associated with the lower "levels" in the conventional biological/ecological hierarchy (Figure 7.3) are easily characterized by scale, there exists a confused association between scale, level, and criteria for system specification. By tradition, if not definition, *populations* are thought of as a collection of individual organisms and *communities* as collections of populations. The *ecosystem* is the community plus the environment, and *landscapes* have been defined as a collection of ecosystems (Forman and Godron 1986). With this conceptual background, the scale relationships between tangibles of the biological hierarchy (cells to organisms) have been extrapolated to apply to higher "levels," with the implication that higher "levels" in the conventional ecological hierarchy are larger scaled than lower ones. The extrapolation implies, for example, that the community is larger scaled than the population. The inclination to make this extrapolation is reinforced by the fact that questions associated with the topmost "levels" of the conventional hierarchy have historically been addressed at larger spatial scales. Consequently, ecosystems are somehow thought of as being larger scaled than populations and landscapes as larger scaled than ecosystems.

It is easy, however, to come up with examples that contradict this view. The ecosystem of a rotting log is smaller than the forest community, and the surface of that log is a moss landscape (Allen and Hoekstra 1990). The skin of a large vertebrate is the landscape for a community of fleas and lice. As argued by Allen and Hoekstra (1990, 1992), "Conventional levels of organization are not scale-dependent . . ." (Allen and Hoekstra 1990, p. 5). The principles, concepts, and approaches associated with each conventional "level" can be applied readily at many different scales. Biome concepts, for example, can credibly be applied to a frost pocket within a forest (Allen and Hoekstra 1992).

The "levels" in the conventional hierarchy are best thought of as approaches (O'Neill et al. 1986), perspectives, criteria for observation and system specification (Allen et al. 1984, Allen and Hoekstra 1990, 1992), or types (Allen et al. in press). For example, I may choose to view the forested hillside I can see from my office window from an ecosystem perspective; in which case I will likely describe it as a collection of carbon and nitrogen pools (e.g., biomass and soils) with flows among them. From this perspective I do not resolve (I am not interested in) individual trees, and I will assume the species composition is constant and parameterized by rates of biomass accumulation or litter decomposition. I may instead decide to view the hillside as a community in which I am very much interested in species composition. I might elect to describe the hillside as a community type (e.g., oak–hickory) and quantify species composition with a dominance curve, or I might take an individual-based approach and describe the dynamics of the hillside with FORET (Shugart and West 1977) or one of the other forest-gap succession models (Shugart 1984), in which case I will not resolve the spatial location of individual trees. Alternatively, I might adopt a spatially explicit landscape perspective and indeed quantify the location of each tree on the hillside, perhaps to investigate seed dispersal or the movement of squirrels through the landscape and the consequences for their population dynamics. In each case, it is the same material system, just looked at in different ways. These "ways" are the "levels" in the conventional ecological hierarchy (O'Neill et al. 1986, Allen and Hoekstra 1992). They are not "scales," and one perspective is not intrinsically larger scaled than another. In my view, the "levels" in the conventional hierarchy are "perspectives," as in "an ecosystem perspective," or "criteria for system specification." But in the discussion that follows I will also refer to the conventional "levels" as ecological 'types" to avoid confusing them with *level* as in "level of hierarchical organization."

The confusion between level and scale engendered by the conventional ecological hierarchy is even more egregious when the type (e.g., community or ecosystem) or perspective (e.g., community perspective or ecosystem perspective) is naively assumed to be a level in a hierarchically organized system. Because a type is thought of as a collection of other types (e.g., the community is a collection of populations), there is an a priori assumption that the type represents a level of organization. This simplistic view of hierarchical organization is reinforced by extrapolating the classical cell-to-organism example of hierarchical organization to higher order types in the conventional hierarchy. Thus, for example, the sequence of types from individual to population to community is somehow thought to be equivalent to the sequence from cells to tissues to organs. Moreover, because an ecological type is thought, erroneously as we have seen, to represent a specific range of scales, those scales are thought to represent a particular level of organization. For example, ecosystems are generally thought to be large, larger at least than the population or community. Areas large enough to encompass more than one population or community are thus thought to represent the ecosystem "level" in the conventional hierarchy. But, as we have seen, an ecosystem perspective can just as easily be applied to the area occupied by a rotting log. Ironically, discussions of scale and hierarchical organization in ecological systems have reinforced this latter error because it is true that higher levels of organization in a nested hierarchical system do occupy larger spatial scales (extents) than lower levels. It is just that levels of hierarchical organization extracted from the observation set by manipulation of the scale of observation (grain and extent) "do not correspond in any simple way to the traditional levels [types] of biological organization" (O'Neill and King in press)!

## 7.3.3 *Landscape, Landscape Scale, and Landscape Level*

The insidious confusion between level and scale has found its way into landscape ecology. The terms *landscape scale* and *landscape level* appear frequently. They are often used as if they were interchangeable. I hope that it is clear by now that they are not. In this section, I define the differences between these terms, and I define the concept of a *landscape* from the perspective of hierarchy theory.

The common term *landscape scale* is best understood as a convenient shorthand for *the scale of the landscape*. But even this is wanting. Forman and Godron (1986, p. 11) defined the lower limit of landscape size as "a few kilometers in diameter." They argued (p. 11) that "areas of a few meters or hundreds of meters across are at a finer scale than a landscape." This is too restrictive and too anthropocentric a definition of *landscape*. The fundamental themes of landscape ecology are *not* scale dependent or limited only to spatial extents greater than a few square kilometers. The fundamental concepts of landscape can be, and have been, successfully applied to areas much smaller than "a few kilometers in diameter" (e.g., at scales from $0.1\,m^2$ to 1 ha; Wiens and Milne 1989, Crist and Wiens 1994, With and Crist 1995). Questions of how spatial heterogeneity influences biotic and abiotic processes, one of the themes defining landscape ecology (Turner and Gardner 1991), can be addressed at virtually any spatial scale. Having adopted an organism-centered view (Wiens 1985), a landscape can as credibly be viewed with a microscope (mites in a field of epidermis) as from a satellite orbiting the Earth.

Thus, the term *landscape scale* is reasonably a shorthand for *the spatial extent of the landscape under consideration*. As such, it is probably useful, but only in conjunction with and following an explicit definition of the spatial scales (both grain and extent) of that landscape (Bissonette, this volume; Milne, this volume). The term should be avoided if used without qualification or even to imply a certain range of scales (Allen in press). The reasonable range of scales is too large. Note also that this usage leaves out any reference to temporal scale. A full qualification of the landscape scale should also include specification of the temporal grain and extent.

The term *landscape level* should be avoided because it implies that the landscape is, a priori, a level of organization. It should not be assumed that the landscape is a level just because landscape is a type in the conventional hierarchy. Neither should it be assumed that the landscape is a level just because it is larger than and contains either an observed or assumed level of organization. The term *landscape level* might be an appropriate shorthand for the *level of organization revealed by observation at the scale of the landscape* (i.e., over extent S with grain σ), but only after that level and the accompanying hierarchical organization have been extracted from the data by manipulation of the scale of observation. Reference to the landscape level implies that the landscape exists or behaves as a consequence of interactions among components at the next lower level and that it interacts with other landscapes as part of a system at the next higher level. Conceiving of the landscape as a level also implies that the entirety of the landscape is contained within the extent of the landscape. If not, the integrity, the wholeness, of the "landscape level" is violated and indeed could not even be perceived as a level of organization. A landscape is often conceptualized as an assemblage of patches. Do these patches interact to form the landscape? Do the holistic or aggregate properties or behaviors of the landscape (the larger extent) change if the pattern of patches is altered but the composition remains the same? I suspect that in many cases they do not, in others they may. If they do, then it may be appropriate to refer to the level of the landscape. However, it must be shown that the landscape exhibits these holistic system properties. Otherwise, if the landscape is simply the collection or set of patches, it is probably more appropriate to simply refer to the (appropriately qualified) scale of the landscape.

Allen (in press) believes that the terms *landscape level* and *landscape scale* should be avoided "in favor of scale-specific and scale-independent usages" because of the confusion between scale and level, but he prefers *landscape level* to *landscape scale*. His landscape level is a type of organization defined by certain properties. He refers to landscape level as "level of organization," but his levels of organization are defined types rather more like the types in the conventional hierarchy and not levels of system organization revealed by observation. He refers to levels revealed by manipulations of the scale of observation as "scalar levels of observation." His "scalar levels of observation" are my "levels of organization" revealed by "scales of observation." His "landscape level of organization" is my "landscape type," "landscape criteria," or "landscape perspective." I prefer to retain *level of organization* for a hierarchical level of organization revealed by manipulation of observation grain and extent and to redefine the conventional levels of organization as types rather than introduce the new concept: *level of observation*. I also think the implications of a careless use of the term *landscape level* are more egregious than the use of *landscape scale*. I agree that *landscape scale* should always be explicitly qualified, but I think the vagueness of too broad an implicit range of spatial scales is preferable to the implication of a hierarchical organization that may or may not exist. I argue that the term *landscape level* should be used only with great care, and never used to mean "scale of the landscape."

What then is a landscape? A *landscape* is first and foremost a spatial extent. This view of landscape is usually qualified as a spatially heterogeneous area (e.g., Turner and Gardner 1991). Heterogeneity or two dimensions may not be necessary qualifiers. Spatial distribution has implications for interactions among system components even in a homogeneous environment (neighboring objects are expected to interact more frequently and perhaps more strongly than distant objects), and interaction in one-dimensional or three-dimensional space is conceivable. I will concede, however, that for most purposes a landscape is a heterogeneous area. Allen and Hoekstra (1992) identify landscape as one of several criteria of ecological integration: landscape, ecosystem, community, organism, population, and biome and biosphere. I agree that there is a landscape perspective or approach (i.e., the perspective of landscape ecology) and landscape criteria (i.e., the themes and subject matter of landscape ecology), but I think that *landscape* without the qualifiers of "scale" or "level" is best understood as a spatial extent. The landscape may sometimes qualify as a holon or level of organization if revealed by changes in the grain and extent of observation. Recall, however, that if the landscape is to qualify as focal level, the extent of the observation set must be larger than the extent of the landscape if it is to reveal the next higher level of organization. Manipulations of the grain within the landscape extent may reveal lower levels of system organization, in which case, the landscape will always exist as spatial context, either the higher level context in a hierarchical organization or the simple box or frame in which a hierarchical organization is observed.

## 7.3.4 Implications for Landscape Ecology

With this introduction to some fundamentals of hierarchy theory (and I encourage readers to explore for themselves the discussions of hierarchy theory applied to evolutionary and ecological systems, as well as the more general treatments of hierarchy theory cited above), I now consider further the implications of a hierarchical perspective for landscape ecology.

First, note that hierarchy theory presupposes the existence of a system. The application of hierarchy theory to landscape ecology thus presupposes conceptualization of the landscape as a system or that at least it contains a spatially distributed system. If the landscape is simply viewed as the stage, the spatial extent, on which ecological phenomena are played out, many of the insights and tools provided by hierarchy theory do not apply.

Much of the discipline of landscape ecology has nothing, or very little, to do with hierarchy theory. Some of the dominant topics in landscape ecology over the past decade or more (e.g., indices of landscape pattern and percolation models) have everything to do with spatial scale, especially scale of observation (e.g., Turner et al. 1989a, Gardner et al. 1989), but very little to do with hierarchical system organization. Even nominally hierarchical approaches often are more about hierarchical sets or hierarchical ordering than hierarchical system organization as I have presented it here (e.g., hierarchical patch structure, Kotliar and Wiens 1990). This in no way implies either that the landscape studies are wanting or that consideration of hierarchical nesting of patches is inappropriate or even misnamed. It simply means that good landscape ecology can be practiced without explicit reference to hierarchy theory as a theory of system organization. It probably cannot be done without a theory of scale, especially spatial scale. Landscape ecology has both drawn upon and contributed to (e.g., Urban et al. 1989, Wiens 1989) a theory of spatial scale in ecological systems.

On the other hand there are areas where hierarchy theory is beneficial, even necessary. "Scaling-up" and other translations of information across spatial scales often involve translations across levels of organization (e.g., translating plant physiology from leaves to canopies to stands of vegetation). These translations benefit from an explicit consideration of hierarchical organization (Allen et al. 1984, King et al. 1990, 1991, King 1991). Hierarchy theory can also inform considerations of landscape response to disturbance (e.g., Allen and Starr 1982, King 1993). Explorations of the temporal dynamics of the landscape and landscape stability can draw deeply on hierarchy theory (O'Neill et al. 1989). More generally, whenever a spatially distributed dynamical system is involved, hierarchy theory may prove useful.

Consider, as an example, an idealized or hypothetical wildlife species occupying a spatial extent of several thousand hectares. The females of this species establish exclusive territories or home ranges with only minimal spatial overlap. The female territories are very nearly constant in size (approximately 1 ha), and they are vigorously patrolled and defended. In poor years the female territories are sparsely distributed; in good years they are distributed more densely. The species is polygamous; a male's territory/ home range overlaps those of several females. He mates with every female within his territory and defends his females against other males. The sizes of the male territories vary with the number of females in his home range. Some have only one female; others have many. The number of females that a male defends and mates varies from year to year. Breeding occurs once a year. Average litter size at parturition is 6 but varies from 2 to 8 and seems to depend on the quality of habitat within the female's home range. The young remain in the maternal home range until the next breeding season. Then they disperse to seek and establish their own home range. Survivorship to dispersal varies from 20% to 80%, and like litter size, it varies with the habitat quality of the maternal territory. Groups of males and their females form colonies or social groupings. The colonies seem to occupy patches of suitable habitat separated by habitat in which females will not or cannot establish a territory. The colonies generally are separated by several kilometers. Adult mortality is high, and the turnover of home ranges within a colony is rapid. Infrequently, dispersing males will emigrate to another colony, usually when favorable environmental conditions have resulted in a larger than normal number of males. Even more infrequently, surplus females migrate to join other colonies. Rarely, apparently in synchrony with a drought cycle with a return time of many decades to a century, the social structure collapses, colonies are abandoned, and individuals wander about but do not breed until conditions are again favorable. When that occurs, new colonies are established, usually in new locations, with a random reassortment of individuals who have survived the drought.

The complex dynamics of this admittedly artificial middle-number system span a range of temporal and spatial scales, and there are several layers of organization: the metapopulation of colonies, the colonies themselves, the male plus several females forming a cluster within a colony, and the individuals and their home ranges. Resolution of the long-term low-frequency dynamics of colony breakup, population slump, and establishment of new colonies followed by population increase would require an observation set of long temporal extent and large spatial extent (many decades and many square kilometers). The spatial and temporal grain of that observation set could be coarse: perhaps sampling at intervals of 1 km once every few years. The resolution of individual territories would require an observation set of much finer spatial grain (perhaps 100 m). Describing the turnover of the female territories would require an observation set with a temporal grain of about a year; a finer grain would provide little or no additional information. Only an extraordinary observation set of large extent and fine grain would resolve the full range of system frequencies.

If the colony was chosen as the focal level and the scale of the observation set designed to resolve that structure, the infrequent introduction of immigrants would appear as an external forcing, a boundary condition. It might appear stochastic or constant depending upon the temporal scale of the observation set. Only by increasing the extent of the observation set would that behavior be observed and explained as exchange between interacting colonies. Similarly, the occasional collapse of the colony would appear as an episodic stochastic event. Only by increasing the extent of the observation set would that behavior be explained.

Explanation of interannual variations in population size of the colony might be best achieved by manipulating the grain of the observation set to resolve the clusters of male plus females as the next lower level of organization. Details of the spatial distribution of individual female territories could probably be successfully ignored. Certainly the high frequency movements of individuals patrolling their territories and interactions between mother and young could be ignored. Forcing a decomposition of the colony into individuals based on the simplistic or conventional assumption that populations are composed of individuals could lead to an inefficient and perhaps inaccurate description of the system.

Space does not allow further analysis of this hypothetical system, nor is it warranted given its extremely artificial nature. The example does, however, illustrate the issues of scale and system organization that must be addressed in approaching a spatially distributed population. It also illustrates the guidance that hierarchy theory can provide. Analysis of spatially distributed wildlife populations may benefit from an explicit and rigorous application of hierarchy theory and the principles of hierarchical organization. Spatially explicit population models promise to become an increasingly common tool in the management of spatially distributed wildlife populations. Principles of hierarchy theory and hierarchical organization should be employed in the development and implementation of those models.

## 7.4  Lessons Learned

Hierarchy theory is perhaps most important as a heuristic for system specification. The theory provides guidance for approaching the system by selective choice of the grain and extent with which to observe the system. If a landscape, a spatially distributed system, can be described as a hierarchically organized system, hierarchy theory provides an explanatory framework and guidance for further analysis. It also provides guidelines for model development.

Over the past decade, landscape ecology, more than any other field of ecology, has been a test bed and stimulus for the development of an ecological hierarchy theory. Most of the advances have been in a theory of scale and a framework for describing the hierarchical ordering of patch structure. Advances in hierarchy theory as a theory of spatially distributed dynamics have been fewer and slower. Nevertheless, a few guiding principles have emerged that should be viewed as lessons learned.

First, levels of organization should always be extracted from the data and not preimposed by a priori assumptions about what the levels of explanation should be (O'Neill and King in press). The observation set is primary. The investigator's choice of focal level will help define the scale of the initial observations. However, it is only by manipulating the grain and extent of the observations (while keeping criteria and the currency of interactions between components constant) that meaningful and explanatory relationships can be found.

Second, the traditional or conventional ecological hierarchy should be abandoned. Evidence shows that levels extracted from the empirical observation sets do not correspond to the traditional levels (O'Neill and King in press). Reference to the conventional "levels" or attempting to reconcile them with levels extracted from the data only generates confusion. The conventional types are useful only when interpreted as relative scale-independent criteria, perspectives, and approaches.

Third, the concepts of scale and level should be clearly distinguished. The terms *scale* and *level* should never be used synonymously. The term *landscape scale* should only be understood as a shorthand for *scale of the landscape*, and it should always be accompanied by an explicit specification of at least the spatial extent. It should not be used to refer to an implicit scale or range of scales. The term *landscape level* should be avoided unless it has been explicitly shown that a level of organization exists at the scale of the landscape under consideration.

Further development of hierarchy theory as a theory of spatially distributed systems is needed. Those developments will build upon and continue to benefit from the synergy of the relationship between landscape ecology and hierarchy theory as applied to ecological systems.

*7.5 Acknowledgments.* I want to thank Joyce Waterhouse for introducing me to Allen and Starr's (1982) *Hierarchy: Perspectives for Ecological Complexity*. That introduction led to subsequent interactions with Bob O'Neill and with Tim Allen. Those interactions, and the long hours of animated discussion with Bob, Tim, Alan Johnson, and Dean Urban that followed have had an enormous influence on my thinking about hierarchy theory and spatially distributed ecological systems. Support during the preparation of this paper was provided by U.S. Department of Defense Strategic Research and Development Program (SERDP) through the Office of Technology Policy, Office of Health and Environmental Research, U.S. Department of Energy, under contract DE-AC05-96OR22464 with Lockheed Martin Energy Research Corp.

## 7.6 References

Allen, T.F.H. In press. The landscape "level" is dead: persuading the family to take it off the respirator. In D.L. Peterson and V.T. Parker, editors. Scale issues in ecology. Columbia University Press, New York, New York, USA.

Allen, T.F.H., and T.B. Starr. 1982. Hierarchy: perspectives for ecological complexity. University of Chicago Press, Chicago, Illinois, USA.

Allen, T.F.H., and T.W. Hoekstra. 1984. Nested and non-nested hierarchies: a significant distinction for ecological systems. Pp. 175–180 in A.W. Smith, editor. Proceedings of the Society for General Systems Research. I. Systems methodologies and isomorphies. Intersystems Publications, Coutts Library Services, Lewiston, New York, USA.

Allen, T.F.H., R.V. O'Neill, and T.W. Hoekstra. 1984. Interlevel relations in ecological research and management: some working principles from hierarchy theory. General Technical Report RM-110. U.S. Department of Agriculture, Forest Service, Rocky Mountain Forest and Range Experiment Station, Fort Collins, Colorado, USA.

Allen, T.F.H., and T.W. Hoekstra. 1990. The confusion between scale-defined levels and conventional levels of organization in ecology. Journal of Vegetation Science 1:5–12.

Allen, T.F.H., and T.W. Hoekstra. 1992. Toward a unified ecology. Columbia University Press, New York, New York, USA.

von Bertalanffy, L. 1968. General system theory. George Braziller, New York, New York, USA.

Caswell, H., H.E. Koenig, J.A. Resh, and Q.E. Ross. 1972. An introduction to systems science for ecologists. Pp. 3–78 in B.C. Patten, editor. Systems analysis and simulations in ecology. Vol. II. Academic Press, New York, New York, USA.

Crist, T.O., and J.A. Wiens. 1994. Scale effects of vegetation on forager movement and seed harvesting by ants. Oikos 69:37–46.

DeAngelis, D.L., W.M. Post, and C.C. Travis. 1986. Positive feedback in natural systems. Springer-Verlag, Berlin, Germany.

Dunning, Jr., J.B., D.J. Stewart, B.J. Danielson, B.R. Noon, T.L. Root, R.H. Lamberson, and E.E. Stevens. 1995. Spatially explicit population models: current forms and future uses. Ecological Applications 5:3–11.

Forman, R.T.T., and M. Godron. 1986. Landscape ecology. John Wiley and Sons, New York, New York, USA.

Gardner, R.H., R.V. O'Neill, M.G. Turner, and V.H. Dale. 1989. Quantifying scale-dependent effects of animal movement with simple percolation models. Landscape Ecology 3:217–227.

Grobstein, C. 1973. Hierarchical order and neogenesis. Pp. 31–47 in H.H. Pattee, editor. Hierarchy theory: the challenge of complex systems. George Braziller, New York, New York, USA.

Huston, M., D. DeAngelis, and W. Post. 1988. New computer models unify ecological theory. BioScience 38:682–691.

Kerner, E.H. 1957. A statistical mechanics of interacting biological species. The Bulletin of Mathematical Biophysics 19:121–146.

King, A.W. 1991. Translating models across scales in the landscape. Pp. 479–517 in M.G. Turner and R.H. Gardner, editors. Quantitative methods in landscape ecology. Springer-Verlag, New York, New York, USA.

King, A.W. 1993. Considerations of scale and hierarchy. Pp. 19–45 in S. Woodley, G. Francis, and J. Key, editors. Ecological integrity and the management of ecosystems. Lewis Publishers Inc., Chelsea, Michigan, USA.

King, A.W., W.R. Emanuel, and R.V. O'Neill. 1990. Linking mechanistic models of tree physiology with models of forest dynamics: problems of temporal scale. Pp. 241–248 in R.K. Dixon, R.S. Meldahl, G.A. Ruark, and W.G. Warren, editors. Process modeling of forest growth responses to forest stress. Timber Press, Portland, Oregon, USA.

King, A.W., A.R. Johnson, and R.V. O'Neill. 1991. Transmutation and functional representation of heterogeneous landscapes. Landscape Ecology 5:239–253.

Koestler, A. 1967. The ghost in the machine. Macmillan, New York, New York, USA.

Kotliar, N.B., and J.A. Wiens. 1990. Multiple scales of patchiness and patch structure: a hierarchical framework for the study of heterogeneity. Oikos 59:253–260.

Liu, Y., J.B. Dunning, Jr., and H.R. Pulliam. 1995. Potential effects of a forest management plan on Bachman's sparrow (*Aimophila aestivalis*): linking a spatially explicit model with GIS. Conservation Biology 62–75.

O'Neill, R.V. 1988. Hierarchy theory and global change. Pp. 29–45 in T. Rosswall, R.G. Woodmansee, P.G. Risser, editors. Scales and global change. John Wiley and Sons, New York, New York, USA.

O'Neill, R.V. 1989. Perspectives in hierarchy and scale. Pp. 140–156 in J. Roughgarden, R.M. May, and S.A. Levin, editors. Perspectives in ecological theory. Princeton University Press, Princeton, New Jersey, USA.

O'Neill, R.V., D.L. DeAngelis, J.B. Waide, and T.F.H. Allen. 1986. A hierarchical concept of ecosystems. Princeton University Press, Princeton, New Jersey, USA.

O'Neill, R.V., J.R. Krummel, R.H. Gardner, G. Sugihara, B. Jackson, D.L. DeAngelis, B.T. Milne, M.G. Turner, B. Zygmunt, S.W. Christensen, V.H. Dale, and R.L. Graham. 1988. Indices of landscape pattern. Landscape Ecology 1:153–162.

O'Neill, R.V., A.R. Johnson, and A.W. King. 1989. A hierarchical framework for the analysis of scale. Landscape Ecology 3:193–206.

O'Neill, R.V., and A.W. King. In press. Homage to St. Michael: or why are there so many books on scale? In D.L. Peterson and V.T. Parker, editors. Scale issues in ecology. Columbia University Press, New York, New York, USA.

Overton, W.S. 1977. A strategy of model construction. Pp. 53–70 in C.A.S. Hall and J.W. Day, editors. Ecosystem modeling in theory and practice: an introduction with case histories. John Wiley and Sons, New York, New York, USA.

Prigogine, I. 1980. From being to becoming: time and complexity in the physical sciences. W.H. Freeman and Company, New York, New York, USA.

Pulliam, H.R., J.B. Dunning, Jr., and J. Liu. 1992. Population dynamics in complex landscapes: a case study. Ecological Applications 2:165–177.

Rosen, R. 1989. Similitude, similarity, and scaling. Landscape Ecology 3:207–216.

Salthe, S.N. 1985. Evolving hierarchical systems. Columbia University Press, New York, New York, USA.

Serway, R.A. 1986. Physics for scientists and engineers with modern physics. Second edition. Saunders College Publishing, Philadelphia, Pennsylvania, USA.

Shugart, H.H. 1984. A theory of forest dynamics. Springer-Verlag, New York, New York, USA.

Shugart, H.H., Jr., and D.C. West. 1977. Development of an Appalachian deciduous forest succession model and its application to assessment of the impact of the chestnut blight. Environmental Management 5:161–179.

Simon, H.A. 1962. The architecture of complexity. Proceedings of the American Philosophical Society 106:467–482.

Simon, H.A. 1973. The organization of complex systems. Pp. 3–27 in H.H. Pattee, editor. Hierarchy theory: the challenge of complex systems. George Braziller, New York, New York, USA.

Turner, M.G., R.V. O'Neill, R.H. Gardner, and B.T. Milne. 1989a. Effects of changing spatial scale on the analysis of landscape pattern. Landscape Ecology 3:153–162.

Turner, M.G., V.H. Dale, and R.H. Gardner. 1989b. Predicting across scales: theory development and testing. Landscape Ecology 3:245–252.

Turner, M.G., and R.H. Gardner. 1991. Quantitative methods in landscape ecology: an introduction. Pp. 3–14 in M.G. Turner and R.H. Gardner, editors. Quantitative methods in landscape ecology. Springer-Verlag, New York, New York, USA.

Turner, M.G., G.J. Arthaud, R.T. Engstrom, S.J. Hejl, J. Liu, S. Loeb, and K. McKelvey. 1995. Usefulness of spatially explicit population models in land management. Ecological Applications 5:12–16.

Turner, M.G., Y. Wu, W.H. Romme, and L.L. Wallace. 1993. A landscape simulation model of winter foraging by large ungulates. Ecological Modelling 69:163–184.

Urban, D.L., R.V. O'Neill, and H.H. Shugart, Jr. 1987. Landscape ecology. BioScience 37:119–127.

Weinberg, G.M. 1975. An introduction to general systems thinking. John Wiley and Sons, New York, New York, USA.

Wiens, J.A. 1985. Vertebrate responses to environmental patchiness in arid and semiarid ecosystems. Pp. 169–193 in S.T.A. Pickett and P.S. White, editors. The ecology of natural disturbance and patch dynamics. Academic Press, New York, New York, USA.

Wiens, J.A. 1989. Spatial scaling in ecology. Functional Ecology 3:385–397.

Wiens, J.A., and B.T. Milne. 1989. Scaling of "landscapes" in landscape ecology, or landscape ecology from a beetle's perspective. Landscape Ecology 3:87–96.

With, K.A., and T.O. Crist. 1995. Critical thresholds in species' responses to landscape structure. Ecology 76:2446–2459.

# Section 2
## Landscape Metrics

# 8
# Neutral Models: Useful Tools for Understanding Landscape Patterns

Scott M. Pearson and Robert H. Gardner

## 8.1 Introduction

A neutral model is a minimum set of rules required to generate pattern in the absence of a particular process (or set of processes) being studied. The results of the neutral model provide a means of testing the effect of the measured process on patterns that are actually observed (Caswell 1976). If observed patterns do not differ from the neutral model, then the measured process has not significantly affected the observed pattern. Conversely, when results differ from model predictions in a way that is consistent with a particular process, then strong evidence for the importance of this process has been obtained. Several authors have argued that formulation of a proper neutral model is necessary for hypothesis testing, because data often exhibit nonrandom patterns in the absence of the causal mechanisms of interest (Quinn and Dunham 1983). This approach has been discussed extensively in the field of community ecology (e.g., Conner and Simberloff 1984, 1986; Haefner 1988) as well as other areas of biology (Nitecki and Hoffman 1987).

Neutral models are useful in landscape ecology, a field of ecology that emphasizes the complex relationships between landscape pattern and eco-

Scott M. Pearson is an Assistant Professor of Biology at Mars Hill College in western North Carolina. He teaches undergraduate biology courses in ecology and evolution. Landscape ecology principles are taught as part of these courses. His research interests include understanding the effects of spatial pattern of habitat on species density and diversity. He is presently studying the effects of forest fragmentation and historical land use on native species of the Southern Appalachian Mountains.

Robert H. Gardner's research interests involve the detection of scale-dependent phenomena and the development of new approaches for predicting ecological dynamics across temporal and spatial scales. He is currently a Professor at the Applachian Environmental Laboratory, which is part of the University of Maryland System, where he teaches a graduate course in landscape ecology. Gardner currently serves as Editor-in-Chief of the journal *Landscape Ecology*.

logical process (Turner 1989, Gardner and O'Neill 1991). Processes, such as disturbance, can produce landscape patterns by changing the abundance and location of habitat patches (Baker 1992). Likewise, patterns have important effects on ecological processes. For example, habitat fragmentation affects metapopulation dynamics (Holt et al. 1995), gene flow (Ballal et al. 1994), and dispersal (Santos and Telleria 1994). The purpose of this chapter is to demonstrate the usefulness of neutral models to landscape ecology by discussing how neutral models (1) assist the investigator in understanding patterns in spatial data and (2) are useful for generating maps for quantifying the effect of landscape pattern on ecological processes.

## 8.2  A Simple Neutral Model

Neutral models help landscape ecologists understand relationships between measures of spatial pattern and landcover abundance. A simple neutral model designed to explore the effect of changes in the abundance of a habitat on the spatial pattern of landcover (Gardner et al. 1987) was derived from the principles of percolation theory (Stauffer and Aharony 1992). The complex patterns generated by simple random maps and the insights provided by percolation theory were proposed as sufficient to test the importance of landscape process on observed landcover patterns.

To understand the effect of changing habitat abundance on measures of spatial pattern, Gardner et al. (1987) used the following simple neutral model consisting of a few straightforward rules:

1. Generate a map of $100 \times 100$ grid cells.
2. For each cell, randomly assign the presence or absence of landcover to each grid cell with probability $P$. Thus, $P$ represents the proportion of the map occupied by the landcover type of interest. In this chapter, $P$ represents the proportional abundance of suitable habitat; the abundance of unsuitable habitat is $1 - P$.
3. For each map, quantify the spatial pattern of landcover by recording the number, size, and shape of habitat patches.

This model is "neutral' in the sense that there is no process governing the spatial distribution of landcover; i.e., the distribution is random. Figure 8.1 shows examples of maps produced using this model. A set of nine landscapes having values of $P$ ranging from 0.1 to 0.9 were produced. Patch-based metrics (e.g., the number of patches, mean patch size, and the area of the largest patch), as well as the amount of edge measured as the total number of cells along the perimeter of all patches, were recorded.

We defined a *patch* as a cluster of adjacent cells of the same habitat type, allowing a theoretical organism to visit all the cells of the patch without having to cross any cells of unsuitable habitat. The rule used to connect cells and define patches is important because it affects both the number and size

P = 0.4

P = 0.6

P = 0.8

FIGURE 8.1. Simple random maps (100 × 100 cells) having different proportions (P) of suitable (black) habitat.

a.

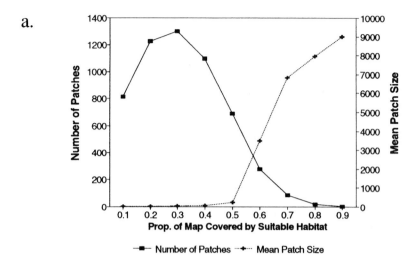

b.

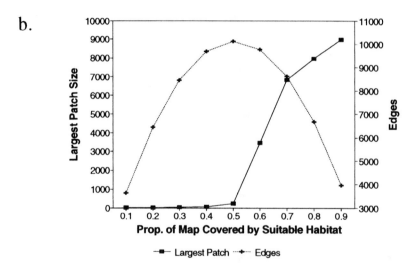

FIGURE 8.2. Landscape metrics for suitable habitat in simple random maps. Units for mean patch size, largest patch size, and edge are number of cells. *Edge* is defined as the number of cells bordering at least one cell of a different cover type. A "four-neighbor" rule was used to identify the patches (i.e., diagonal cells were not considered connected).

of patches. Some applications of percolation theory specify that habitat cells belonging to the same patch must be adjacent on the four cardinal directions. Habitat cells touching on the diagonals (subcardinal directions) are not considered to be connected. This type of rule is called a four-neighbor or nearest-neighbor rule. The analyses and results presented here used the four-neighbor rule. An eight-neighbor rule would allow organisms to move to cells on the diagonal; therefore, connecting cells in the subcardinal directions would be included as part of a habitat patch.

All of the patch-based measures were dependent on $P$. As $P$ increased from its lowest value, the number of patches of suitable habitat increased initially, but patch number began to decline above $P > 0.3$ as small patches began to coalesce (Figure 8.2a). The process of patch coalescence causes mean patch size to increase simultaneously (Figure 8.2a). Likewise, the amount edge increased initially as the number of patches increased, but declined as patches coalesced and the map became filled with suitable habitat (Figure 8.2b). The area of the largest patch increased slowly between $0.1 \leq P \leq 0.5$, then it increased rapidly at higher values of $P$ (Figure 8.2b). This behavior indicated the presence of a critical threshold near $P = 0.59$. A *critical threshold* is a region in parameter space where a small change in the parameter produces a relatively large change in the measured response. In this case, a small change in $P$ in the region $0.5 \leq P \leq 0.6$ produced a large change in the area of the largest patch.

The existence of such critical thresholds is an important contribution of percolation theory to our understanding of spatial patterns. This theory states that if we identify patches with the four-neighbor rule, 50% of very large maps randomly generated with $P = 0.5928$ will have a single, large habitat patch (i.e., the percolating cluster) that will span the entire map (Stauffer and Aharony 1992). If the eight-neighbor rule is used to identify habitat patches, then the critical threshold for $P$ is equal to 0.4072 (see Plotnick and Gardner 1993 for a discussion of critical thresholds for different neighborhood rules). O'Neill et al. (1988) have demonstrated that these patch-definition rules can be used to study the effect of $P$ on landscape connectivity. Pearson et al. (1996) used different neighborhood rules to compare the connectivity of landscapes for organisms with different dispersal capabilities. For example, a species with limited vagility (e.g., a snail) was represented by a four-neighbor movement rule, while a more vagile species (e.g., a mouse) was represented by an eight-neighbor rule. The critical values for landscape connectivity were lower for the eight-neighbor rule than for the four-neighbor rule.

Ecologists and land managers should be aware of the existence of these thresholds because they show that the relationship between habitat abundance and spatial pattern is nonlinear, especially near a critical threshold. This theory implies that small changes in abundance and landscape pattern might have little effect if the habitat is especially abundant or, in some cases, very rare. However, if the habitat is near a threshold level, then a small change in habitat abundance might change a previously connected landscape into a fragmented one. A manager might be caught by surprise when small reductions in habitat (e.g., due to management or economic development), that previously had little effect on a population, suddenly resulted in a population crash. In percolation theory, the threshold values are dependent on the neighborhood rule used to define patches. Therefore, it is important to realize that the "threshold value" must be defined with respect to a given taxon or ecological process because organisms and processes vary in their ability to move across a patchy landscape (O'Neill et al. 1988, Pearson et al. 1996).

The simple neutral model states that in the absence of specific landscape processes that affect the distribution of landcover, the number, size, and shape of patches will change as a function of the fraction of landscape occupied by the landcover type of interest (Gardner et al. 1987). These simple random models also have been used to compare data from actual landscapes. Gardner et al. (1991) found that the spatial pattern of habitat in maps of forests in Georgia (USA) was similar to randomly generated maps when the proportion of forest ($P$) was $\geq 0.6$. At $P < 0.6$, the spatial pattern of forest often differed from that of random maps. They concluded that the many processes that shape landscapes—including human alterations, local variations in topography and soils, the distribution of biota, and the frequency and severity of disturbances—act together to produce "unique" patterns of landcover. Use of random maps as neutral models revealed that the patterns of deforestation in landscapes where more than 40% of the forest had been lost were nonrandom and likely driven by the interactions between land-use choices and environmental characteristics such as topography and soil fertility.

## 8.3  Hierarchical Neutral Landscapes

The use of simple random maps as neutral models is based on the null hypothesis that there are no significant processes affecting the spatial pattern of landcover. In nature, spatial pattern often is structured by many factors, and ecologists may desire a model that is neutral to a specific set of factors while considering the organizing effects of other processes on landscape pattern. Gardner and O'Neill (1991) have described a method of producing maps that vary in levels of contagion; allowing the user to control pattern at a single spatial scale. However, real landscapes often exhibit shifts in spatial patterns as the spatial extent of the landscape being studied increases (O'Neill et al. 1991). For example, broad-scale differences in geology may affect abiotic properties of soils and/or hydrology that in turn, will significantly affect broad-scale changes in vegetation. Thus, the variation in geological patterns may be an important consideration for predicting the broad-scale patterns of vegetation. Ecologists interested in the spatial patterns of disturbance and plant succession also require neutral models that can generate random patterns at fine scales (i.e., the disturbance–succession scale) while allowing users to control the broad-scale variation of landcover types (i.e., geological constraints). Models for generating *hierarchical neutral landscapes* satisfy these needs

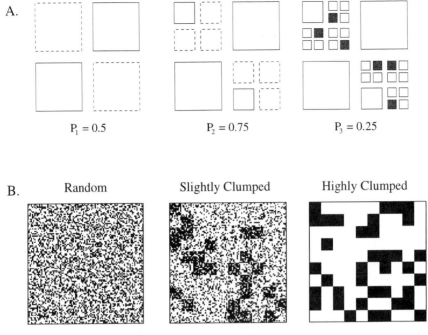

FIGURE 8.3. Method for making a hierarchical neutral map (A; see text) and examples of three maps produced using this method (B). In (A), dashed lines indicate division randomly selected to contain habitat in hierarchical levels 1 and 2. Shaded cells at the finest scale, level 3, contain suitable habitat. Each map in (B) has $P = 0.375$.

by creating random maps with control over pattern at successively finer spatial scales.

Hierarchical neutral models produce scale-dependent patterns by specifying the proportion of habitat present at several hierarchical scales. For example, a $100 \times 100$ cell map may be divided into four equal sized regions ($50 \times 50$ cells). The proportion of these regions that contain suitable habitat is specified as $P_1$ (see Figure 8.3a). Next, each of these regions can be divided into four subregions of $25 \times 25$ cells each. The proportion of these subregions that contain suitable habitat is $P_2$. Finally, specify the proportion of cells within a subregion with suitable habitat ($P_3$). The proportion of cells containing suitable habitat for entire map is the product of these three $P$ values ($P_1 \times P_2 \times P_3$). Altering the relative magnitude of these proportions (or changing the number of hierarchical levels) changes the scale of variation. This technique allows the user to produce random maps that vary in the abundance and dispersion (clumping) of habitat (Figure 8.3b).

Hierarchical neutral maps have been used to better understand the interactions between $P$ and spatial pattern at different scales. Lavorel et al.

(1993) found that spatial structure (e.g., spatial organization) depends on the relationship between spatial pattern and $P$. Patches tend to be larger on hierarchical maps than on random maps when $P >$ critical threshold. When $P <$ critical threshold, patches were smaller on hierarchical maps than on random maps. The relationship between $P$ and spatial structure likewise influences the consequences of landscape change experienced by species requiring the changing habitat. Pearson et al. (1996) used hierarchical maps to demonstrate how $P$ interacts with spatial pattern to affect habitat fragmentation (or connectivity) for different species. For a given value of $P$, landscapes with a fine-scale, random pattern of habitat were more likely to be fragmented than landscapes having some spatial organization (e.g., clumping). This work illustrates how species with different dispersal abilities vary in their sensitivities to a given pattern of habitat loss. With respect to habitat fragmentation, the species with limited dispersal ability were the most sensitive to changes in pattern and abundance of habitat, whereas vagile species were generally more tolerant (see Dale et al. 1994 for a similar analysis of real landscapes). Fractal geometry has been used to describe the hierarchical structure of real landscapes (Palmer 1988; Milne 1991, 1992, this volume). Hierarchial neutral maps based on fractal geometry have been used to study insect movements (Johnson et al. 1992) and species coexistence (Palmer 1992).

## 8.4 Fractal Landscapes

The simple neutral and hierarchical neutral models discussed above produce binary maps composed of cells showing the presence and absence of a single habitat type. Real landscapes frequently have gradients rather than abrupt changes in habitat suitability. That is, habitat quality often varies in a more continuous, rather than discrete, manner. This continuous variation is not represented in a binary map but can be simulated by maps having grid cells of two or more types (e.g., gradient maps). Maps showing semicontinuous variation in habitat suitability can be produced using techniques from fractal geometry. One technique for producing fractal maps employs algorithms based on fractional Brownian motion of a two-dimensional random walk. This technique produces maps with varying amounts of spatial autocorrelation. Consider a sequence of steps $X$, with the value of the current step equal to $X_t$. If each successive step $(X_{t+1} - X_t)$ of the random walk is drawn independently from the Gaussian distribution, the resulting sequence of steps is a Brownian motion. Fractional Brownian motion is produced by introducing the parameter $H$, which controls the correlation between successive steps (Saupe 1988; Plotnick and Prestegaard 1993). $H$ can assume values over the interval $(0, 1)$. When $H = 0.5$, steps are not correlated; when $H < 0.5$, steps are negatively correlated; and when $H > 0.5$, steps are positively correlated. Although the fractional Brownian motion algorithm generates random numbers (see Saupe 1988), a map can be produced by assigning the values obtained from the random walk (i.e., $X_t$) to grid cells by scaling the results to the mean and range of the generated sequence.

This technique was used to study the consequences of varying spatial patterns of $^{137}$Cs contamination in a Tennessee reservoir. While the total amount of contaminant in the reservoir was generally known, its dispersion in space was not. The spatial pattern of contamination could affect the strategies for detecting sites with high contaminant levels ("hot spots") and strategies for cleanup of $^{137}$Cs. The degree of spatial autocorrelation in contaminant levels was unknown, so maps that varied in their levels of autocorrelation were produced in an effort to study the effect of spatial variation on the effectiveness of various sediment sampling schemes. Three values of $H$ (0.01, 0.05, and 0.75) were used to produce three fractal maps of $^{137}$Cs inventory for the reservoir (FRAC01, FRAC50, FRAC75, respectively; Figure 8.4). FRAC01 had the least amount of spatial autocorrelation. That is, cell values in this map were only weakly correlated with adjacent cell values. FRAC50 showed a moderate degree of spatial autocorrelation, and FRAC75 showed a high degree of spatial autocorrelation. These numbers were scaled to a reservoir-wide mean $^{137}$Cs inventory of 225 picocuries per square centimeter for each map.

The degree of spatial autocorrelation affected the level of sampling intensity needed to detect hot spots of contamination. The spatial pattern of contamination could be detected accurately in maps with a large degree of spatial autocorrelation using relatively few samples. As autocorrelation declined, the number of samples required to achieve the same degree of accuracy increased dramatically. This investigation also compared the fractal maps to maps of $^{137}$Cs contamination that incorporated hydrody-

FRAC01                    FRAC50                    FRAC75

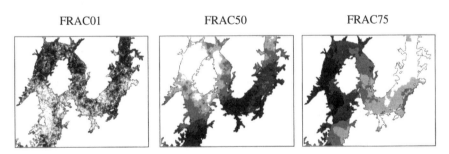

FIGURE 8.4. Neutral maps of semicontinuous values of $^{137}$Cs contamination in a Tennessee reservoir. These fractal maps were produced using a model of fractional Brownian motion in a two-dimensional random walk. The map FRAC01 has little or no spatial autocorrelation. FRAC50 has moderate levels, and FRAC75 has high levels of autocorrelation. Darker areas have lower levels of contamination; light areas are highly contaminated.

namic mechanisms of sediment deposition. $^{137}$Cs in solution is often bound to fine particles, such as clay, by a static charge. Therefore, hydrodynamic mechanisms that influence the particle-size distribution of sediments consequently affect the concentration of contaminants in sediments. Comparison of the fractal maps, which were neutral with respect to sediment deposition, to the sedimentation maps demonstrated that sediment deposition produced patterns quite different from those of the neutral fractal maps. For example, maps produced by the sedimentation models had positive spatial autocorrelation in contaminant levels within zones having similar processes of deposition, but little or no correlation between zones having different sedimentation processes. This investigation documented that sampling strategies that incorporated some stratification with respect to sedimentation zones were most effective in mapping contamination and detecting hot spots that may require cleanup.

## 8.5  Neutral Landscapes and Population Responses to Habitat Fragmentation

Landscape pattern is the result of many ecological processes, and pattern, in turn, directly affects processes. One of the most noticeable effects of pattern on process is the effect of habitat fragmentation on the rate of population growth and the probability of extinction. We found neutral models to be particularly useful in investigating the relationships between habitat loss and fragmentation on biological diversity in human-dominated landscapes of the Southern Appalachian Mountains. The following paragraphs describe results from this research.

Habitat fragmentation is complicated by the fact that both the abundance and spatial pattern of habitat vary among landscapes. For example, consider a landscape covered by 20% suitable habitat. This habitat can be arranged in a few large patches (relatively unfragmented), many small patches (highly fragmented), or in a mixture of large and small patches. Moreover, landscape characteristics can impose structure on habitat patchiness. For example, floodplains provide corridors of undeveloped habitat that maintain habitat connectivity, but an interstate highway may form a barrier to dispersal by bisecting the landscape and exacerbating the isolation of habitat fragments. Topographic patterns affect the arrangement of xeric and mesic habitats. Thus, there are many possible ways that habitat can be spatially arranged in a landscape.

Any set of real landscapes will contain only a limited number of these possibilities. This reality limits the ability of ecologists to investigate the effects of habitat fragmentation empirically because a full range of combinations of $P$ with different spatial patterns usually is not available. Field biologists often are constrained by the small set of "unique" landscapes in which habitat pattern and abundance are observed. Theoretical experiments involving simulation models are useful for testing hypotheses concerning our limited understanding of the effects of landscape pattern on processes. These experiments, in turn, help the investigator put empirical data from a few unique landscapes in a broader conceptual framework.

Changes in landcover result in changes in both the abundance and spatial pattern of habitat. Such changes undoubtedly affect the population dynamics of native species; however, all species using a particular habitat may be differentially affected by these changes. Species that differ in their patterns of survival, fecundity, and dispersal will respond differently to habitat loss and fragmentation. For example, species having large area requirements and limited vagility are impacted more by fragmentation than species that disperse well and can live in small isolated patches. To incorporate the varying responses of species to landcover change, we developed a spatially explicit model that simulates population dynamics on habitat maps. This model can represent species that differ in the three life-history characters listed above (Pearson et al., unpublished manuscript). The purpose of the model was to provide a means of comparing the response of species with different life-history strategies to changes in the abundance and spatial pattern of suitable habitat. Rather than providing predictions about the population dynamics of specific species, the model was used to determine which landscapes could support populations that employ a particular life-history strategy (e.g., high dispersal–high survivorship versus low dispersal–low survivorship). We also determined which life history strategies were most sensitive to landscape changes occurring in the Southern Appalachians.

We used model simulations to rank landscapes in the Southern Blue Ridge province with respect to their ability to support species with a suite of different life histories. Maps derived from remotely sensed images of real landscapes were also used, but these maps encompassed a limited number of combinations of habitat abundance and pattern. Therefore, neutral models for generating maps were used to produce maps that had the same levels of abundance ($P$) but different spatial arrangements of habitat (Table 8.1,

TABLE 8.1. Landscape metrics and population response for six habitat maps generated for the population dynamics model. Percent habitat occupied is the occupation rate after 10 time steps and the mean of 10 replicate simulations

| | Percent suitable | Number of patches | Mean patch size | Landscape shape index | Percent habitat occupied |
|---|---|---|---|---|---|
| Map 5 Actual | 36 | 104 | 3.75 | 9.14 | 23.90 |
| Map 5 Random | 36 | 288 | 1.36 | 23.98 | 0.20 |
| Map 5 Clumped | 36 | 6 | 65.30 | 2.95 | 49.30 |
| Map 8 Actual | 53 | 38 | 17.20 | 7.26 | 44.20 |
| Map 8 Random | 53 | 24 | 24.00 | 25.86 | 1.70 |
| Map 8 Clumped | 53 | 2 | 288.60 | 3.32 | 54.00 |

Map 5                    Map 8

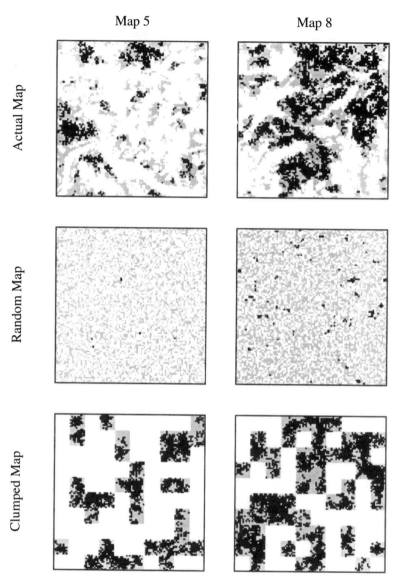

FIGURE 8.5. Actual and generated maps of forest habitat used in a simulation study of population dynamics. Maps in the same column have the same $P$ but different spatial patterns (Map 5, $P = 0.36$; Map 8, $P = 0.53$). The actual maps were derived from actual landcover maps. The random and clumped maps were produced using methods described for hierarchical neutral maps. In all maps, white cells represent unsuitable habitat. Shaded cells represent suitable habitat that is either occupied (black) or unoccupied (grey) by a population. The maps show patterns of occupancy after 10 time steps of a model simulating the growth of populations of an annual plant with moderate fecundity and limited dispersal ability (see Table 1).

Figure 8.5). Landscape metrics, used to quantify these spatial pattern of suitable habitat, were calculated using FRAGSTATS (McGarigal and Marks 1995; see Hargis et al. [this volume] for a further discussion of landscape metrics).

The population dynamics model was used to simulate populations of an annual plant having a life-history strategy of moderate fecundity and limited dispersal, similar to that of *Trillium erectum*, a common mesic forest species in the Southern Appalachian Mountains. At the beginning of each simulation, 33% of cells of suitable habitat were selected at random and marked as occupied. The percentage of suitable cells that were occupied after 10 time steps was recorded and compared for each map (Table 8.1). Rates of occupation greater than 33% indicated growing populations, while rates lower than 33% indicated declining populations.

This life-history strategy was strongly affected by the spatial pattern of habitat. The populations increased on actual Map 8 but declined on actual Map 5 (Figure 8.5). It declined on both random maps and increased on both clumped maps. Analysis of variance confirmed that spatial pattern ($F_{2,56}$ = 1924.0, $r^2$ = 0.95) had a stronger effect than habitat abundance ($F_{2,56}$ = 122.4, $r^2$ = 0.02). Moreover, the occupation rate was correlated with mean patch size across all maps ($r$ = 0.68, $P$ < 0.01; Figure 8.6). The exception to this

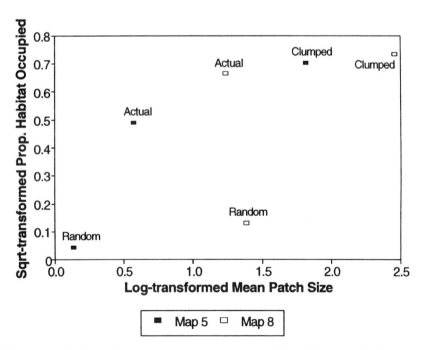

FIGURE 8.6. Relationship between percentage of suitable habitat occupied by population and mean patch size. Occupation rate is square-root transformed, and patch size is log-transformed.

relationship is found with the random Map 8 (Figure 8.6), indicating that patch shape was also important. The patches in this map were more complex (cf. landscape shape index, Table 8.1) than those in the actual and clumped maps. This understanding between the interaction of landscape pattern and population dynamics would not have been possible if the analysis had been limited to the actual maps in which habitat abundance and pattern were confounded. The neutral maps made available additional comparisons that revealed the relatively greater importance of habitat pattern as manifest in patch size and shape.

## 8.6 Summary

Neutral models are useful for testing whether a particular hypothesized mechanism is needed to explain an observed pattern. They are useful in landscape ecology, a field that attempts to understand the complex relationships between landscape patterns and ecological processes. The relationship between landscape processes and observed patterns can be tested rigorously only if the expected pattern in absence of the process is known. Real landscapes can be compared to maps produced from neutral models to test hypotheses related to proposed mechanisms (processes) that control habitat abundance and spatial pattern. Neutral models can produce maps that vary in their complexity and control over spatial pattern at different scales. Obviously, pattern also affects processes. Neutral models can be used to produce maps having habitat patterns not observed in the limited number of real landscapes available to a researcher. Maps of real and theoretical landscapes can be used as input to spatially explicit models to help ecologists explore the implications of landscape patterns for ecosystem processes, population dynamics, and management decisions.

*8.7 Acknowledgments.* The original research described in this chapter grew from collaborations with M.G. Turner, R.V. O'Neill, V.H. Dale, and K.R. Cox. C. Hargis and J. Bissonette provided useful suggestions for improving early drafts. Work on $^{137}$Cs contamination was done in collaboration with K.A. Rose, W.W. Hargrove, and D.A. Levine and was sponsored by the Office of Environmental Restoration and Waste Management, U.S. Department of Energy under contract DE-AC05-84OR21400 with Martin Marietta Energy Systems, Inc. The research on the interaction between life-history strategies and landscape pattern was supported by funding from the U.S. Department of State, Man-and-the-Biosphere Program, Temperate Ecosystem Research and from National Science Foundation Grant DEB 9416803.

## 8.8 References

Baker, W.L. 1992. Effects of settlement and fire suppression on landscape structure. Ecology 73:1879–1887.

Ballal, S.R., S.A. Fore, and S.I. Guttman. 1994. Apparent gene flow and genetic structure of *Acer saccharum* subpopulations in forest fragments. Canadian Journal of Botany 72:1311–1315.

Caswell, H. 1976. Community structure: a neutral model analysis. Ecological Monographs 46:327–354.

Connor, E.F., and D. Simberloff. 1984. Neutral models of species' co-occurrence patterns. Pp. 316–331 in D.R. Strong, Jr., D. Simberloff, L.G. Abele, and A.B. Thistle, editors. Ecological communities: conceptual issues and the evidence. Princeton University Press, Princeton, New Jersey, USA.

Connor, E.F., and D. Simberloff. 1986. Competition, scientific method, and null models in ecology. American Scientist 74:155–162.

Dale, V.H., S.M. Pearson, H.L. Offerman, and R.V. O'Neill. 1994. Relating patterns of land-use change to faunal biodiversity in the central Amazon. Conservation Biology 8:1027–1036.

Gardner, R.H., B.T. Milne, M.G. Turner, and R.V. O'Neill. 1987. Neutral models for the analysis of broad-scale landscape pattern. Landscape Ecology, 1:19–28.

Gardner, R.H., and R.V. O'Neill. 1991. Pattern, process, and predictability: the use of neutral models for landscape analysis. Pp. 289–307 in M.G. Turner and R.H. Gardner, editors. Quantitative methods in landscape ecology: the analysis and interpretation of landscape heterogeneity. Springer-Verlag, New York, New York, USA.

Gardner, R.H., M.G. Turner, R.V. O'Neill, and S. Lavorel. 1991. Simulation of the scale-dependent effects of landscape boundaries on species persistence and dispersal. Pp. 76–89 in M.M. Holland, P.G. Risser, and R.J. Naiman, editors. Ecotones: the role of landscape boundaries in the management and restoration of changing environments. Chapman and Hall, New York, New York, USA.

Haefner, J.W. 1988. Assembly rules for Greater Antillean Anolis lizards: competition and random models compared. Oecologia 74:551–565.

Holt, R.D., G.R. Robinson, and M.S. Gaines. 1995. Vegetation dynamics in an experimentally fragmented landscape. Ecology 76:1610–1624.

Johnson, A.R., B.T. Milne, and J.A. Weins. 1992. Diffusion in fractal landscapes: simulations and experimental studies of tenebrionid beetle movements. Ecology 73:1968–1983.

Lavorel, S., R.H. Gardner, and R.V. O'Neill. 1993. Analysis of patterns in hierarchically structured landscapes. Oikos 67:521–528.

McGarigal, K., and B.J. Marks. 1995. FRAGSTATS: spatial pattern analysis program for quantifying landscape structure. U.S. Department of Agriculture Pacific Northwest Research Station General Technical Report PNW-GTR-351, Portland, Oregon, USA.

Milne, B.T. 1991. Lessions from applying fractal models to landscape patterns. Pp. 199–235 in M.G. Turner and R.H. Gardner, editors. Quantitative methods in landscape ecology: the analysis and interpretation of landscape heterogeneity. Springer-Verlag, New York, New York, USA.

Milne, B.T. 1992. Spatial aggregation and neutral models of fractal landscapes. American Naturalist 139:32–57.

Nitecki, M.H., and A. Hoffman, editors. 1987. Neutral models in biology. Oxford University Press, New York, New York, USA.

O'Neill, R.V., B.T. Milne, M.G. Turner, and R.H. Gardner. 1988. Resource utilization scales and landscape pattern. Landscape Ecology 2:63–69.

O'Neill, R.V., R.H. Gardner, B.T. Milne, M.G. Turner, and B. Jackson. 1991. Heterogeneity and spatial hierarchies. Pp. 85–96 in J. Kolasa and S.T.A. Pickett, editors. Ecological Heterogeneity. Springer-Verlag, New York, New York, USA.

Palmer, M.W. 1988. Fractal geometry: a tool for describing spatial patterns of plant communities. Vegetatio 75:91–102.

Palmer, M.W. 1992. The coexistence of species in fractal landscapes. American Naturalist 139:375–397.

Pearson, S.M., M.G. Turner, R.H. Gardner, and R.V. O'Neill. 1996. An organism-based perspective of habitat fragmentation. Pp. 77–95 in R.C. Szaro and D.W. Johnston, editors. Biodiversity in managed landscapes: theory and practice. Oxford University Press, New York, New York, USA.

Pearson, S.M., M.G. Turner, and K.R. Cox. Population persistence on fragmented landscapes: interactions between life history strategy and landscape pattern (unpublished manuscript).

Plotnick, R.E., and R.H. Gardner. 1993. Lattices and landscapes. Pp. 129–157 in Lectures on mathematics in the life sciences: predicting spatial effects in ecological systems. Vol. 23. R.H. Gardner, editor. American Mathematical Society, Providence, Rhode Island, USA.

Plotnick, R.E., and K. Prestegaard. 1993. Fractal analysis of geological time series. Pp. 207–222 in N. Lam and L. DeCola, editors. Fractals in geography. Prentice-Hall, Inc., Englewood Cliffs, New Jersey, USA.

Quinn, J.F., and A.E. Dunham. 1983. On hypothesis testing in ecology and evolution. American Naturalist 122:602–617.

Santos, T., and J.L. Telleria. 1994. Influence of forest fragmentation on seed consumption and dispersal of Spanish juniper. Biological Conservation 70:129–134.

Saupe, D. 1988. Algorithms for random fractals. Pp. 71–136 in H. Peitgen and D. Saupe, editors. The science of fractal images. Springer-Verlag, New York, New York, USA.

Stauffer, D., and A. Aharony. 1992. Introduction to percolation theory. Second edition. Taylor and Francis, London, UK.

Turner, M.G. 1989. Landscape ecology: the effect of pattern on process. Annual Review of Ecology and Systematics 20:171–197.

# 9
# Understanding Measures of Landscape Pattern

CHRISTINA D. HARGIS, JOHN A. BISSONETTE, and JOHN L. DAVID

## 9.1 Introduction

A major emphasis in landscape ecology is the study of landscape pattern, including pattern dynamics, ecological processes that influence pattern, and effects of pattern on organisms. In order to investigate relationships between landscape pattern and ecological processes, it is often helpful to describe the patterns in quantifiable terms, and a variety of spatial metrics have been developed for this purpose (Whitcomb et al. 1981, Forman and Godron 1986, O'Neill et al. 1988, Turner 1990, Milne 1991, Gustafson and Parker 1992, Li and Reynolds 1993, Plotnick et al. 1993, McGarigal and Marks 1995). Although these metrics enable ecologists to address landscape-level questions in a more rigorous fashion, they often are applied without a clear understanding of the strengths and limitations of each measure or how various measures are interrelated.

Christina D. Hargis has recently completed her doctorate in Wildlife Ecology at Utah State University, Logan, Utah, where she investigated the effects of forest fragmentation and landscape pattern on the American marten. She is currently Assistant National Wildlife Ecologist for the Forest Service.

John A. Bissonette is Leader of the Utah Cooperative Fish and Wildlife Research Unit and Professor in the Department of Fisheries and Wildlife at Utah State University. His research and teaching emphasize landscape ecology and the influence of large-scale processes on wildlife populations and habitat.

John L. David has recently completed his B.S. in Computer Science at the University of New Mexico in Albuquerque and has directed his computer skills toward ecological applications. In addition to the programming for this chapter, he assisted in developing a geo-referenced bibliographic data base for the Geography Department at Utah State University, and built an interface between a mountain pine beetle phenology model and long-term weather data for the Forest Service Intermountain Research Station. He works as a crew leader for wetland restoration for the Forest Service, and is commencing a graduate program in ecosystem restoration.

The goal of this chapter is to provide a basic foundation for understanding measures of landscape pattern, using six metrics for in-depth discussion and illustration: edge density, contagion, mean nearest-neighbor distance, mean proximity index, perimeter-area fractal dimension, and mass fractal dimension. We have chosen these measures because all can be applied to the simplest form of landscape pattern, e.g., pattern arising from two classes of a single landscape attribute. Measures of landscape heterogeneity and diversity involving more than two classes will not be addressed.

We illustrate the range of attainable values for each measure by first using randomly generated landscape patterns. We then demonstrate how the range of values is altered on more realistic landscapes due to the effects of patch size, shape, and placement. We close with a discussion of the strengths and limitations of each measure.

## 9.2  Characteristics of Landscape Pattern

Landscape pattern is the product of a perceived discontinuity in biotic or abiotic factors, as viewed from the standpoint and scale of a particular organism, population, or metapopulation (Wiens 1976, Kotliar and Wiens 1990, Lord and Norton 1991). Although landscape pattern is typically viewed as differences in vegetation and topography at the scale of several kilometers, patterns can emerge at any scale and as a result of any discontinuity, including differences in food availability, temperature, moisture, soil properties, and other attributes critical to the well-being of plants and animals. In all cases, each discontinuity is manifest as a patch (Wiens 1976), the basic building block of landscape pattern.

The lower limit of patch size is the *grain*, or smallest scale at which an organism detects discontinuities (O'Neill et al. 1986, Kotliar and Wiens 1990). From the human standpoint, grain is equivalent to map resolution (Milne 1991) because it represents the smallest scale at which we portray discontinuities in biotic and abiotic factors in map form. In our examples, we use raster-format maps, in which the underlying structure is a two-dimensional grid consisting of pixels that represent the resolution of the map. If no unit of measure is assigned, we can equate a pixel with the grain of any organism, in a generic sense. The simplest form of landscape pattern occurs when each pixel or grain is assigned one of two possible classes of a single landscape attribute such as two vegetation cover types, two classes of food resource, or two thermal classes. In our examples, we have created two hypothetical cover types represented in black and gray.

The countless variations in pattern of two-class landscapes arise from four phenomena: the proportional representation of each class, aggregation of each class into patches, frequency distribution of patch size, and spatial distribution of patches. These phenomena influence landscape pattern in the following ways: the proportional representation of each class generally determines which class forms the landscape matrix; aggregation affects size, shape, and perimeter-to-area ratio of each patch; the frequency distribution of patch size creates landscape texture; and the spatial distribution of patches determines whether patches are clustered or dispersed.

The influence of proportional representation on landscape pattern can be demonstrated by representing the proportion of black and gray pixels on our landscapes as $P_b$ and $P_g$, respectively. If we begin with a landscape entirely consisting of the black cover type and gradually increase $P_g$ with random placement of gray pixels, aggregations of gray pixels will begin to form and increase in size at an exponential rate until a single cluster spans the entire landscape. The $P_g$ at which a single aggregation of gray pixels spans the map can be predicted by percolation theory, a branch of physics that investigates the flow of particles or energy through a porous lattice of grid cells (Stauffer 1985). Assuming that flow can occur through only one of two possible cell types, the critical probability ($P_c$) of a substance percolating through an entire lattice occurs when the porous cells are connected in one continuous cluster. On an infinitely large lattice comprising two randomly interspersed cell types, $P_c \approx 0.5928$ when clusters are formed through cell sides only, and $P_c \approx 0.4072$ if clusters can form through diagonal (corner) connections as well (Stauffer 1985) (Figure 9.1). In landscape ecology, $P_c$ is approximated by $P_g$ (or $P_b$) at the proportion when a single cluster or

FIGURE 9.1. A simulated landscape of random pixels at $P_g = 0.40$, showing percolation through a cluster that spans the map through nearest neighbor and next nearest neighbor connections.

patch of that class spans the landscape. This can occur at variable proportions within an approximate range of 0.50 to 0.65, depending on the degree of aggregation within each landscape class (Gardner and O'Neill 1991).

Regardless of the actual value of $P_g$ at the point of percolation, this proportional value represents a critical shift in landscape pattern that aids in understanding the behavior of several landscape measures. At this juncture, the role of landscape matrix is passed from one class to another, frequently resulting in an abrupt change in the numeric values derived from landscape metrics, particularly for edge density, contagion, mean proximity index, and mass fractal dimension. This will be discussed in more detail as each metric is presented.

The second component, aggregation of pixels within each class, creates the distinctive patch shapes and sizes that characterize each landscape. Greater aggregation is associated primarily with larger patch areas and smaller perimeter–area ratios, but it can also be associated with rounder patch shapes and simpler edges. Aggregation thus influences the numeric values of metrics that quantify patch area, patch shape, patch edge, or perimeter–area ratio, as well as any metrics that incorporate these components in a summary index, such as mean proximity index and contagion. The influence of pixel aggregation will be stressed as we discuss each selected measure.

The third component of landscape pattern, patch size frequency distribution, creates landscape texture (Mandelbrot 1983, Plotnick et al. 1993). Landscapes with a narrow range of patch sizes appear more uniform in texture than landscapes where a variety of patch sizes are present. Patch size frequency distribution affects all metrics that incorporate patch area, perimeter–area ratios, and interpatch distances in the calculations. A broad patch size frequency distribution increases the variance in values derived from these measures.

The fourth component, spatial distribution of patches, places the other components of landscape pattern in a spatially explicit context that gives each landscape its singular identity. This aspect of landscape pattern is of great interest to ecologists, yet it is the most elusive to quantify. Even general notions of spatial distribution, such as clustered versus dispersed patches, are difficult to measure. Metrics that do not include interpatch distances provide no information on spatial context, whereas metrics that utilize interpatch distances are misleading when the distances are not placed in the context of the entire landscape. This is an important limitation that will receive more attention later in the chapter.

A final aspect that affects all metrics is the role of human intervention in portraying landscape pattern. Unless gradient maps are used, boundaries drawn between different map classes are simple lines that do not adequately represent a broad or complex ecotone. Because patch boundaries are often arbitrarily chosen, landscape metrics that quantify these inexact delineations should be viewed as approximate rather than absolute. Mapping errors in depicting patch boundaries, as well as classification errors, can be greater than the differences in numeric values of edge, patch size, and other patch attributes when two or more landscape patterns are compared (Cherrill and McClean 1995).

Moreover, patches tend to be simplified during the mapping process. Patch edges, already smoothed by the resolution of the map, are further simplified by the cartographer. The result is shorter edge lengths and greater aggregation of pixels than actually occur, affecting the calculation of most measures. Measures of fractal dimension are made suspect by smoothing, since the scaling relationship between patch area and edge length is truncated at scales approaching the resolution of the map.

Raster-format maps create both opportunities and difficulties in measuring landscape pattern. Some measures are easier to compute on raster-format maps, and a few measures, such as contagion, require this format. However, raster maps distort landscape pattern into blocky shapes with stair step edges, an additional source of error that must be considered when interpreting landscape metrics. In particular, true edge lengths are overestimated on raster maps (McGarigal and Marks 1995). Furthermore, users of raster maps must decide whether pixels that touch corners are considered connected or not. Pixels that share a common edge are considered nearest neighbors, while pixels that touch only at the corners, forming a diagonal connection, are referred to as next nearest neighbors. The decision whether next nearest neighbors are included in the calculations affects the outcome of measures that incorporate patch area and interpatch distance.

## 9.3 Simulated Landscape Patterns

We will use 11 series of landscape patterns to illustrate the behavior of selected landscape measures, with each landscape $101 \times 101$ pixels in extent. Each series consists of successive increases in the proportion of gray at 0.10 increments, from 0.05 to 0.95 of the map, or to the maximum proportion allowable under simulation constraints, discussed below. Each series consists of five unique landscapes at each proportional interval, and because there are 7 to 10 intervals per series, there are 35 to 50 landscapes within each series.

In one series, landscape pattern is created by randomly placing gray pixels on a black surface until the desired proportion of gray ($P_g$) is reached. These will be referred to as random pixel maps, and can be viewed as a neutral model of landscape pattern (Gardner and O'Neill 1991). In all other series, landscape pattern is created by using preexisting gray patches that conform to one of three basic patch types: (1) small rectangular patches; (2) small, irregular-shaped patches; and (3) large, irregular-shaped patches. The first two patch types both range in size from 1 to 25 pixels, with edge lengths of 1 to 5 pixels, and both have a mean patch size of $9 \pm 6.4$ pixels.

Large, irregular patches are digitized images of 109 actual timber clear-cut harvest patches from the Uinta Mountains of northern Utah and henceforth will be referred to as clearcut patches. Actual patch sizes range from 0.6 to 36 ha, which corresponds to 7 to 400 pixels ($x = 115 \pm 76$ pixels) in our raster images. All three patch types are internally homogeneous; that is, none of the patches contains any other cover types within its boundaries.

The small, rectangular patches and small, irregular patches differ from one another only in patch shape. The two data bases containing these patches also have the same patch size frequency distribution to minimize the effect of this attribute on the values derived from each landscape metric. Clearcut patches differ from small, irregular patches primarily in patch size, and from rectangular patches in both size and shape. Because of the larger size of the clearcut patches, the patch size frequency distribution of this data base also differs from the other two data bases.

We will use the landscape series just described to illustrate the effects of patch size and shape on the range of values obtained from each landscape measure. In addition, a comparision of random pixel maps with landscapes built from preconstructed patches will be used to illustrate the effect of aggregation on the numeric values of the various metrics. Random pixel maps have no aggregation other than that created by randomly formed clusters of like pixels. Of the three patch types used in our simulations, landscapes built from small, irregular patches have the least inherent aggregation, while landscapes built from large clearcuts have the most.

The level of aggregation within a landscape pattern is influenced by the way that the proportional representation of a landscape class increases. A landscape class can increase through the gradual enlargement of one or more patches, the addition of new patches that abut and coalesce into larger patches, or the addition of new patches that are isolated or buffered from all other patches. In general, the gradual enlargement of a single patch will result in the greatest aggregation of a landscape class, unless the patch is extremely convoluted or linear in shape. In our simulations, we increase the proportional representation of the gray landscape class in three ways: (1) enlarging patches, (2) abutting patches, and (3) buffered patches. Because our patch sizes are fixed, we cannot cause patches to enlarge, but we can simulate enlargement by specifying that added patches can overlap an existing patch on the landscape. We allow enlargement of several patches simultaneously per landscape. Abutting patches were created by specifying that each subsequent patch could share boundaries with, but not overlap, an existing patch. To create buffered patches, each successive patch was placed a minimum of one pixel from an existing patch.

We have generated nine landscapes series from all combinations of the three patch types and three methods of increasing the proportion of gray from 0.05 to 0.95 (Figure 9.2a,b,c). Due to constraints imposed by the abutting and buffered patch rules, not all landscape series achieve $P_g = 0.95$. The landscape series of clearcut patches under the abutting-patch rule is truncated at 0.70 because clearcut patches are too large to fit in the remaining matrix, unless patch overlap is allowed. All landscape series with buffered patches are unable to achieve high proportions of gray because of the map space needed to maintain buffers. $P_g$ does not exceed 0.40 for landscapes built with clearcut patches and is limited to 0.35 and 0.30 for rectangular patch and small, irregular-patch landscapes, respectively (Figure 9.3). Because of the narrower range of $P_g$ for landscapes with buffered patches, we have increased the proportion of gray at smaller increments (0.05 rather than 0.10) while maintaining five landscapes per increment.

Although the patch size frequency distribution is generally the same within each landscape series, we wish to emphasize that the buffered-patch rule alters the patch size distribution of landscapes near or at the maximum $P_g$. As landscapes approach this maximum, smaller patches are used to fill the remaining space, and mean patch size of the resulting landscapes is smaller than that of other landscapes.

The nine landscape series just described all contained patches that were spatially dispersed because the patches were placed at random. In order to illustrate the effectiveness of landscape metrics in distinguishing clumped versus dispersed spatial distributions of patches, we generated an additional series in which aggregated patterns were formed (Figure 9.4). Using clearcut patches, we specified that all patches could be no further than three pixels apart and created a series of 40 landscapes consisting of five landscapes at each 0.05 level of $P_g$ from 0.05 to 0.40. We will refer to this series in Section 9.5 when we compare landscapes with aggregated patches to the original series of buffered, clearcut landscapes in which patches were spatially dispersed.

## 9.4 The Behavior of Selected Measures of Landscape Pattern

We used the FRAGSTATS spatial pattern analysis program ver. 2.0 (McGarigal and Marks 1995) to calculate edge density, contagion, mean nearest neighbor distance, proximity index, and perimeter–area fractal dimension. Algorithms used in these calculations are listed in Appendix C of the FRAGSTATS manual (McGarigal and Marks 1995). Mass fractal dimension was calculated using software developed by B. Milne and T. Keitt at the University of New Mexico, and incorporated into the Khoros® image processing environment as an add-on toolbox.

Our landscapes were generated without reference to scale, but in order to express the landscape metrics in actual numeric values, we set the pixel

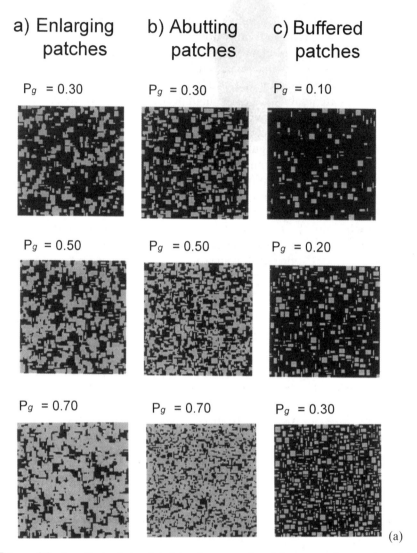

FIGURE 9.2. A subset of simulated landscapes using (a) rectangular patches, (b) small, irregular-shaped patches, and (c) timber clearcut patches. Each column represents a different method of increasing the proportional representation of the gray landscape class.

length at 30 m. Many ecologists have access to satellite imagery and digital elevation models (DEM) with 30-m resolution, and we chose this resolution so that the values we derived from our simulations could be compared more readily with those obtained from actual landscapes. At this scale, a simulated landscape of $101 \times 101$ pixels represents a map extent of $9 \text{km}^2$, a size useful for analysis of a small watershed.

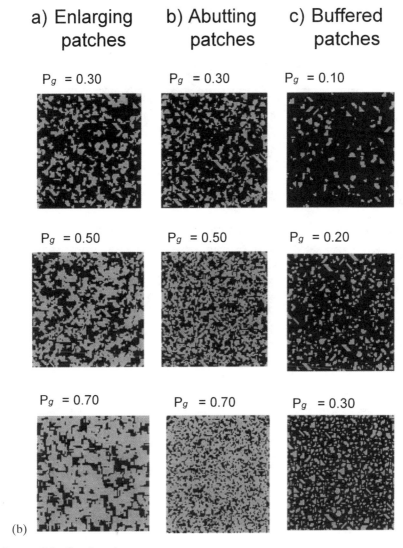

## a) Enlarging patches     b) Abutting patches     c) Buffered patches

$P_g = 0.30$     $P_g = 0.30$     $P_g = 0.10$

$P_g = 0.50$     $P_g = 0.50$     $P_g = 0.20$

$P_g = 0.70$     $P_g = 0.70$     $P_g = 0.30$

(b)

FIGURE 9.2. *Continued*

The combination of map resolution and extent are factors to consider before applying spatial statistics. If map extent is small relative to the resolution, there may be insufficient vector points or raster pixels to run the algorithms used in generating some of the landscape measures, particularly measures of fractal dimension. Also, small maps are more prone to effects caused by map border, since a greater proportion of patches will be truncated at the edges of the map. In statistical tests comparing the range of values derived from maps of varying sizes, we found that values calculated on maps with $64 \times 64$ pixels often deviated from those obtained on maps of

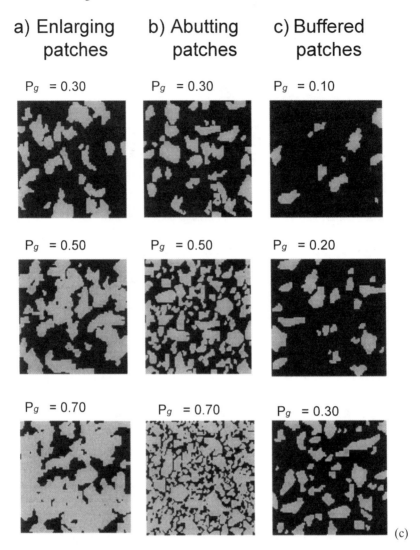

FIGURE 9.2. *Continued*

101 × 101 pixels, while larger map extents (256 × 256 and 512 × 512) yielded similar values for the same landscape pattern. This suggests that maps with 101 × 101 pixels may approach the minimum size needed for achieving consistent values of the same landscape pattern.

## 9.4.1 Edge Density

*Edge density* is the total length of patch edge per unit area, and can be calculated for each class or an entire landscape (McGarigal and Marks

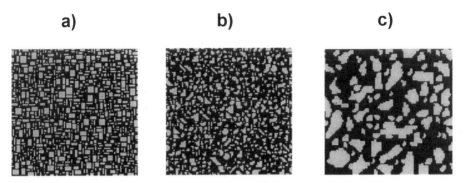

a)          b)          c)

FIGURE 9.3. Maximum density of the gray landscape class under the buffered patch rule: (a) rectangular patches, $P_g = 0.35$; (b) small, irregular patches, $P_g = 0.30$; (c) clearcut patches, $P_g = 0.40$.

1995). For a two-class landscape, class and landscape level yield the same numeric values. This measure is sensitive to map resolution, with fine resolution yielding greater edge length. Therefore, landscapes must be at the same resolution in order to use this measure comparatively.

On random pixel landscapes, edge density forms a parabolic curve with increasing levels of $P_g$ (Figure 9.5a). Initially, edge density increases rapidly because every gray pixel placement contributes four units of edge, but the rate of increase declines when pixels are placed adjacent to existing ones, adding only three or two units of edge per pixel. At $P_g = 0.50$, edge density declines because individual edges are absorbed as pixels ar added.

The same general parabolic shape is apparent on simulated landscapes with increasing $P_g$, especially when the increase is through patch enlargement (Figure 9.5b), but the magnitude of values is considerably less, reflecting the influence of pixel aggregation on this measure. The more

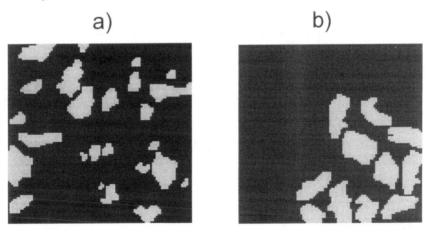

a)          b)

FIGURE 9.4. Examples of simulated landscapes at $P_g = 0.20$ where patches are (a) dispersed or (b) clumped.

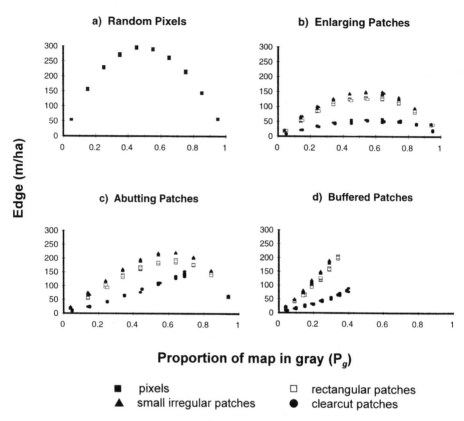

FIGURE 9.5. Response curves for edge density over increasing proportions of gray cover type, showing differences caused by the method of increasing gray (four graphs) and differences due to patch size and shape (symbols within each graph).

aggregation within a class (larger patches or rounder shapes), the fewer units of edge that will be added with the increase in that class. In our simulations, highest edge density values are derived from landscapes built from small, irregular patches, which have the least amount of aggregation due to both size and shape. Second in value are landscapes built from rectangular patches, and the lowest values are associated with clearcut patches because of their large size. The range of values for clearcut landscapes is so low (5 to 50m/ha) that there is little differentiation between landscapes of widely differing $P_g$ (Figure 9.5b).

Edge density values are somewhat higher on landscapes with abutting rather than enlarging patches (Figure 9.5c), reflecting the lower aggregation brought about by interstices between patches that do not abut completely (Figure 9.2a,b,c). On clearcut landscapes with abutting patches, edge density never decreases because each successive patch placement creates irregular-shaped interstices with high edge density that cannot be absorbed by the placement of additional patches once the interstices are smaller than the clearcut patches. On landscapes with buffered patches, edge density for all three patch types continually increases, because buffers restrict patches from coalescing into larger shapes.

The treatment of next nearest neighbors does not change edge density values as calculated by FRAGSTATS. If two patches share a diagonal connection and are treated as one patch, edge density for this patch will be the same as the combined edge density of both patches individually. Diagonal connections change patch area, but they have no bearing on edge density because they do not change the number of pixels that are adjacent to pixels of a different class.

Ultimately, however, patch size influences the effectiveness of this measure because of the low range of values produced by large-patch landscapes. Edge density could be an unsatisfactory measure for differentiating landscapes characterized by large patches with simple edge configurations, a typical pattern in timber harvest areas, agricultural lands, and other managed landscapes. It is intuitive that edge density is influenced by patch area, but the potentially narrow range of values is often overlooked when this measure is applied. Of course, greater discrimination is possible if maps are available at finer resolution.

Another notable difficulty in using edge density as a descriptor of landscape pattern is that two landscapes can have widely differing proportions of a particular cover type, yet have similar edge density values because of the parabolic shape of the edge density curve. For example, landscapes containing small, irregular-shaped patches will have approximately 50 m/ha of edge when the proportion of gray cover type occupies either 0.15 or 0.85 of the landscape (Figure 9.5b). This could be problematic when looking for correlations between edge density and ecological phenomena with increasing representation of a class. Over short ranges of increasing proportions, however, edge density has a near-linear relationship with proportional representation of a class, and the slope is determined by patch size and irregularity of patch edge.

## 9.4.2 Contagion

*Contagion* is an index designed to quantify the degree of aggregation found within landscape classes, originally formulated by O'Neill et al. (1988) and later modified by Li and Reynolds (1993). Requiring raster-format landscapes for calculation, it is the probability that two, randomly chosen, adjacent pixels belong to the same class. In FRAGSTATS, calculations involve the product of two probabilities: (1) the probability that a randomly selected pixel belongs to a given class, which is equivalent to the proportional representation of each class, and (2) the conditional probability that, given a pixel is of one class, an adjacent pixel is of a differing class (McGarigal and Marks 1995).

Contagion index values range between 0% and 100% of the maximum aggregation possible, with maximum aggregation occurring when a landscape is entirely occupied by a single class. Contagion is inversely related to edge density, as seen by the inverse parabolas derived from contagion index values with increasing $P_g$ (Figure 9.6). Rogers (1993) defines *contagion* as a measure of the frequency of occurrence of all possible edge types, because adjacent pixels belong to two differing classes only at patch edges. In our landscape simulations, edge density and contagion had high inverse correlation for all landscape series (Tables 9.1, 9.2, 9.3). Differences between the two measures are due primarily to rounding errors created by the use of logarithmic values in the calculation of contagion.

Because both measures are related to pixel aggregation, the influences of patch size and shape also apply to contagion. Larger patch size and simpler

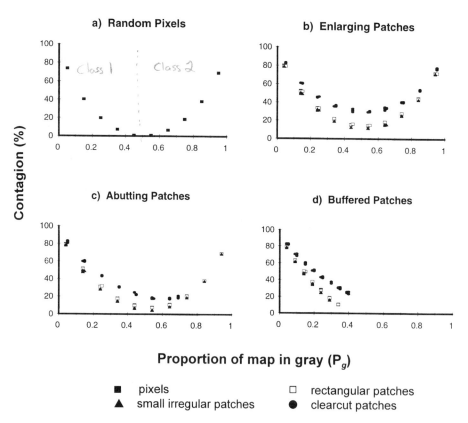

FIGURE 9.6. Response curves for contagion over increasing proportions of gray cover type, showing differences caused by the method of increasing gray (four graphs) and differences due to patch size and shape (symbols within each graph).

TABLE 9.1. Correlation matrices of landscape metrics applied to rectangular-patch landscapes under three types of disturbance growth, with all $|r| > 0.80$ highlighted

### Enlarging patches

|  | $P_1$ | Edge dens. | Contag. | Near. neight. dist | Prox. index | P-A fractal dim. | Mass fractal dim. |
|---|---|---|---|---|---|---|---|
| $P_1$ | 1.00 | 0.28 | −0.13 | −0.74 | 0.71 | 0.64 | −0.81 |
| Edge dens. |  | 1.00 | −0.98 | −0.61 | 0.32 | 0.90 | 0.10 |
| Contag. |  |  | 1.00 | 0.55 | −0.17 | −0.82 | −0.25 |
| Near. neigh. dist. |  |  |  | 1.00 | −0.36 | −0.79 | 0.58 |
| Prox. index |  |  |  |  | 1.00 | 0.54 | −0.51 |
| A-p fractal dim. |  |  |  |  |  | 1.00 | −0.30 |
| Mass fractal dim. |  |  |  |  |  |  | 1.00 |

### Abutting patches

|  | $P_1$ | Edge dens. | Contag. | Near. neight. dist | Prox. index | P-A fractal dim. | Mass fractal dim. |
|---|---|---|---|---|---|---|---|
| $P_1$ | 1.00 | 0.47 | −0.19 | −0.77 | 0.42 | 0.65 | −0.82 |
| Edge dens. |  | 1.00 | −0.83 | −0.62 | 0.61 | 0.95 | −0.13 |
| Contag. |  |  | 1.00 | 0.53 | −0.45 | −0.85 | −0.14 |
| Near. neigh. dist. |  |  |  | 1.00 | −0.21 | −0.78 | 0.64 |
| Prox. index |  |  |  |  | 1.00 | 0.55 | −0.22 |
| A-p fractal dim. |  |  |  |  |  | 1.00 | −0.37 |
| Mass fractal dim. |  |  |  |  |  |  | 1.00 |

### Buffered patches

|  | $P_1$ | Edge dens. | Contag. | Near. neigh. dist. | Prox. index | P-A fractal dim. | Mass fractal dim. |
|---|---|---|---|---|---|---|---|
| $P_1$ | 1.00 | 0.98 | −0.99 | −0.83 | 0.99 | 0.53 | −0.88 |
| Edge dens. |  | 1.00 | −0.97 | −0.78 | 0.99 | 0.49 | −0.83 |
| Contag. |  |  | 1.00 | 0.89 | −0.99 | −0.61 | 0.93 |
| Near. neigh. dist. |  |  |  | 1.00 | −0.84 | −0.82 | 0.91 |
| Prox. index |  |  |  |  | 1.00 | 0.52 | −0.88 |
| A-p fractal dim. |  |  |  |  |  | 1.00 | −0.66 |
| Mass fractal dim. |  |  |  |  |  |  | 1.00 |

edge configurations result in higher contagion values at the same level of $P_g$. As with edge density, however, contagion for a given landscape is the same whether or not diagonal pixels are viewed as connected.

Use of both contagion and edge density to quantify landscape pattern is redundant, and choice of measure depends on the nature of the investigation. For example, habitat ecotones are best quantified by edge density (McGarigal and McComb 1995), whereas the relationship between pixel aggregation and percolation theory (Gardner and O'Neill 1991) may seem intuitively easier to understand in terms of contagion rather than edge density. The level of contagion increases the range of probability for $P_c$. Maps with low contagion due to numerous small patches tend to achieve a percolating cluster at proportions closer to 0.65 than the predicted probability of 0.5928 for random pixels maps (Gardner and O'Neill 1991). Also, maps with contagion >0.90 will consist of very large patches, none of which may span the map (Gardner and O'Neill 1991).

TABLE 9.2. Correlation matrices of landscape metrics on landscapes with irregular-shaped patches and three types of disturbance growth, with all |r| > 0.80 highlighted

Enlarging patches

| | $P_1$ | Edge dens. | Contag. | Near. neigh. dist. | Prox. index | P-A fractal dim. | Mass fractal dim. |
|---|---|---|---|---|---|---|---|
| $P_1$ | 1.00 | 0.25 | −0.12 | −0.71 | 0.75 | 0.42 | −0.91 |
| Edge dens. | | 1.00 | −0.99 | −0.61 | 0.31 | 0.94 | 0.14 |
| Contag. | | | 1.00 | 0.57 | −0.18 | −0.92 | −0.26 |
| Near. neigh. dist. | | | | 1.00 | −0.40 | −0.81 | 0.53 |
| Prox. index | | | | | 1.00 | 0.33 | −0.55 |
| A-p fractal dim. | | | | | | 1.00 | −0.08 |
| Mass fractal dim. | | | | | | | 1.00 |

Abutting patches

| | $P_1$ | Edge dens. | Contag. | Near. neight dist. | Prox. index | P-A fractal dim. | Mass fractal dim. |
|---|---|---|---|---|---|---|---|
| $P_1$ | 1.00 | 0.42 | −0.16 | −0.76 | 0.43 | 0.61 | −0.89 |
| Edge dens. | | 1.00 | −0.94 | −0.55 | 0.73 | 0.91 | −0.03 |
| Contag. | | | 1.00 | 0.48 | −0.55 | −0.85 | −0.19 |
| Near. neigh. dist. | | | | 1.00 | −0.23 | −0.82 | 0.64 |
| Prox. index | | | | | 1.00 | 0.55 | −0.11 |
| A-p fractal dim. | | | | | | 1.00 | −0.30 |
| Mass fractal dim. | | | | | | | 1.00 |

Buffered patches

| | $P_1$ | Edge dens. | Contag. | Near. neigh. dist. | Prox. index | P-A fractal dim. | Mass fractal dim. |
|---|---|---|---|---|---|---|---|
| $P_1$ | 1.00 | 0.99 | −0.99 | −0.82 | 0.99 | 0.68 | −0.95 |
| Edge dens. | | 1.00 | −0.98 | −0.79 | 0.99 | 0.65 | −0.93 |
| Contag. | | | 1.00 | 0.88 | −0.99 | −0.75 | 0.97 |
| Near. neigh. dist. | | | | 1.00 | −0.83 | −0.88 | 0.90 |
| Prox. index | | | | | 1.00 | 0.72 | −0.95 |
| A-p fractal dim. | | | | | | 1.00 | −0.78 |
| Mass fractal dim. | | | | | | | 1.00 |

TABLE 9.3. Correlation matrices for landscape metrics applied to landscapes with clearcut patches, under three types of disturbance growth, with all $|r| > 0.80$ highlighted

| | | | Enlarging patches | | | | |
|---|---|---|---|---|---|---|---|
| | $P_1$ | Edge dens. | Contag. | Near. neigh. dist. | Prox. index | P-A fractal dim. | Mass fractal dim. |
| $P_1$ | 1.00 | 0.40 | −0.12 | −0.71 | 0.41 | 0.35 | −0.91 |
| Edge dens. | | 1.00 | −0.94 | −0.64 | 0.42 | 0.85 | −0.04 |
| Contag. | | | 1.00 | 0.54 | −0.27 | −0.85 | −0.23 |
| Near. neigh. dist. | | | | 1.00 | −0.27 | −0.79 | 0.53 |
| Prox. index | | | | | 1.00 | 0.29 | −0.25 |
| A-p fractal dim. | | | | | | 1.00 | −0.07 |
| Mass fractal dim. | | | | | | | 1.00 |

| | | | Abutting patches | | | | |
|---|---|---|---|---|---|---|---|
| | $P_1$ | Edge dens. | Contag. | Near. neigh. dist. | Prox. index | P-A fractal dim. | Mass fractal dim. |
| $P_1$ | 1.00 | 0.99 | −0.93 | −0.74 | 0.75 | 0.92 | −0.99 |
| Edge dens. | | 1.00 | −0.91 | −0.70 | 0.77 | 0.91 | −0.99 |
| Contag. | | | 1.00 | 0.87 | −0.50 | −0.92 | −0.92 |
| Near. neigh. dist. | | | | 1.00 | −0.35 | −0.80 | 0.73 |
| Prox. index | | | | | 1.00 | 0.67 | −0.75 |
| A-p fractal dim. | | | | | | 1.00 | −0.92 |
| Mass fractal dim. | | | | | | | 1.00 |

| | | | Buffered patches | | | | |
|---|---|---|---|---|---|---|---|
| | $P_1$ | Edge dens. | Contag. | Near. neigh. dist. | Prox. index | P-A fractal dim. | Mass fractal dim. |
| $P_1$ | 1.00 | 0.99 | −0.99 | −0.77 | 0.93 | 0.73 | −0.98 |
| Edge dens. | | 1.00 | −0.97 | −0.73 | 0.93 | 0.67 | −0.98 |
| Contag. | | | 1.00 | 0.84 | −0.92 | −0.80 | 0.96 |
| Near. neigh. dist. | | | | 1.00 | −0.78 | −0.85 | 0.72 |
| Prox. index | | | | | 1.00 | 0.68 | −0.91 |
| A-p fractal dim. | | | | | | 1.00 | −0.70 |
| Mass fractal dim. | | | | | | | 1.00 |

## 9.4.3 Mean Nearest-Neighbor Distance

*Mean nearest-neighbor distance* defines the average edge-to-edge distance in meters between a patch and its nearest neighbor of the same class. This measure yields absolute values and requires maps of similar resolution for comparisons between landscapes.

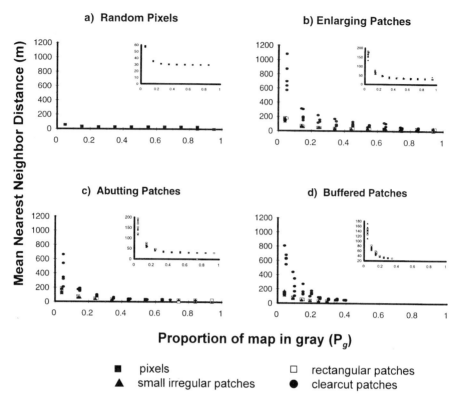

**Proportion of map in gray (P$_g$)**

■   pixels                 □   rectangular patches
▲   small irregular patches     ●   clearcut patches

FIGURE 9.7. Response curves for mean nearest neighbor distance over increasing proportions of gray cover type, showing differences caused by the method of increasing gray (four graphs) and differences due to patch size and shape (symbols within each graph).

Mean nearest-neighbor distance values become exponentially shorter with increasing $P_g$, regardless of the patch type or method of increasing the proportion of gray (Figure 9.7). All landscape series produce similar-shaped response curves, although this is not apparent when all are graphed on the same ordinate scale because of low values derived from landscapes with small patches (Figure 9.7). However, the similarity is evident when landscapes with small patches are graphed so their maximum values fill the ordinate axis (Figure 9.7, inserts).

In all landscape series, when $P_g$ exceeds 0.20, the range of values is so narrow that discrimination among landscapes is difficult. However, the high variance in values at $P_g < 0.20$ suggests that this metric may be useful in differentiating interpatch distances of landscapes when the proportional representation of the two class types are the most dissimilar (high $P_b$ and low $P_g$, or the reverse).

Although this measure does not use patch area in the calculations, it is sensitive to patch size because few, large patches can produce greater distances and more variation between distances than many small patches at the same $P_g$. This is evident in the values derived from larger, clearcut patches compared to either of the small patch types (Figure 9.7). For example, several of the landscapes containing clearcuts achieved $P_g = 0.05$ with only four patches; in one landscape, the patches were widely dispersed, giving a mean nearest-neighbor distance value of 1078 m, whereas another landscape had a mean nearest-neighbor distance value of 631 m. In contrast, two landscapes built from small, irregular patches required 45 patches to achieve $P_g = 0.05$, and these landscapes had smaller and more similar mean nearest neighbor distance values of 133 m and 154 m.

Nearest neighbor distances are affected by the choice of including next nearest neighbors as connections. If patches can be connected diagonally, the landscape pattern will have fewer, larger patches than if diagonal connections are ignored, and interpatch distances will be greater and more variable in the same manner as described above. Our graphed results were derived from including next nearest neighbor connections. Notice that with random pixel maps, this yielded only three possible values: at $P_g = 0.05$, average distance between nearest neighboring patches was two pixels (60 m); for values between 0.15 and 0.85, the distance was one pixel (30 m); and at $P_g = 0.95$, there was no mean nearest-neighbor distance because only one patch was present (Figure 9.7a, insert).

McGarigal and Marks (1995) recommend that the standard deviation be reported with mean nearest-neighbor distance, because it adds information on patch dispersion. A small standard deviation implies a fairly uniform spatial distribution of patches, whereas a large value indicates clumping patterns.

## 9.4.4  Mean Proximity Index

Proximity or isolation indices (Whitcomb et al. 1981, Gustafson and Parker 1992) are designed to measure the isolation of a patch within a complex of patches, given a specified search radius. The original "isolation coefficient" (Whitcomb et al. 1981) is calculated as the inverse of the sum of each patch area divided by the nearest edge-to-edge distance between it and the patch being indexed. A proximity index developed by Gustafson and Parker (1992) is calculated as the sum of the ratio of patch size to interpatch distance between each patch and its nearest neighbor for all patches within the search radius (Gustafson and Parker 1992). In FRAGSTATS, proximity index is calculated as the sum of each patch size divided by the square of the nearest edge-to-edge distance between it and the patch being indexed. This last version is the calculation used on our landscapes. The proximity index of Gustafson and Parker (1992) can be standardized across a variety of landscape extents as long as the same map resolution and search radius are used, but the algorithm used in FRAGSTATS apparently cannot be scaled.

*Mean proximity index* is the mean value derived from the proximity indices of all patches within a landscape, and the highest values occur on landscapes where patches of the same class are large and in close proximity. This measure yields absolute values, with the upper limit determined by the search radius and minimum distance between patches (McGarigal and Marks 1995).

In order to use mean proximity index comparatively among landscapes, all must be at the same resolution and use the same search radius, although overall extent may vary. Maximum values are determined by the size of the specified search radius and the minimum distance between patches (McGarigal and Marks 1995). We used a search radius of 10 pixels to allow comparison of our results with those of Gustafson and Parker (1992), who examined the behavior of this measure on random pixel and random clump maps, the latter being equivalent to our rectangular-patch landscapes. They

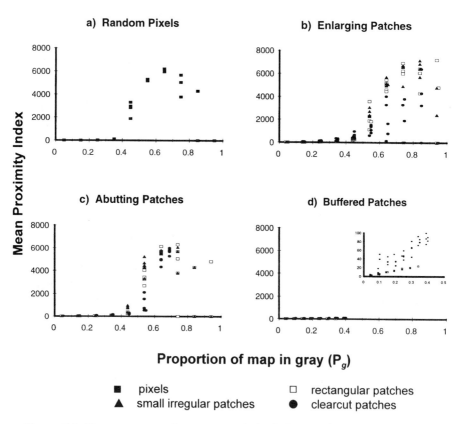

FIGURE 9.8. Response curves for mean proximity index over increasing proportions of gray cover type, showing differences caused by the method of increasing gray (four graphs) and differences due to patch size and shape (symbols within each graph).

found a characteristic sigmoid shape in the response curve of this measure with increasing proportions of a cover type, which we found in our applications of this measure as well (Figure 9.8). Mean proximity index values gradually increase until approximately $P_g = 0.40$, when a dramatic shift in the slope occurs, producing greatly increased values until approximately $P_g = 0.80$, after which a decline in value occurs on some landscapes.

The abrupt increase in values at $P_g = 0.40$ is due to percolation of the increasing landscape class, when the presence of a spanning cluster or patch reduces the nearest neighbor distance among all patches (Gustafson and Parker 1992). At this point, the numerator of the proximity index equation, patch area, increases substantially while the denominator, interpatch distance, decreases. After percolation, small changes in interpatch distances result in large differences in proximity index values at high $P_g$, as seen by the increase in variability at high proportions of the dominant landscape class, especially on landscapes with large patches. Gustafson and Parker (1992) reported a distinct fork of high and low index values at high proportions of the indexed landscape class, depending on whether the nearest pixel was on the diagonal or a full pixel away.

When percolation is inhibited by the requirement that all patches be surrounded by buffers, mean proximity index increases linearly with increasing $P_g$ (Figure 9.8d, insert). Over this range, the rate of increase and average values of the mean proximity index are greater when patches are large, as seen in the steeper slope of large, clearcut patches.

Mean proximity index is best used when the landscape class of interest is below the critical probability for percolation. Above this, the wide range of values caused by slight differences in spatial arrangement of patches may be difficult to interpret ecologically. Moreover, the class of interest is then the landscape matrix, and a measure of patch isolation is less applicable because of increased connectivity. Mean proximity index is most applicable when patches occur in low densities and under different degrees of isolation, such as in species distribution gap analysis and the study of spatial patterns of metapopulations. Spetich et al. (1997) used the proximity index to quantify the relative isolation of old-growth forest patches in Indiana. Hargis and Bissonette (1997) found a significant correlation between mean proximity index of clearcut harvest patches and capture rates of American martens (*Martes americana*), indicating that martens avoid landscapes where unforested patches are large and closely spaced.

## 9.4.5  Perimeter-Area Fractal Dimension

Perimeter–area fractal dimension provides information on irregularity of patch configurations and can be calculated a number of ways (Lovejoy 1982, Burrough 1986). In FRAGSTATS, it is computed as 2 divided by the slope of $\log(P)$ on $\log(A)$, where $P$ and $A$ are the perimeter (m) and area (m²) of each patch in the class or landscape (McGarigal and Marks 1995). The authors refer to this measure as the "double-log fractal dimension," although all fractal dimension measures are based on log-log relationships between a quantity of interest and changes in the scale of the ruler with which the quantity is measured (Mandelbrot 1983, Voss 1988).

The term *perimeter* in McGarigal and Marks (1995) is used to describe the outermost occupied pixels of a patch, whereas in the physics literature, these pixels are known as a special form of edge called the hull, and the term *perimeter* is reserved for the unoccupied pixels adjacent to the hull (Voss 1984, Grossman and Aharony 1987). In our present discussion, we will use the term *perimeter–area fractal dimension* in reference to the measure calculated by FRAGSTATS, but will refer to the outermost occupied pixels of a patch as the edge rather than the perimeter.

With increasing proportions of the gray cover type, perimeter–area fractal dimension produces a parabolic curve similar to that produced by edge density, reflecting the similarity between these measures in their relationship to patch edge length (Figure 9.9). A major difference is that edge density changes with map resolution and extent, while perimeter–area fractal dimension is invariant over all scales. The theoretical limits of all perimeter–area fractal dimension measures are between one and two, with higher values indicating greater complexity of patch edge (Lovejoy 1982, Mandelbrot 1983). On computerized images of landscapes, the attainable range is somewhat narrower, because edges are rarely convoluted to the extent that they produce values greater than 1.6.

Perimeter–area fractal dimension is sensitive to the treatment of next nearest neighbors, yielding greater values when patches connect on the diagonal than when diagonal connections are ignored, because a single patch containing a small, connecting isthmus has greater irregularity than either of the two separate patches produced by breaking the isthmus. On random maps, fractal dimension ranges from 1.4 to 1.7 when diagonal connections are allowed, and from 1.3 to 1.5 when diagonal connections are ignored. The values we have illustrated (Figure 9.9) were all obtained from calculations that included next nearest neighbors.

In our simulated landscapes, patch size and shape, and the method in which additional patches are added have interesting effects on perimeter–area fractal dimension. When patches gradually enlarge with increasing $P_g$, landscapes built from small patches have more complex perimeters and therefore higher fractal dimension values than landscapes built from large patches (Figure 9.9b). However, when each additional patch is a buffered distance from all previous patches, landscapes with large patches have the highest fractal dimension values. Note that the fractal dimension value for all landscapes with buffered patches show little change over increasing $P_g$, especially landscapes built from small rectangles (Figure 9.9d). This is because each additional patch does not alter the configuration of existing patches due to the presence of buffers. As a result, the fractal dimension of the landscape reflects the average fractal dimension of all original patches

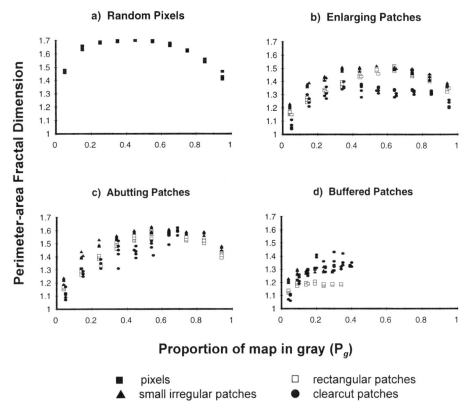

FIGURE 9.9. Response curves for perimeter–area fractal dimension over increasing proportions of gray cover type, showing differences caused by the method of increasing gray (four graphs) and differences due to patch size and shape (symbols within each graph).

drawn from the patch data base. Landscapes built from clearcut patches show the greatest variation in fractal dimension, because the clearcut patch data base contains the greatest variation in patch perimeters of all three patch types.

The nearly flat response curves produced from buffered-patch landscapes indicate that perimeter–area fractal dimension is a poor measure of landscape pattern when all patches exhibit similar irregularity. This measure can differentiate landscapes containing simple patches from landscapes containing irregular-shaped patches, but cannot differentiate landscapes containing different numbers or proportions of patches with similar shapes. An example of this limitation is found in Ripple et al. (1991), where a comparison of two landscapes containing patches of similar shapes yielded perimeter–area fractal dimension values of 1.26 and 1.28, although the proportion of disturbance represented by the patches increased from 8.5% to 23%.

## 9.4.6 Mass Fractal Dimension

Mass fractal dimension can be used to quantify the complexity of an area containing internal heterogeneity rather than the complexity of the area's border. As such, it is useful for quantifying the shape of a landscape matrix caused by patch placement; in other words, the patches are viewed as holes within an otherwise two-dimensional space. The theoretical limits of this measure are between 0 and 2, with the maximum value derived from a landscape containing a single class that entirely fills the two-dimensional space.

Mass fractal dimension represents the scaling relationship between the number of pixels of a given class within a sample of the landscape and the size of the box defining the sample. A range of box sizes is used to delineate sample sizes, from 3 pixels on a side to a maximum of approximately $\frac{1}{3}$ of the landscape. For each box size, the mean number of pixels of the selected class is determined by centering the box on every pixel of that class in the landscape and counting the number of pixels of that class included in the box sample. Mass fractal dimension is equal to the slope derived from regressing the log of the mean number of pixels for each box size on the log of the box lengths (Voss 1988, Milne 1991).

An alternative form of calculating this measure, known as the box dimension, is derived by placing a grid over the entire landscape and counting the mean number of pixels of the selected class in each grid cell, and repeating this procedure using grids of increasingly finer resolutions. Both methods yield comparable values (Voss 1988).

On random pixel maps, mass fractal dimension produces a negative, curved slope over increasing $P_g$, with a minimum value of approximately 0.8 when $P_g = 0.95$. Similar slopes are generated for all other landscape series, regardless of patch size or shape (Figure 9.10). Thus, mass fractal dimension is more responsive to changes in the proportional representation of a cover type than to changes in patch size or shape.

Simplifications of landscape pattern produced during the mapping process yield images that are not truly fractal, and the application of fractal measures is questionable. Our simulated landscapes lack fractal properties of actual landscapes, as do many classified images in which smoothing and renormalization functions have been applied. One of the properties of fractal geometry is that the quantity being measured and the ruler length of measurement have a linear relationship when both are logarithmically transformed (Voss 1988). If this condition is not met, the object of interest may not be fractal. Images generated by simulations or map smoothing functions may not produce the appropriate linear relationship.

Also, landscape extent and resolution affect the ability to derive an accurate scaling relationship between box size and mean number of pixels (B. Milne, University of New Mexico, personal communication). Large landscape extents or fine resolution are needed to produce a reasonable

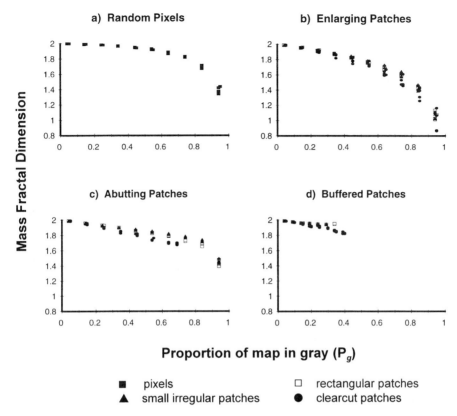

FIGURE 9.10. Response curves for mass fractal dimension over increasing proportions of gray cover type, showing differences caused by the method of increasing gray (four graphs) and differences due to patch size and shape (symbols within each graph).

number of box sizes for generating an accurate slope. Also, small map extents or coarse grain may produce a slope that appears linear, when in fact it is simply a linear segment of a curve, a situation that indicates the image is not truly fractal.

Even if the images were fractal, differences between small-patch landscape patterns would be apparent over only the smallest box sizes, and if map resolution allows only two or three box sizes before the sample box exceeds the size of the average patch, the slope derived from these points is questionable. When larger box sizes are added to the regression line, any interesting differences are lost in the averaging process (R. Voss, Florida Atlantic University, personal communication).

Therefore, although mass fractal dimension has theoretical potential as a measure of landscape pattern, its utility is limited by mapping procedures that remove fractal structure and by small map extents and pixel resolution that are inadequate to observe the scaling relationship. Nevertheless, mass fractal dimension is a useful conceptual tool for understanding scaling relationships between organisms of different sizes and habitat patterns at different scales (see both Milne and Ritchie, this volume). Using landscape simulations, Milne et al. (1992) used fractal geometry to characterize resource abundance at various scales, and investigated allometric relationships between three sizes of herbivores and fractally distributed resources.

## 9.5  The Challenge of Measuring Spatial Distribution of Patches

Although the measures just described provide useful information on the size, shape, and distance relationships among patches on a landscape, none is able to distinguish overall landscape pattern caused by a unique spatial distribution of patches. To demonstrate this, we generated a landscape series comprising clearcut patches with clumped spatial distributions (Figure 9.4) and compared the values derived for each measure with those obtained from dispersed patches. Edge density, contagion, perimeter–area fractal dimension, and mass fractal dimension all yield similar values regardless of patch distribution (Figure 9.11). Mean proximity index values differ, but the values are misleading, because landscapes with few, clumped patches produce values that would be expected at high densities of the selected class when the interpatch distance is low. Likewise, nearest neighbor distance values on landscapes with few, clumped patches are similar to those obtained on landscapes with numerous patches. Although both of these measures quantify interpatch distances, neither measure includes the distance from the outermost patch in a cluster to the edge of the landscape, and it is this distance that places patches in the context of the entire landscape and differentiates clumped versus dispersed patterns of patches.

## 9.6  Correlation of Landscape Measures

Each measure described in this chapter quantifies a different aspect of landscape pattern, but correlations among measures are fairly high. This is partly because the separate components that constitute landscape pattern are fundamentally related and changes in any one component affects one or more other aspects (Li et al. 1993). The size and shape of patches, perimeter–area relationships, and interpatch distances are all controlled by the amount of aggregation within each class, and therefore any differences in class aggregation between two or more landscapes will be manifest in several aspects of landscape pattern. Moreover, all measures are derived from a small set of parameters related to patch size, shape, and interpatch distances, and measures that share one or more parameters in their calcula-

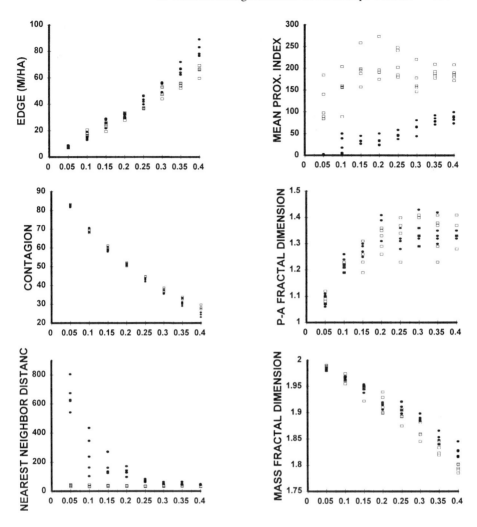

**Proportion of map in gray (P$_g$)**

□ = aggregated patches    ● = dispersed patches

FIGURE 9.11. Response curves for six landscape metrics on landscapes with clumped or dispersed patches.

tions will respond simultaneously with changes in the values of the shared parameters. For example, if two landscapes differ only in the amount of edge, this affects edge density, contagion, and perimeter–area fractal dimension simultaneously.

Correlation matrices for landscape measures applied to our simulated landscapes (Tables 9.1, 9.2, and 9.3) indicate high inverse correlation between edge density and contagion, as discussed previously. Correlation coefficients between perimeter–area fractal dimension and each of these measures is also quite high, reflecting the role of edge length in deriving all three measures. Proximity index generally has the lowest correlation with all other measures.

In our simulations, correlations between $P_g$ and each measure were fairly high, especially when buffered patches were used to increase a class proportion (Tables 9.1, 9.2, and 9.3). This relationship is perhaps stronger than would occur on actual landscapes, due to the random and somewhat uniform dispersal of patches of the same size and shape with each incremental increase in $P_g$ in our simulations. However, landscapes derived from the same ecological or anthropogenic process may also exhibit high correlations between class proportions and other measures. This could be advantageous to the ecologist, because class proportions, which are often easy to estimate or calculate, could be used to estimate more complex measures of landscape pattern.

## 9.7 Summary

A meaningful interpretation of landscape metrics is possible only when the limitations of each measure are fully understood and the range of attainable values is known. It is also helpful to know whether a particular measure provides a unique contribution to our understanding of landscape structure or simply echoes a change in the proportional representation of a landscape class.

Each of the measures presented quantifies a single component of landscape pattern rather than a comprehensive, quantified summary, and none of the measures is able to differentiate spatial patterns of patch dispersion. We have shown that edge density and contagion are both based on edge and are inversely redundant. Mean nearest neighbor distance may be an effective measure when the proportional representation of the selected landscape class is <0.20, but is a poor discriminator at proportions greater than that. Mean proximity index is a good indicator of patch isolation at low proportions of the selected landscape class, but should not be used if the selected class spans the map or is the landscape matrix. Perimeter–area fractal dimension is most useful in differentiating landscapes with high contrast in patch shapes, but does not effectively discriminate between landscapes comprising patches of similar shapes, a situation likely to occur when patches are derived from the same process. Mass fractal dimension is limited in ability to differentiate landscape patterns when applied to small maps, and it is highly correlated with the proportional representation of each landscape class. The graphs accompanying this chapter illustrate the behavior of each measure on simulated landscapes and may be used by ecologists to aid in interpretation of values derived from actual landscapes.

Source code and documentation for the LSF software package is available at the following website: http://www.arc.unm.edu/lsf.

*9.8 Acknowledgments.* This project was carried out within the Utah Cooperative Fish and Wildlife Research Unit, U.S.G.S.–Biological Resources Division, using landscape simulation software developed and owned by J. David. We appreciate use of the FRAGSTATS spatial pattern analysis program developed by K. McGarigal and B. Marks, Oregon State University. We thank B. Milne and T. Keitt, University of New Mexico, for use of their software for calculating mass fractal dimension, and B. Milne and R. Voss for consultation on the behavior of this measure. Landscape simulation software was generated within the Khoros® software development environment. Khoros is the registered trademark of Khoral Research, Inc. Simulations were run on an IBM RS6000/370 network at the Albuquerque Resource Center, University of New Mexico from a remote connection at Utah State University. Use of the supercomputer at the Albuquerque Resource Center was sponsored in part by the Phillips Laboratory, Air Force Materiel Command, USAF, under cooperative agreement number F29601-93-2-0001. The views and conclusions of the authors do not represent official policies or endorsements of Phillips Laboratory or the U.S. Government.

## *9.9 References*

Burrough, P.A. 1986. Principles of geographic information systems for land resources assessment. Oxford University Press, New York, New York, USA.

Cherrill, A., and C. McClean. 1995. An investigation of uncertainty in field habitat mapping and the implications for detecting land cover change. Landscape Ecology 10:5–21.

Forman, R.T.T., and M. Godron. 1986. Landscape ecology. John Wiley and Sons, New York, New York, USA.

Gardner, R.H., and R.V. O'Neill. 1991. Pattern, process, and predictability: the use of neutral models for landscape analysis. Pp. 289–307 in M.G. Turner and R.H. Gardner, editors. Quantitative methods in landscape ecology. Springer-Verlag, New York, New York, USA.

Grossman, T., and A. Aharony. 1987. Accessible external perimeters of percolation clusters. Journal of Physics A: Mathematical Gen. 20:L1103–L1201.

Gustafson, E.J., and G.R. Parker. 1992. Relationships between landcover proportion and indices of landscape spatial pattern. Landscape Ecology 7:101–110.

Hargis, C.D., and J.A. Bissonette. 1997. The influence of forest fragmentation and prey availability on American. marten populations: a multi-scale analysis. In G. Proulx, H.N. Bryant, and P. M. Woodard, editors. Martes: taxonomy, ecology, techniques, and management. Provincial Museum of Alberta, Edmonton, Alberta, Canada.

Kotliar, N.B., and J.A. Wiens. 1990. Multiple scales of patchiness and patch structure: a hierarchical framework for the study of heterogeneity. Oikos 59:253–260.

Li, H., and J.F. Reynolds. 1993. A new contagion index to quantify spatial patterns of landscapes. Landscape Ecology 8:155–162.

Li, H., J.F. Franklin, F.J. Swanson, and T.A. Spies. 1993. Developing alternative forest cutting patterns: a simulation approach. Landscape Ecology 8:63–75.

Lord, J.M., and D.A. Norton. 1990. Scale and the spatial concept of fragmentation. Conservation Biology 2:197–202.

Lovejoy, S. 1982. Area–perimeter relation for rain cloud areas. Science 216:185–187.

Mandelbrot, B. 1983. The fractal geometry of nature. W.H. Freeman and Company, New York, New York, USA.

McGarigal, K., and B. Marks. 1995. FRAGSTATS: Spatial analysis program for quantifying landscape structure. U.S. Department of Agriculture Forest Service General Technical Report PNW-GTR-351.

McGarigal, K., and W.C. McComb. 1995. Relationships between landscape structure and breeding birds in the Oregon coast range. Ecological Monographs 65:235–260.

Milne, B.T. 1991. Lessons from applying fractal models to landscape patterns. Pp. 199–235 in M.G. Turner and R.H. Gardner, editors. Quantitative methods in landscape ecology. Springer-Verlag, New York, USA.

Milne, B.T., M.G. Turner, J.A. Wiens, and A.R. Johnson. 1992. Interactions between the fractal geometry of landscapes and allometric herbivory. Theoretical Population Biology 41:337–353.

O'Neill, R.V., D.L. DeAngelis, J.B. Waide, and T.F.H. Allen. 1986. A hierarchical concept of ecosystems. Princeton University Press, Princeton, New Jersey, USA.

O'Neill, R.V., J.R. Krummel, R.H. Gardner, G. Sugihara, B. Jackson, D.L. DeAngelis, B.T. Milne, M.G. Turner, B. Zygmunt, S.W. Christensen, V.H. Dale, and R.L. Graham. 1988. Indices of landscape pattern. Landscape Ecology 1:153–162.

Plotnick, R.E., R.H. Gardner, and R.V. O'Neill. 1993. Lacunarity indices as measures of landscape texture. Landscape Ecology 8:201–211.

Ripple, W.J., G.A. Bradshaw, and T.A. Spies. 1991. Measuring forest landscape patterns in the Cascade Range of Oregon, USA. Biological Conservation 57:73–88.

Rogers, C.A. 1993. Describing landscapes: indices of structure [thesis]. Simon Fraser University, Burnaby, British Columbia, Canada.

Spetich, M.A., G.R. Parker, and R.J. Gustafson. 1997. Spatial and temporal relationships of old-growth and secondary forests in Indiana. Natural Areas Journal 17:118–130.

Stauffer, D. 1985. Introduction to percolation theory. Taylor and Francis, London, UK.

Turner, M.G. 1990. Spatial and temporal analysis of landscape patterns. Landscape Ecology 4:21–30.

Voss, R.F. 1984. The fractal dimension of percolation cluster hulls. Journal of Physics A: Mathematical Gen. 17:L373–377.

Voss. R.F. 1988. Fractals in nature: from characterization to simulation. Pp. 21–70 in H.O. Peitgen and D. Saupe, editors. The science of fractal images. Springer-Verlag, New York, New York, USA.

Whitcomb, R.F., C.S. Robbins, J.F. Lynch, B.L. Whitcomb, M.K. Klimkiewicz, and D. Bystrak. 1981. Effects of forest fragmentation on avifauna of the eastern deciduous forest. Pp. 125–205 in R.L. Burgess and D.M. Sharpe, editors. Forest island dynamics in man-dominated landscapes. Sprnger-Verlag, New York, New York, USA.

Wiens, J.A. 1976. Population responses to patchy environments. Annual Review of Ecology and Systematics 7:81–120.

# Section 3
## Applications and Large Scale Management

# 10

# The Role of Moose in Landscape Processes: Effects of Biogeography, Population Dynamics, and Predation

R. Terry Bowyer, Victor Van Ballenberghe, and John G. Kie

## 10.1 Introduction

Moose (*Alces alces*) are a dominant feature of Holarctic landscapes. Their massive size (Schladweiler and Stevens 1973, Peterson 1974, Franzmann et al. 1978, Saether 1985, Schwartz et al. 1987), herbivorous diet (Peek 1974), and wide distribution (Peterson 1955, Franzmann 1981) make them a pivotal organism in understanding the dynamics of boreal ecosystems in which they live. An increasing body of evidence suggests these large herbivores play a crucial role in determining the structure and function of the ecosystem they inhabit. Moreover, we contend that the role that moose and other large herbivores play in ecosystem processes has been neglected by many ecologists and that future advances in ecosystem science will require integrating the behavior and population ecology of large mammals into the existing paradigms of landscape ecology.

Our understanding of how moose interact with their environment, however, is inchoate. Even though there has been substantial effort expended to determine how moose and other herbivores affect their food supply and how plants are adapted to herbivory (Bryant et al. 1991), how the life-history strategies and population dynamics of moose interact in this system

R. Terry Bowyer is a Professor of Wildlife Ecology in the Institute of Arctic Biology, and is Head of the Department of Biology and Wildlife at the University of Alaska, Fairbanks. He received a B.S. and M.S. from Humboldt State University, and a Ph.D. from the University of Michigan. He has conducted research on moose in Maine and Alaska over the past 12 years.

Victor Van Ballenberghe is a Research Wildlife Biologist with the U.S. Forest Service in Anchorage, Alaska. He received a B.S. from State University of New York at Oneonta, and his M.S. and Ph.D. from the University of Minnesota. He has been conducting research on moose in Denali National Park for the past 17 years.

John. G. Kie is the Team Leader for the Starkey Project of the U.S. Forest Service in LaGrande, Oregon. He received his B.S., M.S., and Ph.D. from the University of California at Berkeley. He has conducted research on moose in Alaska for the past 5 years.

have yet to be adequately studied. The purpose of this paper is to review how moose interact with landscapes in which they live and to suggest avenues for future investigations to better address and understand these interactions.

## 10.2 Evolutionary Constraints and Biogeography

How moose interact with their environment is the result of a complex interplay of phylogenetic constraints, biogeography, and adaptations to living in a boreal forest ecosystem. Thus, the role of moose in the dynamics of these landscapes cannot be appreciated fully without an understanding of their evolutionary history, especially in the New World.

Moose are relatively recent colonists of North America. Modern moose (*Alces alces*) arrived in Beringia probably no earlier than about 12,000 years ago via the Bering land bridge (Guthrie 1990a, 1990b). Progenitors of modern moose once were thought to have colonized the New World much earlier, and *Alces* was believed to have evolved from *Cervacles*, which roamed both Beringia and areas south of the ice sheet during the last full glacial in North America (Peterson 1955, Kelsall and Telfer 1974). Recent morphological data, however, do not support that view (Azzaroli 1985, Churcher and Pinsof 1987, Guthrie 1990a), and it is likely that *Cervacles* was long extinct by the time *Alces alces* arrived in the New World (Guthrie 1990a, Bowyer et al. 1991).

Four subspecies of moose currently are recognized in North America: *A. a. gigas*, *A. a andersoni*, *A. a. shirshi*, and *A. a. americana* (Peterson 1955, Franzmann 1981, Coady 1982). Although Alaskan moose (*A. a. gigas*) were hypothesized to have been separated from the other subspecies during the Wisconsin Glaciation, dates for *Alces alces* in Alaska of generally <12,000 years ago (Guthrie 1990a) make this idea unlikely. Indeed, a more probable scenario for the colonization of the New World by this large cervid involves the presence of an ice-free corridor (Burns 1990, Catto and Mandryk 1990) in western North America towards the end of the Wisconsin Glaciation (Figure 10.1). Thus, progenitors of modern moose, which were somewhat larger than present-day *A. a. gigas* (Guthrie 1984), crossed the land bridge and rapidly dispersed across North America.

Moose are well adapted to living in boreal and subarctic environments (Kelsall 1969, Kelsall and Telfer 1974, Coady 1982, Van Ballenberghe 1993). Their long legs help them cope with deep snow, and their massive size buffers them against low temperatures that characterize these land-scapes. The lower extreme of the thermal-neutral zone (the temperature below which moose must expend energy beyond that necessary for basal metabolism to maintain body heat) has never been measured in moose, but lies somewhere below −30°C (Renecker and Hudson 1986). Moose are

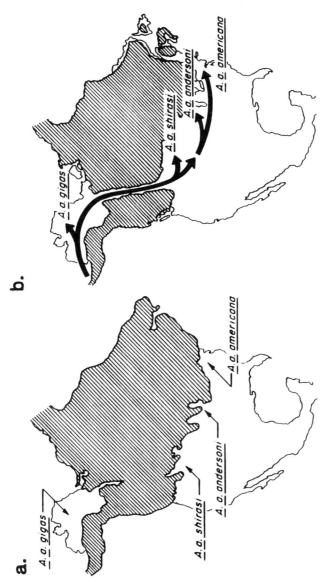

FIGURE 10.1. (a) Traditional view (Peterson 1955) of moose subspeciation in North America caused by the Wisconsin ice-sheet (shaded) isolating moose in Beringia (*A. a. gigas*) from those to the south. (b) The ice-free corridor hypothesis for subspeciation of moose in North America (from Bowyer et al. 1991).

susceptible to heat stress (Renecker and Hudson 1990), however, and this may have helped limit the southern distribution of this ungulate.

Moose are well equipped with dentition to utilize woody vegetation (browse) as the mainstay of their diet (Peterson 1955). Snow tends to cover most plants except tall shrubs and trees in boreal environments for at least one-half of each year (up to nine months in Alaska). Moose can flourish even where only browse is available year-round (Miquelle and Van Ballenberghe 1989, Van Ballenberghe et al. 1989). Nonetheless, moose also are adapted to foraging on aquatic plants during summer, ostensibly to obtain sodium (Belovsky and Jordan 1981).

Despite filling a similar ecological niche, the subspecies of moose in North America are quite different in a number of characteristics. Alaskan moose are distinct in their larger body size (Franzmann 1981), more massive antlers with a "butterfly" configuration (Gasaway et al. 1987), and in their conspicuously marked pledge (Bowyer et al. 1991) than are other subspecies. One of the smaller and most darkly colored subspecies in North America is *A. a. americana*, which has the most easterly distribution. The other subspecies are intermediate in external features—these observations, along with other differences in cranial morphometrics among the subspecies (Peterson 1955), support the pattern of subspeciation depicted in Figure 10.1. Moreover, Alaskan moose differ from other North American *Alces* in one other fundamental way—they exhibit a different mating system.

Alaskan moose often occur in more open habitats than other subspecies (Molvar and Bowyer 1994), tend to occur in larger groups (Peek et al. 1974), and exhibit a harem mating system (Molvar and Bowyer 1994) similar to that of North American elk (*Cervus elaphus*; Bowyer 1981). Other subspecies of moose in North America possess a tending-bond system of mating more typical of North American deer (*Odocoileus*; Hirth 1977, Bowyer 1986). This collection of distributional, historic, morphological, and behavioral data suggests that moose have undergone extremely rapid evolution since their arrival in the New World not much more than 10,000 years ago. These data further indicate that the subspecies have, to some extent, adapted to regional areas of the boreal environments they inhabit. Similarly, woody plants would be expected to have coevolved deterrents to browsers (Bryant et al. 1989), given strong selection, sufficient time, and the necessary genetic variability.

Consequently, plant–animal interactions must be viewed in both an evolutionary and biogeographic framework. For instance, much of the area where the boreal forest is distributed today was covered by ice when moose colonized the New World. The exception was the glacial refugium in Beringia (Kurtén and Anderson 1980). Thus, moose in eastern North America have interacted with the trees and shrubs on which they feed for a comparatively short period of time, whereas large herbivores, including some browsers, have coevolved with plants in Beringia for millennia

(Guthrie 1990b). Presumed differences in the defense systems of plants against herbivores, especially secondary compounds, between Alaska and eastern North America, then, should be evident. Bryant, Tahvainen et al. (1989), and Bryant, Swihart et al. (1994) have documented this pattern for snowshoe hares (*Lepus americana*) and birch (*Betula*), and we hypothesize an analogous situation for moose and their forage.

Moose in eastern North America consume large amounts of conifers in winter. For instance, Ludewig and Bowyer (1985) reported that winter diets of moose in Maine were composed of 73% conifers and dominated by balsam fir (*Abies balsamea*), but also included white spruce (*Picea glauca*). Diets of moose in Alaska, however, contain little white spruce or other conifers in any season (Peek 1974, Oldemeyer 1983, Risenhoover 1989, Van Ballenberghe et al. 1989). Moose inhabiting the midcontinent do consume conifers in winter (mostly *Abies balsamea*), but select hardwoods where the appropriate species are available (Peek 1974). Thus, white spruce, which dominates better drained sites in the northwestern boreal forest, is little affected by foraging moose, whereas this same species (and other conifers as well) are likely influenced by herbivory by moose and white-tailed deer (*Odocoileus virginianus*; Ludewig and Bowyer 1985) in the east. We hypothesize that such dietary differences relate to the defense of these plants via secondary compounds, as Bryant, Tahvanainen et al. (1989), and Bryant, Swihart et al. (1994) documented for hares and birch. Such biogeographic effects have been largely overlooked in studies of plant–animal interactions and mostly ignored in studies of ecosystem structure and function. This is an area clearly in need of additional research to help improve our understanding of variation in the life-history parameters of moose and why they vary across landscapes.

Moose have the ability, under certain conditions, to modify strongly the environments in which they live. Several important studies (Pastor, Naiman et al. 1988, Pastor and Naiman 1992, Pastor, Dewey et al. 1993) demonstrated that herbivory by moose altered patterns of plant succession in the midwestern United States. In that system, foraging by moose presumably caused a shift in species composition from hardwoods to conifers (Risenhoover and Maass 1987). Exclosures that prevented feeding by moose were dominated by hardwoods that were preferred as forage, whereas areas where moose were abundant had mostly conifers (Pastor et al. 1993). This change in species composition was accompanied by lower rates of nutrient cycling in conifer-dominated habitats (Pastor et al. 1993) and likely affected an array of other invertebrates and vertebrates that rely on hardwoods, conifers, or edge for their principal habitats. But, contrary to studies of "herbivore optimization" elsewhere (McNaughton 1984, 1985, 1988, Ruess and McNaughton 1987, Frank and McNaughton 1993), moose in Pastor's system appeared to affect rates of nutrient cycling negatively (Pastor et al. 1993). This outcome needs to be viewed in a broad context. Moose in Pastor's system eat conifers, and the availability of palatable evergreen forage may help keep population densities of moose high enough to alter plant succession. Shifts in succession from hardwoods to conifers may have played a more important role in affecting nutrient cycling than fecal and urine inputs from moose. Thus, plant–animal interactions are not independent of the dynamics of either plant or animal populations. This is a theme to which we will return later. For now, it is sufficient to note that outcomes from studies of the role of large mammals on ecosystem process represent a single point in the dynamics of both plants and animals nested within a geographic setting in which both have evolved. Moreover, primary productivity is strongly correlated with biodiversity in grassland ecosystems (Tilman et al. 1996), and we suspect a similar relationship exists in the boreal forest.

Studies of moose in Alaska in plant communities that are not undergoing rapid succession (treeline stands of white spruce with willow [*Salix*] understories) strongly suggest that moose have a positive effect on rates of nutrient cycling, presumably in part because of the deposition of urine and fecal material (Molvar et al. 1993). Rates of nitrogen mineralization were higher in an area of relatively high density ($5.60 \mu g \, \frac{N}{g \, soil} \times \frac{-1}{day}$) of moose compared with a low-density ($0.48 \mu g \frac{N}{g \, soil} \times \frac{-1}{day}$) area (Molvar et al. 1993). We caution, however, that even in the high-density area, moose were regulated by predators (Gasaway et al. 1992, Van Ballenberghe and Ballard 1994) and that a negative effect likely would have been obtained at very high densities of moose. The point we wish to emphasize is that future research needs to disentangle the effects of plant succession from herbivore optimization (Hilbert et al. 1981, Hik and Jefferies 1990) if we are to gain a more complete understanding of the role of large mammals in ecosystem processes.

Finally, we believe it is essential to begin thinking about large herbivores as more than merely consumers of plants and conveyors of urine and feces. Their interactions with their environment are complex and go far beyond these outcomes. Obvious effects, such as trampling of vegetation (Packer 1953, Pegau 1970) or compaction of soil (Packer 1963) by large mammals have received too little attention. Less well-known behaviors such as the digging of rutting pits by moose (Miquelle 1991) or their scent marking of trees (Bowyer et al. 1994, Figure 10.2) may have considerable effects on the landscapes moose inhabit, but have yet to be considered in studies of ecosystem dynamics. Indeed, selecting specific trees for scent marking (Figure 10.3) may have effects on stand structure via differential mortality of young trees (Bowyer et al. 1994).

We reiterate that studies of ecosystem processes need to be viewed with respect to both the evolutionary history and biogeography of plants and animals. The scale for understanding such processes is continental (or larger), and future insights into such landscape dynamics will likely hinge on our ability to incorporate these notions into our research.

FIGURE 10.2. Female (left) and male Alaskan moose scent marking (rubbing) poll-sized white spruce (from Bowyer et al. 1994).

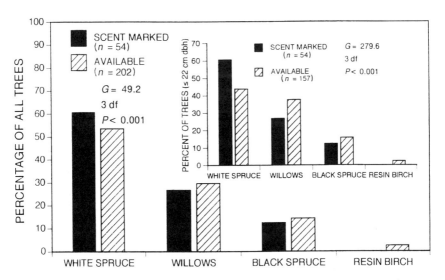

FIGURE 10.3. Selection of trees by Alaskan moose for scent marking (from Bowyer et al. 1994).

## 10.3 Population Dynamics and Sexual Segregation

Fertilization of plants via inputs of herbivore feces and urine can enhance the quality of browse (McKendrick et al. 1980, Day and Detling 1990). Large mammalian herbivores tend to defecate at feeding sites (Etchberger et al. 1988), and nitrogen in their feces increases with forage quality (Leslie and Starkey 1985, Hodgman and Bowyer 1986), at least temporarily resulting in a positive feedback. Heavy browsing, however, can have negative effects on quality of plants to herbivores including shorter, stout stems and inducing defense systems that lower the palatability of forage (Bryant 1981, Bryant, Chapin et al. 1983, Bryant and Chapin 1986, Bryant, Reichardt et al. 1992). The literature on this topic is large, complex, and far beyond the scope of this paper. Nonetheless, the density of ungulates and how they are distributed across the landscape will play a critical role in plant–animal interactions. Our goal is to discuss the role of ungulate population dynamics in these interactions and to suggest ways in which incorporating population dynamics can provide a more complete understanding of this process. Furthermore, we hope to evaluate how this process might vary across landscapes.

Populations of ungulates typically are regulated by density-dependent mechanisms (McCullough 1979). This concept does not exclude the roles of predation or severe weather in limiting populations of large herbivores. Rather, as McCullough (1979) noted, if predators or winter conditions fail to limit such populations, then intraspecific competition for food will do so. Thus, density-dependent effects are mostly mediated by nutrition (Simkin 1974, Saether and Haagenrud 1983, Schwartz and Hundertmark 1993). At the level of the population, such effects typically involve declines in physical condition, ovulation rates, litter size, and survivorship of young or even adults with increasing population size (McCullough 1979). Likewise, increases in age at first reproduction, birth intervals, and hence, mean generation time typically accompany increasing populations as they approach $K$.

Species of large herbivores differ in their responses to density, but there is little question that density dependence plays an important role in their dynamics; this observation holds for moose (Edwards and Ritchey 1958, Pimlott, 1959, 1961, Coady 1982). Because the physical condition of moose is a function of their food supply, and food supply is related to the number of animals competing for this forage, the population dynamics of this herbivore must play a key role in plant–animal interactions, and by extension, to landscape processes. Such a spatially explicit approach has been used to understand the dynamics of animal populations (Pimm and Gilpin 1989) but has yet to be incorporated into models of plant–animal interactions. The herbivore-optimization model (Hik and Jefferies 1990) clearly notes the importance of animal abundance, but does not yet incorporate the dynamics of herbivore populations. Clearly, both spatial and temporal models are required to fully understand how large mammals affect ecosystem processes.

The plant communities inhabited by moose are inherently patchy (Miquelle et al. 1992). Thus, the density of moose is not distributed evenly across the landscape because moose select those patches of habitat in which they are most likely to meet their nutritional or other needs. For instance, some populations of moose are migratory (Le Resche 1974, Van Ballenberghe 1977, Sweanor and Sandegren 1988, Histøl and Hjeljord 1993), creating patchy distributions of these large mammals across the landscape on a seasonal basis. Obviously, severe weather can have adverse effects on populations of moose and other large mammals (Coady 1974, Peterson and Allen 1974, Mech et al. 1987). Moreover, deep snow impedes movements, can cause moose to exhaust energy reserves in winter, and has the potential to limit populations. Snow also can alter the distribution of moose across the landscape and thereby influence plant–animal interactions and, hence, the effects of herbivore optimization. In addition, snow may protect low-growing shrubs from herbivory during winter, even when the density of moose is high. Moose in such a population would likely be distributed in areas with shallower snow, where heavy browsing might initiate strong negative feedbacks between moose and their forage at those sites where moose concentrated. A low-density population of moose, however, might interact far differently with their forage under identical conditions of snow cover. For example, levels of browsing in an area with lower snow depths might result in positive nutritional feedbacks on the population of moose. Similarly, variability in snow cover across the landscape in the same year would result in different levels of browsing and thereby spatial variation in plant–animal interactions, even in homogenous vegetation types. Effects of slope steepness, aspect, ruggedness, and a host of other factors would have consequences even with other variables held constant. It is a daunting prospect to consider just the effects of snowfall in the dynamics of moose and the plants on which they feed. Year-to-year variation in snow depth and structure are likely to have profound effects on moose populations and the manner in which moose use patches of habitat across the landscape. Both snow and population density of moose have consequences on the ecosystem dynamics of such areas.

We see an opportunity to wed modern global positioning (GPS; Moen et al. 1996) and geographic information systems (GIS; Nicholson and Mather 1996) with remote sensing to understand better how moose distribute themselves across the landscape under a variety of environmental conditions. A better understanding of the dynamics of landscapes also will require the integration of models of herbivore optimization with the dynamics of moose and plant populations. We believe this is an essential first step for future research in this area.

Perhaps no other life-history characteristic affects the distribution of moose upon the landscape more than the segregation of the sexes outside of rut. Sexual segregation is a phenomenon common to most ungulates (Bowyer 1984, Bowyer et al. 1996, Main et al. 1996, Bleich et al. 1997), even if the causes of segregation remain uncertain in some species. What is certain is that sexual segregation creates a patchy distribution of moose for much of the year (Miquelle et al. 1992; Figure 10.4), and that this variation in local density has important implications for understanding plant–herbivore interactions. For instance, note the concentration of female groups in the East End and male groups around Jenney Creek in Denali National Park, ALASKA, during winter and at calving; moreover, note that

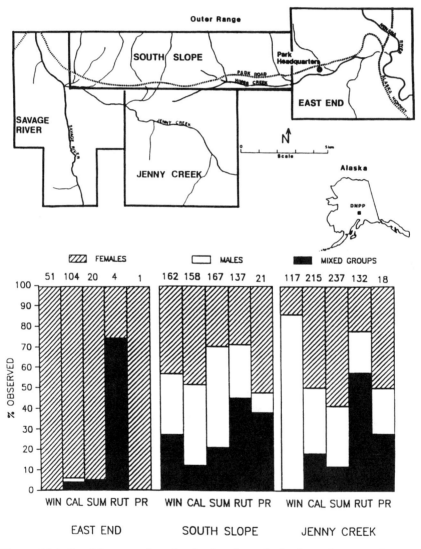

FIGURE 10.4. Spatial segregation of male, female, and mixed-sexed groups of moose through the seasons (winter, calving, summer, rut, and post-rut) at Denali National Park and Preserve, Alaska (from Miquelle et al. 1992).

these areas are generally >5 km apart (Figure 10.4). Males and females may select habitats or forage differentially while the sexes are segregated (Miquelle et al. 1992). Thus, a model of foraging dynamics that fails to consider the needs of both sexes is likely to fall short of predicting reality. Additionally, the scale at which sexual segregation is evaluated is likely to have a profound effect upon our interpretation of this process (Bowyer et al. 1996). Because adult sex ratios often favor females in ungulates (Peterson 1955) and males often have larger home ranges than do females (Hundertmark, in press), males typically occur at low densities when segregated compared with females. These life-history patterns of moose and other ungulates, when considered in relation to the patchy distributions of habitat and forage, determine levels of foraging in particular areas and the response of plants to these levels of herbivory. We suggest that the extreme sexual dimorphism exhibited by ungulates (Ralls 1977) and the differing life-history strategies of the sexes (Main et al. 1996) have profound effects on ecosystem processes and on interpreting hypotheses concerning herbivore optimization. Further advances in these and related plant-herbivore studies will require a more nearly complete understanding of how and why the sexes of ungulates segregate.

The scale over which moose segregate sexually is moderate (Figure 10.4) compared with other ungulates (Bowyer et al. 1996). Nonetheless, selecting the scale at which to measure segregation may be problematical (Bowyer et al. 1996). For instance, by varying the size of the sampling unit or the time interval over which data are collected, it is possible to obtain almost any degree of sexual segregation (Figure 10.5). The effects of scale in understanding this and other phenomena has received too little attention in the literature. Moreover, spatial separation of the sexes causes localized differences in the population, with males typically occurring at lower density (Bowyer 1984), with implications for plant–herbivore interactions. Similar problems exist in understanding how moose "decide" to use a particular patch of habitat, plants within that patch, or stems of an individual plant. A vast body of literature concerns models that address questions related to the use of patches and their varying quality (Stephens and Krebs 1986). Far less attention, however, has been directed toward identifying the scale at which moose (or other large mammals) make decisions. The question of what scale is relevant to a moose is likely to change with the nature of the question being asked. For example, a relatively small patch might be sufficient to meet the short-term needs or a female and her neonate, but be far too small to support her nutritional requirements for lactation later in summer. Nevertheless, answering such conundrums may be necessary to interpret fully the role of moose in ecosystem processes. Failure to select the correct scale can lead to the misinterpretation of data (Morris 1987, Wiens 1989, Powell 1994, Bowyer et al. 1996).

Although our knowledge of the seasonal dynamics of moose home ranges will be improved by recent advances in methodology (Worton 1989, 1995,

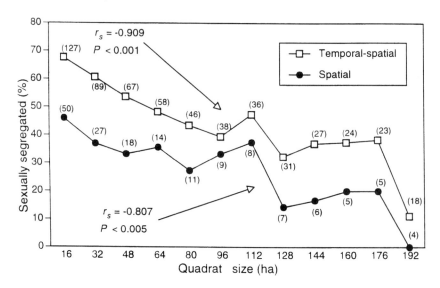

FIGURE 10.5. Reduction in degree of sexual segregation in black-tailed deer with increasing quadrat size (from Bowyer et al. 1996). The spatial scale is for five weeks combined, whereas the temporal-spatial scale is analyzed by week. Similar problems of scale likely cloud interpretation of spatial separation of the sexes for most ungulates.

Kie et al. 1996), we suspect it will be sometime before those studying foraging dynamics, behavioral ecology, and ecosystem processes can integrate their views to produce a more cohesive theory; much exciting work remains to be accomplished (see Ritchie, this volume).

## 10.4  Use of Patches Across the Landscape

Despite the aforementioned problems associated with patch size, scale, and their importance in understanding how moose are distributed upon the landscape, some generalizations are possible. First, it is obvious that a patch must be sufficiently large to support the needs of moose for forage for there to be more than transitory use of that patch. Moreover, a moose must have a sufficient number of such patches within its home range to meet its dietary needs for survival and successful reproduction (McNab 1963, Ford 1983, Swihart et al. 1988), if a moose population is to persist. The sex and age class of moose also affect the size of home ranges (Cederlund and Sand 1994). Even this simplistic representation of the importance of patches to moose, however, offers challenges to our understanding of landscape ecology. For instance, would the value of such patches to moose be similar at high and low population density? Clearly, per capita availability of forage to moose within a patch would decline with increasing population size, but the relative contributions of abundance and quality of forage in relation to this process remain uncertain. Would the correct outcome be predicted by optimization models (i.e., give-up times on foraging within a depleted patch), would moose switch to smaller patches, or move to patches of suboptimal habitat (sensu Fretwell 1972)? There is a vast body of theory, but little empirical evidence.

How might interactions between the biomass of forage available to moose and its quality effect population dynamics of herbivores (White 1983)? Large herbivores likely alter the forage they consume in several important ways. First, browsing reduces the number of growing points so that regrowth of remaining stems is greater than for plants that were unbrowsed. Likewise, browsing releases stems from apical dominance and, in consequence, lateral stems grow longer (Bergstrom and Danell 1987, Molvar et al. 1993). Indeed, moose may obtain larger stems by repeatedly browsing them in subsequent years, thereby reducing the effort necessary to meet their nutritional needs (Vivas et al. 1991). For instance, dry biomass of willow stems that had not been browsed for two years was about 0.6 g per leader of current annual growth; stems browsed in the previous winter, but not the current one, had 0.8 g of available growth. Current annual growth of stems browsed in both winters, however, was nearly 4.0 g (Bowyer and Bowyer, 1997). Browsing also may alter the carbon-nutrient balance of the plant, thereby reducing secondary defense compounds, at least at moderate levels of herbivory (Bryant et al. 1991). Litter from browsed plants decomposes more rapidly than litter from unbrowsed plants (Flanagan and Van Cleve 1983), thereby increasing nutrient availability to the browsed plant. Finally, large herbivores may stimulate plant growth through inputs of urine and feces that fertilize the plant (McKendrick et al. 1980). The net effect is plants with longer stems and larger leaves that are more palatable to moose (Molvar et al. 1993). So, at the same time increasing populations of moose reduce per capita availability of stems on which to feed, they improve the quality of forage over a wide range of population densities.

Within a patch, then, there is likely variation in the availability and quality of plants for moose. How such variation relates to the use of that patch by moose or in relation to other patches is largely unknown. Moreover, how trade-offs between biomass and quality are likely to alter the importance of a patch to moose requires more study.

We have already discussed the complicating effects of snow cover on the distribution of moose upon the landscape and will not reiterate it here. Imagine, however, how a patchy distribution of snow superimposed over an environment with variation in biomass and quality of forage might affect the selection of a patch (or the plants within it) by moose.

There are many other site-specific variables that likely affect the quality of forage for moose (e.g., slope, aspect, litter depth, soil moisture) that we acknowledge but will not expand upon here. One additional complication we will address, however, is shading. Willows growing in sunny areas had higher cell-wall contents than those growing in the shade (Molvar et al. 1993). Indeed, both nitrogen content and digestibility were lower for plants growing in the sun compared with their counterparts in the shade (Molvar et al. 1993). These outcomes suggest that plants growing in direct sunlight may be of less value to moose than shaded ones; Hjeljord et al. (1990) reached similar conclusions. Bryant and Chapin (1986) suggested that such differences occurred because growth of plants in sunny areas outstripped the availability of nutrients in the soil, as a result of higher rates of photosynthesis and subsequent carbon fixation. This process also affects carbon-nutrient balance within the plant and may result in more carbon-based secondary compounds. What is less obvious, however, is that population density of moose has a strong interaction with the effects of shading. For instance, Molvar et al. (1993) reported that when level of browsing was treated as a covariate, leaves and stems were larger in the sun than in the shade where moose density was high, but smaller where moose density was low. A likely explanation is that increased levels of nutrients from urine and feces of moose resulted in greater growth for plants in the sun on the high-density area. Presumably, differences in forage quantity and quality from these interactions would effect the distribution of moose within and among patches. In this instance, moose density (via inputs of urine and feces) may have offset the detrimental effects of sunlight on the palatability of forage (Molvar et al. 1993).

## 10.5  Predation

It is naive to believe that only forage quantity and quality will influence the distribution and density of moose. There is clear evidence that large mammalian predators (e.g., wolves, *Canis lupus*, and bears, *Ursus*) are capable of holding moose populations at low levels (Gasaway, Stephenson et al. 1983, Gasaway, Boertje et al. 1992, Van Ballenberghe 1987, Messier 1994, Van Ballenberghe and Ballard 1994). Where populations of moose are held at relatively low density (Van Ballenberghe and Ballard 1994), negative feedbacks from the overbrowsing of range are unlikely. In many areas of Alaska, then, a natural and abundant fauna of predators plays a major role in the functioning of ecosystem processes. Additionally, McLaren and Peterson (1994) suggested the "top-down" regulation of the Isle Royale ecosystem, whereby wolves limit the number of moose, which in turn affects rates of growth for balsam fir. There is debate over the best model to explain population regulation of moose by their predators, but there is little question that moose are held at low densities by predators for long periods of time (Gasaway et al. 1992, Van Ballenberghe 1987, Van Ballenberghe and Ballard 1994). We suggest that the extirpation of wolves and brown bears from much of their previous range holds the potential to bias our understanding of ecosystem dynamics in many environments and that multiple ungulate prey and plants with well-developed secondary compounds for defense against large herbivores might alter our interpretation of landscape processes in these systems.

Predators do more than help to determine the number of prey. Moose respond in several important ways to the risk of predation. Ungulates typically respond to predation by increasing group size (Hirth 1977). In general, large groups are less vulnerable to attack because of more ears, noses, and eyes with which to detect a predator or the alarm signal of a conspecific (Bowyer et al. 1991). Moreover, the more animals in a group, the lower the probability of an individual being selected as prey (*sensu* Hamilton's [1971] selfish herd). Thus, risk of predation holds the potential to alter group size. Indeed, there are numerous studies of large herbivores, including moose (Molvar and Bowyer 1994) that document an increase in group size with increasing distance from cover (a measure of predation risk). Thus, ungulates foraging in more risky habitats would be expected to form larger groups. Such groups would concentrate foraging activities in specific areas as well as the deposition of urine and feces. These concentrated effects of herbivory are no doubt much different from those of the same number of moose spread more evenly through a patch or across the landscape. Group size, however, is not a component of any foraging model with which we are familiar. Moreover, this information will be difficult to obtain via telemetry and remote sensing unless all animals are telemetered.

Group size and distance from cover also affect the amount of time an animal spends foraging and the selectivity with which it feeds (Edwards 1983; Figure 10.6). Molvar and Bowyer (1994) documented that as moose ventured farther from escape cover and encountered smaller stems, the bites they took were larger and more variable, as evidenced from the willows on which they fed (Figure 10.7). Likewise, moose in the open lowered their feeding efficiency (i.e., spent less of their active time feeding; Figure 10.6). There may be some areas that individuals do not exploit that are of high quality because risk of predation is too great. For instance, females with young use habitats differently and spend more time surveying for predators than do females without young (Miquelle et al. 1992, Molvar and Bowyer 1994). Variables such as group size are seldom considered in studies of ecosystem dynamics, yet have the potential to be major determinants in the patchy distribution of large herbivores across the landscape.

## 10.6 Conclusions

We foresee major breakthroughs in our understand of landscape ecology via studies of plant–animal interactions. We suggest that the time is at hand to meld new technology with a more integrated approach to understanding

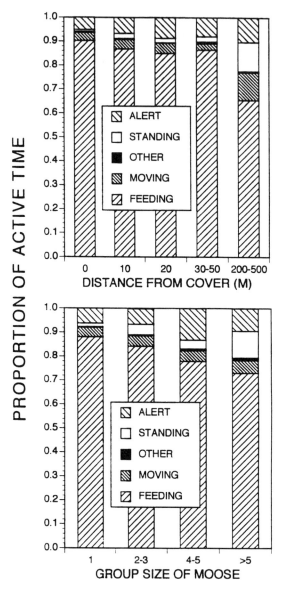

FIGURE 10.6. Proportion of active time spent in various behaviors by Alaskan moose relative to distance from escape cover (above) and group size (below) (from Molvar and Bowyer 1994).

theories related to herbivore optimization, population dynamics of herbivores, the effects of predation, and even their biogeography and evolution. We recognize, however, that this is a tall order and will not be accomplished easily.

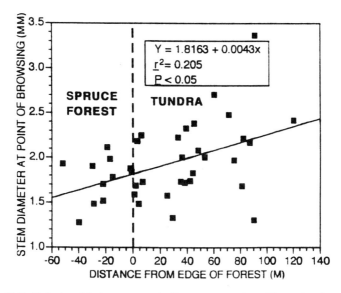

FIGURE 10.7. Relationship between twig diameter at point of browsing by moose on stems of diamondleaf willow and distance from the edge of the forest. Negative values indicate moose were foraging within the forest (from Molvar and Bowyer 1994).

We applaud ongoing attempts to model the interaction of large herbivores with the landscapes in which they live. We caution only that realistic constraints such as differences in sex and age play a role in such efforts and that the population dynamics of the animals be given the same level of attention as the cycling of their nutrients.

## 10.7 References

Azzaroli, A. 1985. Taxonomy of Quarternary Alcini (Cervidae, Mammalia). Acta Zoologica Fennica 170:179–180.

Belovsky, G.E., and P.A. Jordan. 1981. Sodium dynamics and adaptations of a moose population. Journal of Mammalogy 61:613–621.

Bergstrom, R., and K. Danell. 1987. Effects of simulated winter browsing by moose on morphology and biomass of two birch species. Journal of Ecology 75:533–544.

Bleich, V.C., R.T. Bowyer, and J.D. Wehausen. 1997. Sexual segregation in mountain sheep: resources or predation? Wildlife Monographs 134:1–50.

Bowyer, R.T. 1981. Activity, movement, and distribution of Roosevelt elk during rut. Journal of Mammalogy 62:574–582.

Bowyer, R.T. 1984. Sexual segregation in southern mule deer. Journal of Mammalogy 65:410–417.

Bowyer, R.T. 1986. Antler characteristics as related to social status of male southern mule deer. Southwestern Naturalist 31:289–298.

Bowyer, R.T., J.L. Rachlow, V. Van Ballenberghe, and R.D. Guthrie. 1991. Evolution of a rump patch in Alaska moose: an hypothesis. Alces 27:12–23.

Bowyer, R.T., V. Van Ballenberghe, and K.R. Rock. 1994. Scent marking by Alaskan moose: characteristics and spatial distribution of rubbed trees. Canadian Journal of Zoology 72:2186–2192.

Bowyer, R.T., J.G. Kie, and V. Van Ballenberghe. 1996. Sexual segregation in black-tailed deer: effects of scale. Journal of Wildlife Management 60:10–17.

Bowyer, J.W., and R.T. Bowyer. 1997. Effects of previous browsing on the selection of willow stems by Alaskan Moose. Alces 33:11–18.

Bryant, J.P. 1981. Phytochemical deterrence of snowshoe hare browsing by adventitious shoots of four Alaskan trees. Science 213:889–890.

Bryant, J.P., F.S. Chapin, and D.R. Klein. 1983. Carbon/nutrient balance of boreal plants in relation to herbivory. Oikos 40:357–368.

Bryant, J.P, and F.S. Chapin, III. 1986. Browsing-woody plant interactions during boreal forest plant succession. Pp. 213–225 in K. Van Cleve, F.S. Chapin, III, P.W. Flanagan, L.A. Viereck, and C.T. Dyrness, editors. Forest ecosystems in the Alaskan taiga. Springer-Verlag, New York, New York, USA.

Bryant, J.P., J. Tahvanainen, M. Sulkinosa, R. Julkunen-Tiito, P. Reichardt, and T. Green. 1989. Biogeographic evidence for evolution of chemical defense by boreal birch and willow against mammalian browsing. American Naturalist 134:20–34.

Bryant, J.P., F.D. Provenza, J. Pastor, P.B. Reichardt, T.P. Clausen, and J.T. de Toit. 1991. Interactions between woody plants and browsing mammals mediated by secondary metabolites. Annual Review of Ecology and Systematics 22:431–436.

Bryant, J.P., P.B. Reichardt, and T.P. Clausen. 1992. Chemically mediated interactions between woody plants and browsing mammals mediated by secondary metabolites. Journal of Range Management 45:18–24.

Bryant, J.P., R.K. Swihart, P.B. Reichardt, and L. Newton. 1994. Biogeography of woody plant chemical defense against snowshoe hare browsing: comparison of Alaska and eastern North America. Oikos 51:385–394.

Burns, J.A. 1990. Paleontological perspectives on the ice-free corridor. Pp. 61–66 in L.D. Agenbroad, J.I. Mead, and L.W. Nelson, editors. Megafauna and man: discovery of America's heartland. University of Northern Arizona, Flagstaff, Arizona, USA.

Catto, N., and C. Mandryk. 1990. Geology of the postulated ice-free corridor. Pp. 80–85 in L.D. Agenbroad, J.I. Mead, and L.W. Nelson, editors. Megafauna and man: Discovery of America's heartland. University of Northern Arizona, Flagstaff, Arizona, USA.

Cederlund, G., and H. Sand. 1994. Home-range size in relation to age and sex in moose. Journal of Mammalogy 75:1005–1012.

Churcher, S.C., and J.D. Pinsof. 1987. Variation in the antlers of North American Cervalces (Mammalia; Ceridae): review of new and previously recorded specimens. Journal of Vertebrate Paleontology 7:373–397.

Coady, J.W. 1974. Influence of snow on behavior of moose. Naturaliste Canadien 101:417–436.

Coady, J.W. 1982. Moose. Pp. 902–922 in J.A. Chapman and G.A. Feldhamer, editors. Wild mammals of North America: Biology, management, and economics. John Hopkins University Press, Baltimore, Maryland, USA.

Day T.A., and J.K. Detling. 1990. Grassland patch dynamics and herbivore grazing preference following urine deposition. Ecology 71:180–188.

Edwards, J. 1983. Diet shifts in moose due to predator avoidance. Oecologia 60:185–189.

Edwards, R.Y., and R.W. Ritchey. 1958. Reproduction in moose populations. Journal of Wildlife Management 22:261–268.

Etchberger, R.C., R. Mazaika, and R.T. Bowyer, 1988. White-tailed deer, *Odocoileus virginianus*, fecal groups relative to vegetation biomass and quality in Maine. Canadian Field-Naturalist 102:671–674.

Flanagan, P.W., and K. Van Cleve. 1983. Nutrient cycling in relation to decomposition and organic-matter quality in taiga ecosystems. Canadian Journal of Forest Research 13:795–817.

Ford, R.G. 1983. Home range in a patchy environment: optimal foraging predictions. American Zoologist 23:314–326.

Frank, D.A., and S.J. McNaughton. 1993. Evidence for the promotion of aboveground grassland production by native large herbivores in Yellowstone National Park. Oecologia 96:157–161.

Franzmann, A.W. 1981. Alces alces. Mammalian Species 154:1–7.

Franzmann, A.W., R.E. LeResche, R.A. Rausch, and J.L. Oldemeyer. 1978. Alaskan moose measurements and weights and measurement–weight relationships. Canadian Journal of Zoology 56:298–306.

Fretwell, S.D. 1972. Populations in a seasonal environment. Princeton University Press, Princeton, New Jersey, USA.

Gasaway, W.C., R.O. Stephenson, J.L. Davis, P.K.E. Sheperd, and O.E. Burris. 1983. Interrelationships of wolves, prey, and man in interior Alaska. Wildlife Monographs 84:1–50.

Gasaway, W.C., D.J. Preston, D.J. Reed, and D.D. Roby. 1987. Comparative antler morphology and size of North American moose. Swedish Wildlife Research Supplement 1:311–325.

Gasaway, W.C., R.D. Boertje, D.V. Grangaard, D.G. Kellyhouse, R.O. Stevenson, and D.G. Larsen. 1992. The role of predation in limiting moose at low densities in Alaska and Yukon and implications for conservation. Wildlife Monographs 120:1–59.

Guthrie, D.R. 1984. Alaskan megabucks, megabulls, and megarams: the issue of Pleistocene gigantism. Carnegie Museum of Natural History Special Publication 8:482–509.

Guthrie, D.R. 1990a. New dates on Alaskan Quaternary moose, *Cervalces-Alces*—archaeological, evolutionary, and ecological implications. Current Research in the Pleistocene 7:111–112.

Guthrie, D.R. 1990b. Frozen fauna of the mammoth steppe: the story of blue babe. University of Chicago Press, Chicago, Illinois, USA.

Hamilton, W.D. 1971. Geometry for the selfish herd. Journal of Theoretical Biology 31:295–311.

Hik, D.S., and R.L. Jefferies. 1990. Increases in the net above-ground primary production of a salt-marsh forage grass: a test of the predictions of the herbivore-optimization model. Journal of Ecology 78:180–195.

Hilbert, D.W., D.M. Swift, J.K. Detling, and M.I. Dyer. 1981. Relative growth rates and the grazing optimization hypothesis. Oecologia 51:14–18.

Hirth, D.H. 1977. Social behavior of white-tailed deer in relation to habitat. Wildlife Monographs 53:1–55.

Histøl, T., and O. Hjeljord. 1993. Winter feeding strategies of migrating and nonmigrating moose. Canadian Journal of Zoology 71:1421–1428.

Hjeljord, O., N. Hovik, and H.B. Pedersen. 1990. Choice of feeding sites by moose during summer: the influence of forest structure and plant phenology. Holarctic Ecology 13:281–292.

Hodgman, T.P., and R.T. Bowyer. 1986. Fecal crude protein relative to browsing intensity by white-tailed deer on wintering areas in Maine. Acta Theriologica 31:347–353.

Hundertmark, K.J. In press. Home range, dispersal, and migration. In A.W. Franzmann and C.C. Schwartz, editors. Ecology and management of North American moose. Smithsonian Institution Press, Washington, D.C., USA.

Kelsall, J.P. 1969. Structural adaptations of moose and deer for snow. Journal of Mammalogy 50:302–310.

Kelsall, J.P., and E.S. Telfer. 1974. Biogeography of moose with particular reference to western North America. Naturaliste Canadien 101:117–130.

Kie, J.G., J.A. Baldwin, and C.J. Evans. 1996. CALHOME: a program for estimating animal home ranges. Wildlife Society Bulletin 24:342–344.

Kurtén, B., and E.E. Anderson. 1980. Pleistocene mammals of North America. Columbia University Press, New York, New York, USA.

Le Resche, R.E. 1974. Moose migrations in North America. Naturaliste Canadien 101:393–415.

Leslie, D.M., Jr., and E.S. Starkey. 1985. Fecal indices to dietary quality of cervids in old-growth forests. Journal of Wildlife Management 49:142–146.

Ludewig, H.A., and R.T. Bowyer. 1985. Overlap in winter diets of sympatric moose and white-tailed deer in Maine. Journal of Mammalogy 66:390–392.

Main, M.B., F.W. Weckerly, and V.C. Bleich. 1996. Sexual segregation in ungulates: new directions for research. Journal of Mammalogy 77:449–461.

McCullough, D.R. 1979. The George Reserve deer herd: population ecology of a K-selected species. University of Michigan Press, Ann Arbor, Michigan, USA.

McKendrick, J.D., G.O. Batzli, K.R. Everett, and J.C. Swanson. 1980. Some effects of mammalian herbivore fertilization on tundra soils and vegetation. Arctic and Alpine Research 12:565–578.

McLaren, B.E., and R.O. Peterson. 1994. Wolves, moose, and tree rings on Isle Royale. Science 266:1555–1558.

McNab, B.K. 1963. Bioenergetics and determination of home range size. American Naturalist 97:133–140.

McNaughton, S.J. 1984. Grazing lawns: animals in herds, plant form, and coevolution. American Naturalist 124:863–866.

McNaughton, S.J. 1985. Ecology of a grazing system: the Serengeti. Ecological Monographs 55:259–294.

McNaughton, S.J. 1988. Mineral nutrition and spatial concentrations of African ungulates. Nature 334:343–345.

Mech, L.D., R.E. McRoberts, R.O. Peterson, and R.E. Page. 1987. Relationship of deer and moose populations to previous winter's snow. Journal of Animal Ecology 56:615–627.

Messier, F. 1994. Ungulate population models with predation: a case study with the North American moose. Ecology 75:478–488.

Miquelle, D.G. 1991. Are moose mice? The function of scent urination in moose. American Naturalist 138:460–477.

Miquelle, D.G., and V. Van Ballenberghe. 1989. Impact of bark stripping on aspen–spruce communities. Journal of Wildlife Management 53:577–586.

Miquelle, D.G., J.W. Peek, and V. Van Ballenberghe. 1992. Sexual segregation in Alaskan moose. Wildlife Monographs 122:1–57.

Moen, R., J. Pastor, Y. Cohen, and C.C. Schwartz. 1996. Effects of moose movements and habitat use on GPS collar performance. Journal of Wildlife Management 60:659–668.

Molvar, E.M., R.T. Bowyer, and V. Van Ballenberghe. 1993. Moose herbivory, browse quality, and nutrient cycling an Alaskan treeline community. Oecologia 94:472–479.

Molvar, E.M., and R.T. Bowyer. 1994. Costs and benefits of group living in a recently social ungulate: the Alaskan moose. Journal of Mammalogy 75:621–630.

Morris, D.W. 1987. Ecological scale and habitat use. Ecology 68:362–369.

Nicholson, M.C., and T.N. Mather. 1996. Methods for evaluating Lyme disease risks using geographic information systems and geospatial analysis. Journal of Medical Entomology 33:711–720.

Oldemeyer, J.L. 1983. Browse production and its use by moose (*Alces alces*) and snowshoe hares (*Lepus americanus*) at the Kenai Moose Research Center, Alaska, USA. Journal of Wildlife Management 47:486–496.

Packer, P.E. 1953. Effects of trampling disturbance on watershed conditions, runoff, and erosion. Journal of Forestry 51:28–31.

Packer, P.E. 1963. Soil stability requirements for the Gallatin elk winter range. Journal of Wildlife Management 27:401–410.

Pastor, J., R.J. Naiman, B. Dewey, and P.F. McInnes. 1988. Moose, microbes and the boreal forest. BioScience 38:770–777.

Pastor, J., and R.J. Naiman. 1992. Selective foraging and ecosystem processes in boreal forests. American Naturalist 134:690–705.

Pastor, J., B. Dewey, R.J. Naiman, P.F. McInnes, and Y. Cohen. 1993. Moose browsing and soil fertility in the boreal forests of Isle Royale National Park. Ecology 74:467–480.

Peek, J.M. 1974. A review of moose food habits studies in North America. Naturaliste Canadien 101:195–215.

Peek, J.M., R.E. LeResche, and D.R. Stevens. 1974. Dynamics of moose aggregations in Alaska, Minnesota, and Montana. Journal of Mammalogy 55:126–136.

Pegau, R.E. 1970. Effects of reindeer trampling and grazing on lichens. Journal of Range Management 23:95–97.

Peterson, R.L. 1955. North American moose. University of Toronto Press, Toronto, Ontario, Canada.

Peterson, R.L. 1974. A review of the general life history of moose. Naturaliste Canadien 101:9–21.

Peterson, R.O., and D.L. Allen. 1974. Snow conditions as a parameter in moose–wolf relationships. Naturaliste Canadien 101:481–492.

Pimlott, D.H. 1959. Reproduction and productivity of Newfoundland moose. Journal of Wildlife Management 23:381–401.

Pimlott, D.H. 1961. The ecology and management of moose in North America. Terre Vie 2:246–265.

Pimm, S.L., and M.E. Gilpin. 1989. Theoretical issues in conservation biology. Pp. 287–305 in J. Roughgarden, R.M. May, and S.A. Levin, editors. Perspectives in ecological theory. Princeton University Press, Princeton, New Jersey, USA.

Powell, R.A. 1994. Effects of scale on habitat selection and foraging behavior of fishers in winter. Journal of Mammalogy 75:349–356.

Ralls, K. 1977. Sexual dimorphism in mammals: avian models and unanswered questions. American Naturalist 111:917–938.

Renecker, L.A., and R.J. Hudson. 1986. Seasonal energy expenditures and thermoregulatory responses of moose. Canadian Journal of Zoology 64:322–327.

Renecker, L.A., and R.J. Hudson. 1990. Behavioral and thermoregulatory responses of moose to high ambient temperatures and insect harassment in aspen-dominated forests. Alces 26:66–72.

Risenhoover, K.L. 1989. Composition and quality of moose winter diets in interior Alaska. Journal of Wildlife Management 53:568–577.

Risenhoover, K.L., and S.A. Maass. 1987. The influence of moose on the composition and structure of Isle Royale forests. Canadian Journal of Forest Research 17:357–364.

Ruess, R.W., and S.J. McNaughton. 1987. Grazing and the dynamics of nutrient and energy regulated microbial processes in the Serengeti grasslands. Oikos 49:101–110.

Saether, B.-E. 1985. Annual variation in carcass weight of Norwegian moose in relation to climate along a latitudinal gradient. Journal of Wildlife Management 49:977–983.

Saether, B.-E., and H. Haagenrud. 1983. Life history of the moose (*Alces alces*): fecundity rates in relation to age and carcass weight. Journal of Mammalogy 64:226–232.

Schladweiler, P., and D.R. Stevens. 1973. Weights of moose in Montana. Journal of Mammalogy 54:772–775.

Schwartz, C.C., W.L. Regelin, and A.W. Franzmann. 1987. Seasonal weight dynamics of moose. Swedish Wildlife Research Supplement 1:301–310.

Schwartz, C.C., and K.J. Hundertmark. 1993. Reproductive characteristics of Alaskan moose. Journal of Wildlife Management 57:454–468.

Simkin, D.W. 1974. Reproduction and productivity of moose. Naturaliste Canadien 101:517–526.

Stephens, D.W., and J.R. Krebs. 1986. Foraging theory. Princeton University Press, Princeton, New Jersey, USA.

Sweanor, P.Y., and F. Sandegren. 1988. Migratory behavior of related moose. Holarctic Ecology 11:190–193.

Swihart, R.K., N.A. Slade, and B.J. Bergstrom. 1988. Relating body size to the rate of home range use in mammals. Ecology 69:393–399.

Tilman, D., D. Wedin, and J. Knops. 1996. Productivity and sustainability influenced by biodiversity in grassland ecosystems. Nature 379:718–720.

Van Ballenberghe, V. 1977. Migratory behavior of moose in southcentral Alaska. International Congress of Game Biologists 13:103–109.

Van Ballenberghe, V. 1987. Effects of predation on moose numbers: a review of recent North American studies. Swedish Wildlife Research supplement 1:431–460.

Van Ballenberghe, V. 1993. Behavioral adaptations of moose to treeline habitats in subarctic Alaska. Alces supplement 1:193–206.

Van Ballenberghe, V., D.G. Miquelle, and J.G. McCracken. 1989. Heavy utilization of woody plants by moose during summer in Denali National Park, Alaska. Alces 25:31–35.

Van Ballenberghe, V., and W.B. Ballard. 1994. Limitation and regulation of moose populations: the role of predation. Canadian Journal of Zoology 72:2071–2077.

Vivas, H.J., B.-E. Saether, and R. Anderson. 1991. Optimal twig-size selection of a generalist herbivore, the moose *Alces alces*: implications for plant–herbivore interactions. Journal of Animal Ecology 60:395–408.

White, R.G. 1983. Foraging patterns and their multiplier effect on productivity of northern ungulates. Oikos 40:377–384.

Wiens, J.A. 1989. Spatial scaling in ecology. Functional Ecology 3:385–397.

Worton, B.J. 1989. Kernel methods for estimating the utilization distribution in home-range studies. Ecology 70:164–168.

Worton, B.J. 1995. Using Monte Carlo simulation to evaluate kernel-based home range estimators. Journal of Wildlife Management 59:794–800.

# 11
# A Spatial View of Population Dynamics

JERRY L. COOKE

## 11.1 Introduction

The foundation of population studies in wildlife biology and wildlife management rests on the claim that a population can be characterized from local observations (Ludwig and Reynolds 1988). Wildlife managers go through many formal and informal steps in managing the resources under their responsibility, starting with visualizing a population's probable position on a logistics growth curve (Lotka 1925). For example, when an investigator claims: (1) that a species is threatened/endangered and requires special protection, (2) that a species is stable enough for harvest, or (3) that a species is overpopulated in an area and should be reduced to protect important habitats, that claim usually rests on sample-based predictions of population change. The Lotka model permeates what we do in resource management. For example, identification of population phase (e.g., geometric growth, maximum sustained yield, asymptotic growth, carrying capacity) and predicting the effects of population intervention represent a logical framework for doing "thought experiments". Thought experiments constitute much of the trial and error nature of the management process because they are pivotal to interpreting observable results and predicting the outcome of treatments. However, while the traditional principles of population dynamics are the basis of many resource management decisions, treatments are rarely applied to entire populations, but rather to demes that often are thought of, discussed, and treated as though they were populations. Treating demes may be simpler than treating populations, but

Jerry L. Cooke is Program Director for Upland Wildlife Ecology with Texas Parks and Wildlife Department. This program is responsible for all nonmigratory game species in Texas. He has been a field biologist with the department for over 20 years, working mostly in the Chihuahuan Desert of West Texas. Since 1985, his graduate studies have focused on predicting desert mule deer distributions in very large systems using landscape constraints as the only explicit population processes. He also has worked on how to repair population estimates biased by the scales of observations.

may create population changes that are too confounded to predict and may create observable results too complex to interpret easily.

In this chapter I will focus on three pivotal claims of wildlife management: (1) population phase can be determined from local assessments, (2) treatments or intervention can significantly change population behavior, and (3) the impact of a treatment or intervention on a population can be assessed. To do this, I will use a spatially explicit model to create a population whose growth and distribution are known. I will then use survey and harvest procedures recognizable to any wildlife manager. The only relevant differences between these modeled exercises and day-to-day wildlife management will be the precision and accuracy of the harvest treatment and the complete knowledge of what the population is really doing.

## 11.2 Models Neutral to the Claim of Population Processes

One way of matching population changes to lower level processes (i.e., processes from which population characteristics emerge) is through individual-based, spatial models. For example, an investigator may assume that a linkage exists between body condition (i.e., fat reserves) of female deer and the population's in-utero reproduction (e.g., average number of fetuses per female, conception chronology). If this investigator created a spatially explicit, individual-based model in which there was no direct linkage between body condition and reproduction (i.e., the two processes were independent of one another in the model), the model could be used to test the importance of the assumed linkage. Should the spatial results of the model show the expected correlation between body condition and reproduction (e.g., high reproduction concomitant with good body condition and low reproduction with poor body condition), then one would conclude that they probably were not linked directly to each other as might have been assumed from field observations. The model's results could be used to argue that each of the two processes may be linked to a third process that produces correlation between the two, but understanding the relationship is confounded by their co-occurrence. Where a model reproduces an expected pattern in the absence of a process thought to be critical to an event (i.e., model without the conceptual constraint), links between the pattern and process (and the very existence of the process) are challenged under the modeled conditions (Gardner et al. 1987, Cale et al. 1989). Thus, individual-based models may offer a way to study rigorously the processes that shape populations (Prigogine and Stengers 1984).

The goal of this part of the chapter is to determine whether "population" processes are necessary to produce "population" behavior. Specifically, I compare habitat-limited, density-dependent population growth to patterns of change within the population. I will attempt to determine whether spatial structure of a population, normally attributed to disturbance and habitat fragmentation, may be reproduced using a simplistic forage-depletion model in uniform and unperturbed habitats. Furthermore, I will attempt to determine whether a population's status relative to its position on a logistic growth curve can be estimated using only local observations.

## 11.3  A Method of Studying Populations

### 11.3.1  Habitat Simulation

A population of wild animals seldom is distributed uniformly over a landscape. Distribution patterns of animals like deer and other grazing and browsing animals tend to have heterogeneous densities, and these patterns frequently are attributed to fragmentation of habitats or the dispersion pattern of resources. Heterogeneous densities over a landscape create an observable structure within a population, often associated with differing sex ratios of adults and differing reproductive success. Details of this structure are often the focus of intense survey efforts because wildlife managers believe that understanding physical characteristics provides insight and a basis for predicting the potential behavior of the population as a whole. However, assuming that local events follow the same rules attributed to populations can provide a false confidence that may lead to poor management goals and poor treatment design.

To understand how populations can be spatially structured independent of habitat fragmentation or resource dispersion, I evaluated population growth within landscapes having very simplified features. Each landscape consisted of 900 cells ($30 \times 30$) where any individual cell either contained resources or did not. Resource cells (occupied cells) were either uniformly or heterogeneously distributed across the landscape. In uniformly distributed landscapes, resource occupancy was either 23% = Low ($L_L$) or 43% = Intermediate ($L_I$). Heterogeneous landscapes were characterized by having one quarter of the area low in resources, two quarters intermediate in resources, and one quarter where every cell contained resources. I termed this landscape mixed ($L_M$). A mixed landscape had the same total resource availability as intermediate, but with a different spatial arrangement (Figure 11.1). The initial value of all resource cells was chosen to provide the complete nutritional goal of the first individual to forage there (i.e., a value equal to 2). To simulate forage removal by an individual in its habitat, I deducted a resource value of 0.1 from a cell after each individual feeding bout, reducing the value to subsequent foragers until no resources remained. The habitats were renewed each day coinciding with the time step of model. The habitat design was intentionally simplistic so that population expansion became a function of breeding, reproduction, and survival while spatially compressing change into an easily observable extent.

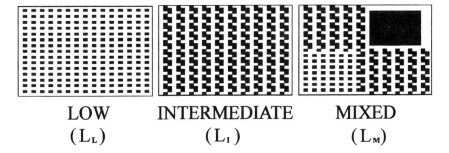

LOW
($L_L$)
INTERMEDIATE
($L_I$)
MIXED
($L_M$)

FIGURE 11.1. The landscapes used in this study contained cells with and without resources arranged in either LOW ($L_L$) or INTERMEDIATE ($L_I$) resource densities. One landscape contained resources in a MIXED ($L_M$) arrangment.

## 11.3.2 Summary of a Behavioral Null Model

I constructed the animal foraging model to resemble a population of desert mule deer in a landscape of approximately 60,000 ha. The population at the beginning of each simulation was 50 individuals having the composition and physical characteristics of the herd found on the Black Gap Wildlife Management Area, Texas (Cooke 1985). The initial location of each animal in the population was chosen randomly, but once assigned, these initial locations were identical for each simulation.

In the model, individuals moved to obtain "food resources," which were unclassified as to type and which have no units. Before each individual began a simulated day's feeding, I calculated a foraging goal (see section 11.4.1) that simulated the basic needs of the animal for food resources. The goal was based on the individual's sex (male goals were higher than those of females), age (young animals' goals were higher than those of older adults), weight (heavy animals needed to attain a higher goal than light animals), and reproductive status (pregnant females needed to achieve higher goals than nonpregnant females, Cooke 1993). The direction and maximum distance an individual foraged was chosen at random (i.e., uniform random movement), but the daily movement could not exceed the mean maximum ability of a mule deer to range (i.e., 3 cells). Individuals foraged in the same order each day (i.e., as they were listed in the object file with new individuals appended to the end).

One way of evaluating population performance is through interpretation of changes in individual condition and reproduction. Body condition ("fat reserve," represented by the familiar categories of excellent = 4, good = 3, fair = 2, poor = 1, and dead = 0) was simulated to reflect how successfully an individual satisfied its foraging goals and the effort expended in moving to an adjacent cell. Conception was coincidental to foraging in the model. Females bred during a limited season when they encountered a mate in the same cell. However, a female in poor condition (body condition <1) could not breed, and pregnancy was terminated when a female could not meet her foraging goals for a critical number of days (Cooke 1993). This will be more fully discussed later under section 11.4.3, Reproduction Rules.

Changes in population size reflected births and deaths over the landscape. Individuals lived a maximum life span of nine years, and death could result earlier from poor body condition (i.e., when body condition = 0). Each population was allowed to grow and expand for 25 years.

## 11.4 Movement Model Description

### 11.4.1 Foraging Goals

The unitless foraging goal for each individual was calculated using the variables of body weight, sex, reproductive condition, and age. The first component of the goal was growth, which, was interpolated from table values (Cooke 1985) using the individual's age. The table values ($C_{age}$) were determined from weight changes between age classes for females (Cooke 1985); as a convention, I multiplied the interpolated value by 1.05 for males ($C_{sex}$). I added the individual's weight divided by 200 (i.e., a maximum weight) to the goal to simulate maintenance of the body mass ($C_{size}$). If a female was pregnant, an additional 0.2 was added to the final foraging goal to simulate fetal growth and placental maintenance ($C_{preg}$). The algorithm was simply:

$$\left(C_{age} \cdot C_{sex}\right) + C_{size} + C_{preg} = \text{foraging goal.} \tag{11.1}$$

Examples:

30 pound buck fawn: $(1.1 \cdot 1.05) + 0.15 + 0 = 1.305$
100 pound 3-year-old buck: $(1 \cdot 1.05) + 0.5 + 0 = 1.550$
150 pound 6-year-old buck: $(1 \cdot 1.05) + 0.75 + 0 = 1.800$
150 pound 7-year-old buck: $(0.9 \cdot 1.05) + 0.75 + 0 = 1.695$
100 pound 3-year-old doe: $(1 \cdot 1) + 0.5 + 0 = 1.500$
100 pound 3-year-old pregnant doe: $(1 \cdot 1) + 0.5 + 0.2 = 1.700$

Hence, the model calculated a foraging goal for each individual deer, based on its age, sex, body mass, and if female, its pregnancy status, at the beginning of each day before foraging began.

### 11.4.2 Movement Rules

Individuals were allowed to forage freely, based on uniform random movement rules, over the landscape (60,000 ha, 64.75 ha/cell) with no "home range" limitation. Distance and direction were chosen independently. Distance was selected by taking a uniform random number (i.e., a 16-digit decimal from the continuous range 0 to 1) for an $X$ and $Y$ displacement and multiplying it by 3 (i.e., maximum mean distance that can be covered in a day: 1.5 km.). For each axis, a second random number (i.e., 16-digit decimal from the continuous range 0 to 1) was used to determine the displacement's sign (if [random]) <0.5 then [+], else [−]). The selection of distance and direction from an individual's location identified a target cell toward which an individual would move. An individual began at rest and traveled along the route toward the target cell until either (1) the foraging goal was met, suggesting its movement grain was determined by resource availability (Addicott et al. 1987), or (2) the target cell was reached (i.e., movement grain determined allometrically, Odum 1959), whichever occurred first. An individual obtained resources equal to the value of the cell that it occupied in each step, up to a maximum of 3 per movement bout. Movement costs were increased by 0.2 for each cell the animal moved to until the target position was attained. Body condition (i.e., $0 \leq$ body condition $\leq 4$) and body mass were changed each day by the difference between the resource goal plus the costs of movement and resources obtained:

$$C_b = C_b - \left\{ \delta \times \left[ \frac{4 - C_b}{4} \right] \times I_{age} \right\}$$
(11.2)

where: $C_b$ = Body condition, $\delta$ = difference between resource goal and resources actually obtained, and $I_{age} = 1$, unless $A < 5$, then $I_{age} = \frac{A}{5}$, and where $A$ = the animal's age

$$W = W + \left\{ \left[ \frac{\delta \times W}{200} \right] \Big/ 10 \right\}$$
(11.3)

where: $W$ = weight.
I reduced the resource value in a cell by 0.1 after an individual fed there, and increased the age of each individual 1 day at the end of the bout.

## 11.4.3 Reproduction Rules

Breeding occurred at the end of a feeding bout if a mate was encountered, but was restricted to the time period between the Julian dates of 352 and 020. Twins were conceived after each breeding, but the sexes were chosen at random. A critical feature of mule deer population dynamics (and to a lesser extent the population dynamics of white-tailed deer as well) is fetal sex ratios that are strongly linked to doe nutrition (Verme 1965, Pittman 1987). If nutrition is poor during gestation, fewer fawns are born. This relationship between nutrition and fawn production more significantly impacts the number of female fawns than the number of male fawns (Verme 1965, Pittman 1987). "Counters" were used to determine whether sufficient resources were obtained by the female to support her fetuses during gestation. At conception, the date was recorded for the doe, and two gestation counters were initiated: one to record elapsed time in days from breeding and one to record number of days in which the female's resource goal was met. When the difference between the gestation counters exceeded 17 and

22, female and male fetuses, respectively, were lost. When gestation had progressed 197 days, the fawns were born with equal initial weights, at the age of 1 day, with a body condition of 1, and at the location of their mother.

## *11.4.4 Death Rules*

Individuals died when they were nine years old or when an individual's body condition reached zero. The only other form of death was for fetuses through poor nutrition during gestation, as described above.

## 11.5  Population Modeling Results

A single 25-year simulation was run for all habitats to assess population behavior as a series of single events, similar to a manager's field assessments of populations. For the purposes of this exercise, conducting multiple iterations of these simulations would not be instructive. To claim that these results would occur at some percentage probability, as would be expected from viewing multiple simulations, is argumentative and entirely a function of the model's assumptions and not useful for my purposes here. The simulations used are intended to demonstrate that a result can occur; I leave it to the reader to determine its application to his/her situation.

In these simulations, all populations generally exhibited sigmoid growth. However, none demonstrated asymptotic behavior. Populations responding to $L_I$ and $L_M$ ($P_I$ and $P_M$ respectively) ultimately reached over 5700 and 5000 individuals respectively, while populations responding to $L_L$ reached about 2500 (Figure 11.2). When comparing population growth between the

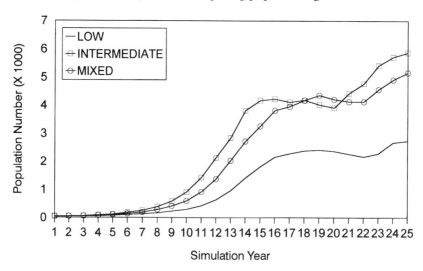

FIGURE 11.2. Populations on all habitat types initially demonstrated sigmoid growth, despite the fact that the model included no birth or death rules linked directly to population density. The ultimate population size in these simulations was apparently determined by the amount of resources available on the landscape.

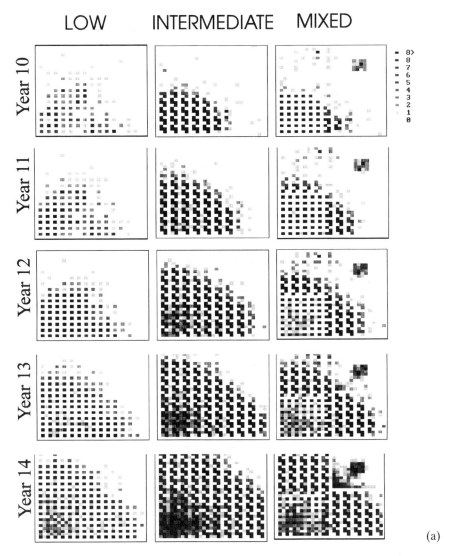

(a)

FIGURE 11.3. Distribution of population, expressed in total number of animals per cell, on LOW ($L_L$), INTERMEDIATE ($L_I$) and MIXED ($L_M$) habitats for years 10–24 of the simulations (*continued*).

three habitats, the plots of population sizes were near mirror images throughout the simulations. However, $P_I$ began its increase earlier and fluctuated more extremely. Average body condition of adults (i.e., age ≥1 year), on all habitats, in year 25 was 3.65 (range = 3.63 to 3.67).

Despite the pattern of resource dispersion across the landscapes ($L_L$, $L_I$, $L_M$), the spatial pattern of reproduction was similar and heterogeneously distributed on all habitats. In these simulations, the populations (Figure 11.3) expanded through time from a single focus ($P_L$, $P_I$) or two foci ($P_M$).

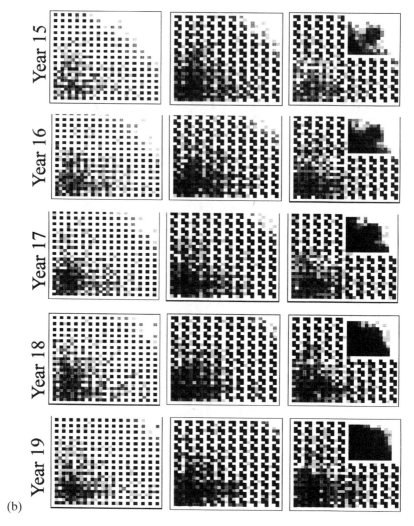

(b)

FIGURE 11.3. *Continued*

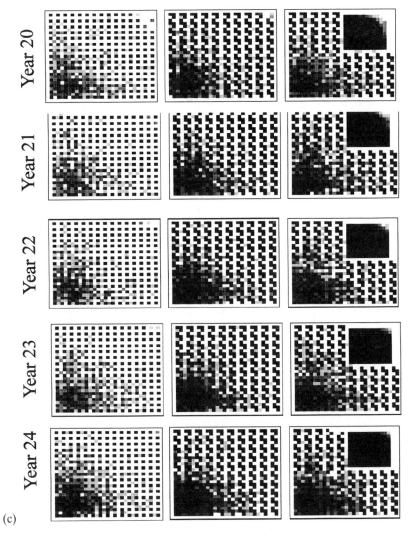

FIGURE 11.3. *Continued*

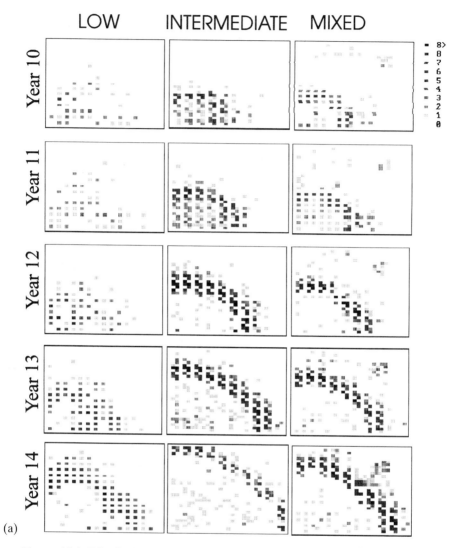

(a)

FIGURE 11.4. Distribution of individuals 1 year fo age (herd increment) expressed in total number of animals per cell, on LOW ($L_L$), INTERMEDIATE ($L_I$) and MIXED ($L_M$) habitats for years 10–24 of the simulations.

The expansion across the landscapes resulted form the survival of offspring on the periphery of the adult concentrations as shown by the maps of yearling animals (Figure 11.4), despite the fact that most (i.e., ≈ 70%) fawns were conceived and born within the adult concentrations. The distribution of yearlings tracks the combination of resources and breeding partners. As

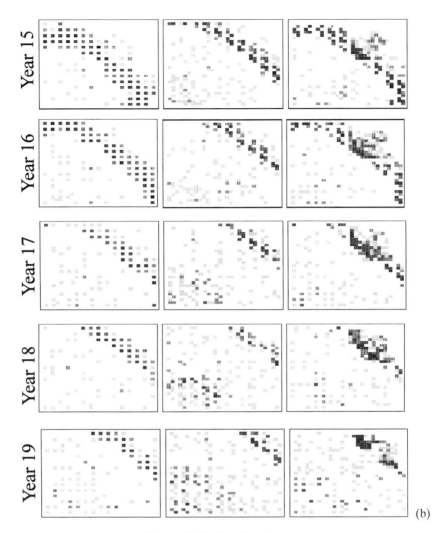

(b)

FIGURE 11.4. *Continued*

the simulations developed, yearlings originated at a population focus and then moved as a wave outward in successive years. Eventually, as deaths decreased numbers of individuals in the concentrations, new waves of yearlings formed, repeating the wave progression as before, creating a secondary pulse in the final years of the simulation.

While these results represent only a single simulation on each habitat type, they do not support the notion that a simple system will produce simple behavior. Populations in these simulations developed complex spa-

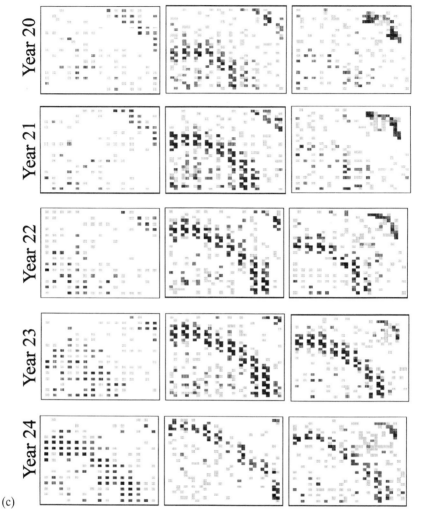

FIGURE 11.4. *Continued*

tial structure despite simple, uniform habitats and simple movement rules for individuals. Complex patterns confound the interpretation of observations, often resulting in complex theories to explain how they come about, where a simple explanation may have sufficed. The results of these simulations are important because they show that complex spatial patterning of populations on landscapes need not be the result of catastrophic habitat disturbance, season dynamics, or complex population processes.

## 11.5.1 A Special Note About Spatial Reproduction Patterns

Reproduction (i.e., conception, fawning, and recruitment) in the model was spatially structured, in a manner different from that suggested by the logistic-population model. The logistic-population model predicts only average growth and death for a population. The highest population-growth rates were at the edge of population foci where body conditions also were best. These results are consistent with the traditional density-dependent notion that nutritionally limited reproduction should be best where there are better resources. The explicit breeding constraint within the model (i.e., females could not breed if in poor body condition: $\leq 1$) was never triggered during any of these simulations. So despite the logistic growth demonstrated by the populations, conception was not constrained by nutrition, but only indirectly by differential mobility across the landscapes. Conception was less frequent among individuals using poorer habitats because greater mobility resulted in lower density, and, therefore, those individuals encountered breeding partners less frequently. This suggests that landscape pattern and movement grain may result in an explicit breeding constraint similar to that previously attributed to poor body condition.

These results are similar to the ecological studies cited by May (1994) where, in several instances, conflicting conclusions about a system could be drawn from observations at two different scales. By observing the simulation results in this study at a population level of organization, population processes seem to be consistent with those predicted by the logistic-population model. However, by observing individual behavior in the same simulations at local scales, it is clear that different processes may be of primary importance. For wildlife managers, management goes awry when treatments are based on local observations assumed to have the same properties and processes as the population as a whole. For example, managers who hope to detect changes in *ecological density* (i.e., relationship between population size and resource availability) by monitoring conception rates in areas of high density could be easily misled if landscape pattern and movement grain were primarily responsible for the patterns observed rather than a simple density dependence.

## 11.5.2 Using the Model Within a Management Context

### 11.5.2.1 Estimating Populations

To demonstrate the impact of complex spatial events on management decisions, I applied two simple, harvest methods (explained in the next section) and then used two standard survey procedures to evaluate the results. Initially, the unharvested populations were estimated using these survey procedures to differentiate the potential impact of harvest on survey estimates from problems that may arise within routine, unharvested systems.

Survey procedure 1 (PLOT) consisted of counts within 90 randomly selected cells (i.e., a 10% sample) to sample the landscape, determine the
mean cell density, and expand the mean density to a population estimate for
each year. The second survey procedure (TRANS) sampled at the same
intensity, but consisted of 9 left-to-right transects of 10 cells each. Again the
mean cell density was expanded to estimate each population in each year.

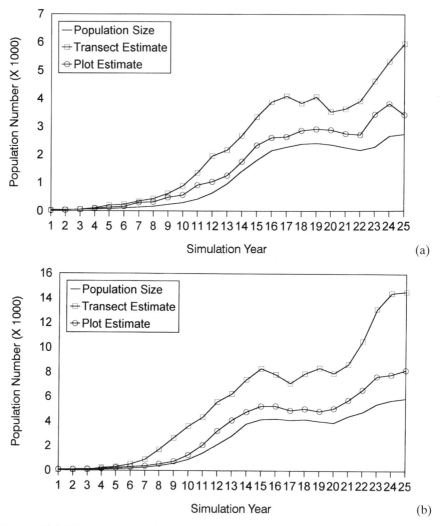

FIGURE 11.5. The way populations grew on the three habitats—(a) LOW (L$_L$), (b)
INTERMEDIATE (L$_1$), (c) MIXED (L$_M$)—could easily be misinterpreted from
estimates made from plot or transect surveys. Estimates of population size ranged
from 2 to 3 times larger than the actual population size. For the populations that
grew during these simulations, overestimates made from either survey method
would likely lead a manager to take inappropriate management actions.

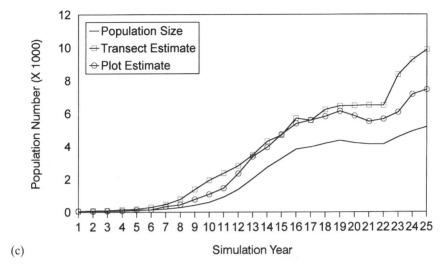

(c)

FIGURE 11.5. *Continued*

The true numerical growth of the unharvested populations was different from the estimates obtained from either PLOT or TRANS survey procedures (Figure 11.5). The greatest differences between actual and estimated numbers occurred in $L_M$ habitats surveyed by TRANS. In this instance, the estimate continuously increased in a near straight line, whereas the true population growth was distinctly sigmoid. These overestimates occurred because the populations violated the basic assumptions of parametric methodology (independence of individuals in a population resulting in random clumping and a random distribution of individuals, Cooke 1993). In these simulations, the procedures used to move individuals were distinctly independent, but the resulting distribution patterns were not Poisson despite the relatively uncomplicated resource patterns on the landscapes. The clumping of individuals across the landscapes created the overestimates (Cooke 1993). This is important, because living individuals in biological populations also are attracted to each other by strong social and behavioral needs unlike the "individuals" in the simulations, and living individuals also are constrained by the variance in resource availability within their nor-mal range. In other words, some biological populations may exhibit even greater tendencies toward clumping than those observed here. Also, aggression between individuals may force very different dispersion patterns from those exhibited in these models. Territoriality, colonial nesting, spacing within feeding or floating flocks of birds, can lead to underestimation because the dispersion of individuals will be more uniform than assumed in the procedure selected for making estimates of population size (Cooke 1993).

Both overestimation and underestimation can have a confounding affect on successful assessment and management of a system. For example, drought influenced white-tailed deer females in South Texas to concentrate in limited resource areas while nursing their fawns during summer and fall of 1980 (Cooke 1993). These animals were surveyed across six South Texas counties using a stratified grid of aerial transects, where 8 km separated the east-west and north-south transects. Despite the survey design or sampling effort, clumping of females resulted in an overestimate of their numbers. Concurrently on the same range, white-tailed deer males were broadly dispersed. Their more uniform distribution resulted in an underestimate of their numbers. Each bias was caused by the lack of randomness, which was a pivotal assumption of the estimate procedure used (Cooke 1993) and which, coincidentally, was identical to that used in these simulations. Assessment and management become more problematic when different subunits of a population (e.g., males, females, and young) have different seasonal needs that can lead to different dispersion patterns on the same landscape, even in years with normal rainfall. Estimating ratios that compare subunits may be more problematic than estimating population size (Cooke 1993) despite the fact that managers tend to have more confidence in ratios than in other population statistics (Connally 1981).

### 11.5.2.2 Harvest

Populations were harvested from each landscape type using two methods. Harvests were conducted in year 18, when growth reach an asymptote. In the first method, cells were selected at random and the first individual encountered in each selected cell was removed. In the second method, cells were selected randomly, but directed proportionally toward the high-density areas, so that cells that initially had many individuals were selected more frequently than were cells with few individuals. Each population was harvested to leave approximately 900 individuals. One way of distinguishing between different populations dispersed across their landscapes is by comparing the variance to mean ratio ($\sigma^2 : \mu$) of individuals counted within cells. A random dispersion of individuals results in $\sigma^2 : \mu$ ratio that equals 1. Uniform distributions yield $\sigma^2 : \mu$ ratios that are less than 1 while clumped distributions have $\sigma^2 : \mu$ ratios that are greater than 1. Random harvest left the inherent spatial sturcture of the original population ($\sigma^2 : \mu$ of individuals/ cell for unharvested versus randomly harvested populations: $P_L = 3.12$ versus 3.18, $P_I = 4.82$ versus 5.17, $P_M = 3.68$ versus 3.50), while directed harvest tended to even out the cell density over the distribution ($\sigma^2 : \mu$ of individuals/cell for unharvested versus directed harvested populations: $P_L = 3.12$ versus 2.70, $P_I = 4.82$ versus 3.27, $P_M = 3.68$ versus 3.57). Following harvest, each population was allowed to grow for five years, establishing new spatial growth patterns that were induced by the harvest, as demon-

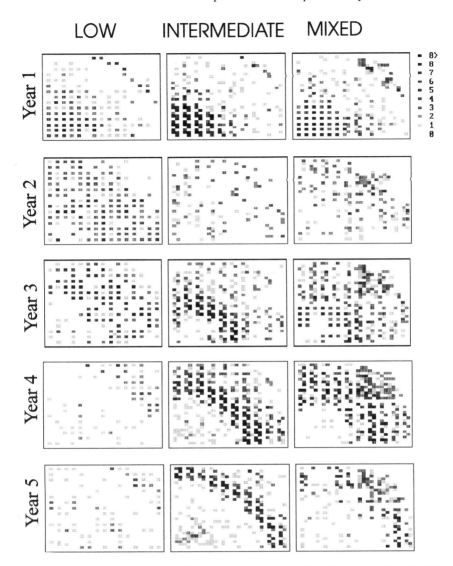

FIGURE 11.6. Distribution of individuals 1 year of age (herd increment) on LOW ($L_L$), INTERMEDIATE ($L_I$) and MIXED ($L_M$) habitats for years 1–5 following a simulated random harvest. This harvest reduced the population to the approximate level equal to year 13 in the original (unharvested) simulations.

strated by the maps of yearling individuals (Figures 11.6 and 11.7). Randomly harvested populations tended to develop the same type of yearling dispersion pattern as seen in the original, unharvested populations' growth (Figure 11.6). However, when harvest was directed toward high density portions of the landscapes, subsequent growth, as demonstrated by the

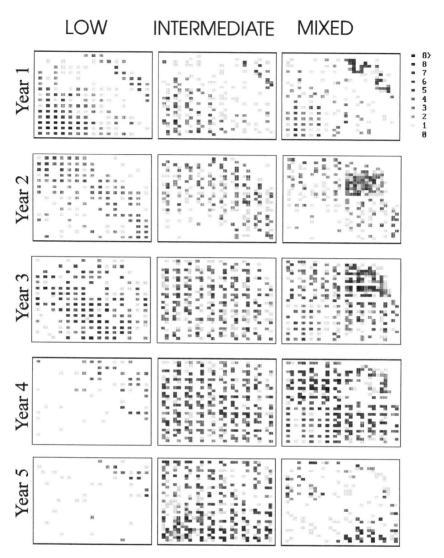

FIGURE 11.7. Distribution of individuals 1 year of age (herd increment) on LOW ($L_L$), INTERMEDIATE ($L_I$) and MIXED ($L_M$) habitats for years 1–5 following a simulated directed harvest. This harvest reduced the population to the approximate level equal to year 13 in the original (unharvested) simulations.

maps of yearling individuals (Figure 11.7), was more generalized over the landscapes. The distribution of yearlings continued to track resources and breeding partners subsequent to both harvest methods because the model's processes were unchanged. However, the spatial optimum of resources and breeding partners was less localized for populations responding to directed harvest.

The individuals in each distribution were counted to get actual population size, and then the population sizes were estimated from the two survey methods described above (Table 11.1). Generally, the largest population sizes were attained by $P_I$ and $P_M$, and these sizes were greater following directed harvest than the sizes attained by the unharvested populations.

A one-way analysis of variance (ANOVA) was used to determine whether the patterns induced by harvest led to different subsequent population growth when compared to the growth from 900 individuals of the unharvested population. No difference was detected between populations from any habitat type ($P \geq 0.05$). The same ANOVA procedures were applied to PLOT estimates (i.e., 90 random cells). PLOT estimates suggested that there were no differences in population growth between habitat types for combined unharvested and harvested populations. However, the comparison using TRANS estimates (i.e., 90 cells arranged in nine groupings of 10 cells each) showed a highly significant difference in the populations' growth in the $L_I$ habitat ($P \leq 0.01$).

This is important from a management perspective because mangers often can limit the number of individuals that may be harvested from a population, but they seldom have much control over how they will be removed in a spatial sense. In these simulations, the numerical size of populations grew

TABLE 11.1. Simulated populations grown on three habitat types ($L_L$, $L_I$, $L_M$) for untreated (ORIG) and following two harvest treatment types (RAN, DIR). Comparisons are of actual population size (ACT) and two estimates using different survey methods (PLOT, TRANS)

| Source | Year | $L_L$ | | | $L_I$ | | | $L_M$ | | |
| | | ORIG | RAN | DIR | ORIG | RAN | DIR | ORIG | RAN | DIR |
|---|---|---|---|---|---|---|---|---|---|---|
| ACT | 1 | 960 | 898 | 899 | 908 | 897 | 894 | 911 | 899 | 897 |
| | 2 | 1411 | 1346 | 1208 | 1424 | 1071 | 1219 | 1370 | 1195 | 1337 |
| | 3 | 1807 | 1905 | 1793 | 2122 | 1622 | 1730 | 2027 | 1784 | 2026 |
| | 4 | 2150 | 2125 | 2217 | 2831 | 2479 | 2585 | 2714 | 2643 | 3023 |
| | 5 | 2275 | 2221 | 2465 | 3807 | 3418 | 3766 | 3252 | 3528 | 3922 |
| | 6 | 2383 | 2347 | 2490 | 4167 | 4208 | 4652 | 3812 | 4193 | 4645 |
| PLOT | 1 | 1260 | 1014 | 1036 | 1271 | 1078 | 929 | 1452 | 1431 | 1185 |
| | 2 | 1708 | 1698 | 1377 | 2082 | 1217 | 1324 | 2338 | 1751 | 1783 |
| | 3 | 2338 | 2349 | 2136 | 3235 | 2007 | 2018 | 3364 | 2669 | 2787 |
| | 4 | 2605 | 2499 | 2808 | 4100 | 3353 | 3193 | 3929 | 3673 | 4047 |
| | 5 | 2637 | 2872 | 3139 | 4782 | 4463 | 4645 | 4720 | 4826 | 5350 |
| | 6 | 2862 | 3022 | 3225 | 5243 | 5414 | 5937 | 5371 | 5766 | 6161 |
| TRANS | 1 | 2168 | 1367 | 1175 | 3620 | 2872 | 812 | 2360 | 1708 | 705 |
| | 2 | 2680 | 1997 | 1367 | 4335 | 2968 | 1164 | 2808 | 1740 | 1004 |
| | 3 | 3342 | 2798 | 2189 | 5584 | 3695 | 1783 | 3460 | 2488 | 1719 |
| | 4 | 3876 | 3278 | 3481 | 6225 | 4922 | 3000 | 4292 | 2990 | 2691 |
| | 5 | 4090 | 3598 | 4154 | 7400 | 6332 | 6268 | 4677 | 4591 | 4688 |
| | 6 | 3833 | 4026 | 4325 | 8307 | 8756 | 9215 | 5702 | 6631 | 7837 |

TABLE 11.2. ANOVA comparisons ($P \leq 0.05$) of populations grown on three habitat types ($L_L$, $L_I$, $L_M$) within the untreated (ORIG) and treatment (RAN, DIR) populations, to determine whether accuracy of survey methods influenced the assessment of treatment efficacy

| | Treatment | | |
| Habitat type | ORIG | RAN | DIR |
| --- | --- | --- | --- |
| LOW | ** | ** | ** |
| INTERMEDIATE | ns | ** | ns |
| MIX | ns | ns | ns |

in a similar way whether their distributions were a result of unharvested population growth or were created artificially by harvest. Knowing whether or not a manager can assess the impact of harvest on a population is perhaps more important to the future of the population than the impact of the harvest itself. Since management is guided by estimates of population size made from surveys, how the impact of harvest may influence estimates is important.

A one-way ANOVA also was used to determine if accuracy of the survey methods could influence the assessment of harvest efficacy (Table 11.2). There were highly significant differences found within harvested and unharvested populations growing on $L_L$ habitat types ($P \leq 0.01$). Also, differences were found within randomly harvested populations growing on the $L_I$ habitat type. No differences were detected within harvested and unharvested populations on the $L_M$ habitat type. These results are interesting because intuitively, populations in complex habitat types should be more difficult to survey and estimate than those in simple or uniform habitats (Connally 1981), but this was not the case for these simulated populations. However, when harvest was applied to $P_M$, the $\sigma^2$:$\mu$ ratio (described above) of individuals/cell was not different between the harvested and unharvested populations, suggesting that the inherent pattern of the populations' distributions were unchanged by the harvest schemes.

## 11.6 Discussion

The model used in this chapter is a simple yet effective way of illustrating some potentially powerful concepts. By viewing the model results through harvest and survey methods familiar to most wildlife managers, I showed that population patchiness need not be caused by habitat patchiness and that animal dispersion over the landscape can have a significant influence over population estimates. These results are important for two reasons: (1) wildlife managers should not assume too much about the dispersion of resources across a landscape from animal distribution or patterns of population density, and (2) estimates of a population's characteristics and size can be changed by factors other than changes in the characteristics and size of the population.

In this chapter, I have shown that population dynamics can be complex in very simple landscapes and our estimates may be poor indicators of the true nature of a population. These results suggest that the observable complexity of a population's distribution may not be a good tool for evaluating the complexity of resource dispersion across a landscape. This illustrates how description in general, and analysis without a distinct spatial component specifically, has in many ways confounded our understanding of many population behaviors, because interrelationships and the sources of variation are masked.

## *11.7 References*

Addicott, J.F., J.M. Aho, M.F. Antolin, D.K. Padilla, J.S. Richardson, and D.A. Soluk. 1987. Ecological neighborhoods: scaling environmental patterns. Oikos 49:340–346.

Cale, W.G., G.M. Henebry, and J.A. Yeakley. 1989. Inferring process from pattern in natural communities. BioScience 39:600–605.

Connally, G.E. 1981. Trends in populations and harvests. Pp. 225–243 in O.C. Wallmo, editor. Mule and black-tailed deer of North America. University of Nebraska Press, Lincoln, Nebraska, USA.

Cooke, J.L. 1985. Theories of distribution and mobility: The desert mule deer of the Black Gap Wildlife Management, Texas. Texas Parks and Wildlife Department, Austin, Texas, USA.

Cooke, J.L. 1993. Assessing populations in complex systems Ph.D. [dissertation]. Texas A&M University, College Station, Texas, USA.

Gardner, R.H., B.T. Milne, M.G. Turner, and R.V. O'Neill. 1987. Neutral models for the analysis of broad-scale landscape pattern. Landscape Ecology 1:19–28.

Lotka, A.J. 1925. Elements of physical biology. Williams & Wilkins Co., Baltimore, Maryland, USA.

Ludwig, J.A., and J.F. Reynolds. 1988. Statistical ecology. John Wiley and Sons, New York, New York, USA.

May, R.M. 1994 The effects of spatial scale of ecological questions and answers. Pp. 1–17 in P.J. Edwards, R.M. May, and N.R. Webb, editors. Large-scale ecology and conservation biology. Blackwell Scientific Publications, London, UK.

Odum, E.P. 1959. Fundamentals of ecology. W.B. Saunders Company, Philadelphia, Pennsylvania, USA.

Pittman, M. 1987. Mule deer reproduction and nutrition. Job Final Report W-109-R Texas Big Game Investigations. Unpublished report on file, Texas Parks and Wildlife Department, Ft. Davis, Texas, USA.

Prigogine, I., and I. Stengers. 1984. Order out of chaos. Bantam Books, New York, New York, USA.

Verme, L.J. 1965. Reproduction studies on penned white-tailed deer. Journal of Wildlife Management 29:74–79.

# 12
# The Importance of Scale in Habitat Conservation for an Endangered Species: The Capercaillie in Central Europe

ILSE STORCH

## 12.1 Introduction

Habitat relationships of any wildlife species include multiple hierarchical levels and spatial scales. While local habitat features such as vegetation structure are most relevant to individuals, large-scale features, e.g., the landscape mosaic, affect populations and metapopulations. In order for a species to persist, its requirements must be met at all scales. Specialized habitat needs and large spatial requirements render the capercaillie *Tetrao urogallus* an excellent example to demonstrate the importance of spatial scale in the conservation of wildlife species.

The capercaillie is the largest of all grouse. It has a polygynous lek mating system (Johnsgard 1983), and its size dimorphism is the most pronounced within the grouse family. At 4 to 5 kg, cocks weigh twice as much as hens (Klaus et al. 1989, Storch unpublished data). Capercaillie are coniferous forest obligates: their winter diet consists almost exclusively of conifer needles, an adaptation to harsh northern winters. The primary habitat of the capercaillie is old-growth boreal and montane forest, which is characterized by coniferous trees, open structure with moderate canopy cover, and rich ground vegetation dominated by ericaceous shrubs such as bilberry *Vaccinium myrtillus*. When this kind of structure is provided, commercially managed forests are used by capercaillie as well. The main distribution range of the capercaillie is the boreal forest of Siberia and Fennoscandia (Figure 12.1). In central Europe, where the landscape is dominated by farmland and deciduous forests, coniferous forests are restricted to

---

Ilse Storch is Chief Scientist and Project Leader with Munich Wildlife Society, a private non-profit organization active in wildlife conservation, management, and research in central Europe. Her interests include research into wildlife-habitat relationships at various scales. Since 1986, the effects of habitat fragmentation on individuals and populations, and their implications for conservation, have been the major focus of her work. Ilse Storch also is a consultant to wildlife conservation programs on several continents, and chairs the IUCN Grouse Specialist Group.

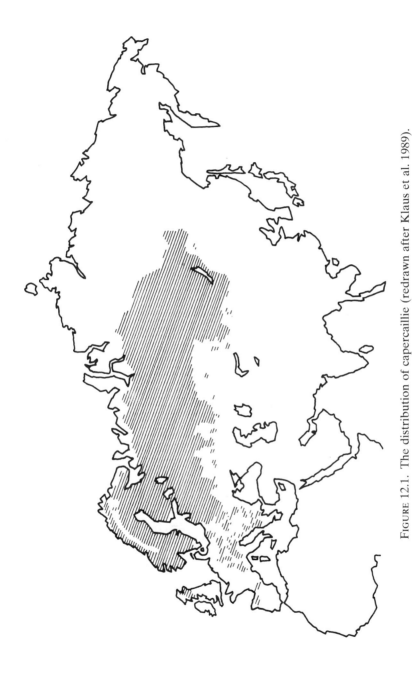

FIGURE 12.1. The distribution of capercaillie (redrawn after Klaus et al. 1989).

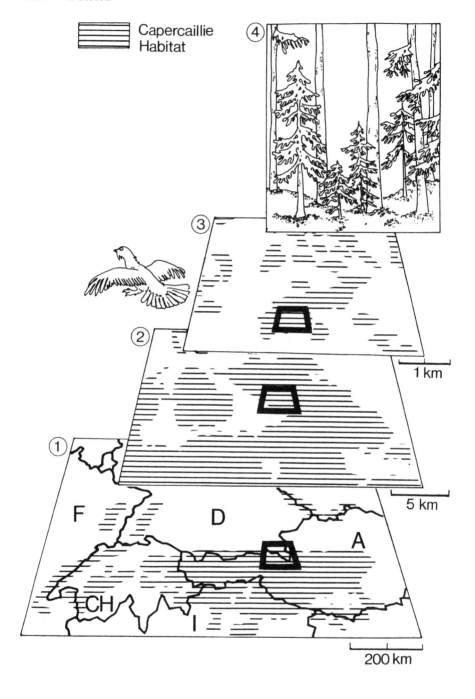

Capercaillie Habitat

① ② ③ ④

1 km

5 km

200 km

F    D    A

CH    I

montane areas. Therefore, at the continental scale, the capercaillie is limited to isolated mountain ranges ($100 \, km^2$ to $50,000 \, km^2$); the largest are the Alps (Klaus et al. 1989). The range of alpine capercaillie extends from the French-Swiss border over parts of Switzerland, southern Germany, northern Italy, western and central Austria where the best populations are located, to northern Slovenia (see Figure 12.2, box 1).

During the past few decades, capercaillie numbers have been declining throughout its range. Loss and deterioration of habitats are assumed to be the major causes of decline (see review in Klaus et al. 1989, Rolstad 1989). In central Europe (i.e., Germany, Switzerland, Austria, northern Italy, Poland, Czechia, Slovakia), many local populations have disappeared, the species has been listed in the Red Data Books, and hunting has been restricted (Austria, Slovakia) or banned in all countries. Hunting bans did not reverse population trends, however.

The capercaillie is a highly valued game bird; its image is a typical element of folklore in its central European strongholds such as the Alps and the Black Forest. The interest of hunters in capercaillie traditionally has been focused on lekking males. In spring, males occupy permanent territories aggregated around leks. As a consequence, capercaillie were generally conceived as sedentary birds with restricted, predictable home ranges. This view prevailed not only among hunters and naturalists but was reflected in popular and scientific writings (e.g., Müller 1974, 1979) and in conservation as well. Thus, capercaillie conservation efforts have been focused on preservation of microhabitat features at forest stand level, and particularly at leks (Popp 1974, Stein 1974, Zeimentz 1974, Weiss 1988).

Smaller scale habitat preferences of capercaillie are generally well understood (Klaus et al. 1989, Storch 1995a). Insight into habitat relationships at the landscape scale came only after Scandinavian researchers conducted the first telemetry studies on capercaillie (Wegge 1985, Wegge and Larsen 1987, Rolstad 1989, Gjerde 1991a). These studies suggested that capercaillie populations were very susceptible to macrohabitat disturbances such as

----

FIGURE 12.2. Central European capercaillie habitats at various levels of spatial scale. (1) Continental scale: montane conifer forest versus lowlands (central Europe), (2) regional scale: forested mountain ranges versus farmland valleys (Bavarian Alps), (3) local scale: old and middle-aged stands versus younger forest stages (Teisenberg), (4) forest stand scale: vegetation structure. At a continental scale, the distribution of montane coniferous forests results in isolated regional distribution ranges. Within distribution ranges, e.g., the Alps, local capercaillie populations live on forested mountain ranges separated by farmland valleys. Within mountain ranges, e.g., Teisenberg, spacing of capercaillie depends on old forest fragmentation patterns. Among old forest stands, capercaillie select according to habitat structure. The capercaillie depends on availability of habitats from local to continental scale. Therefore, conservation efforts are likely to fail unless all scales are considered.

forest fragmentation (Rolstad and Wegge 1989). Although critical for conservation, there were no data and few ideas regarding landscape scale habitat requirements of capercaillie populations in central Europe (e.g., Müller 1974, Scherzinger 1988).

This was the situation in 1988, when the University of Munich and the Munich Wildlife Society began the first central European telemetry study on capercaillie on the Teisenberg mountain range in the Bavarian Alps. The study aimed to document habitat and range use of capercaillie in relation to habitat characteristics at various levels of scale. We believed that hunters and conservationists were correct and that local forest structure mattered. However, we also hypothesized that this was only part of the story. If capercaillie were susceptible to landscape-scale habitat features in Scandinavia, it was likely that they were susceptible in central Europe as well. Thus, it seemed that a vital aspect of capercaillie habitat requirements had been neglected. Therefore, the main objective of the Teisenberg study was to discover whether both local forest structure and landscape mosaic played an important role in habitat relationships of capercaillie in central Europe.

In this paper, I want to use the Teisenberg capercaillie study as an example to illustrate that habitat relationships must be understood on all relevant levels of scale as a vital prerequisite for habitat conservation for endangered species.

## 12.2  The Teisenberg Study

### 12.2.1  Approach, Study Area, and Methods

The Teisenberg study was designed to provide a description of preferred habitat structures, as well as information on spatial requirements of capercaillie, in order to develop guidelines for conservation in the Bavarian Alps and other central European areas. The focus was on the effects of forest fragmentation patterns on capercaillie: we wanted to understand how the distribution of favorable habitats affected range use and spacing of individuals and populations. The Teisenberg population should serve as a model to shape our understanding of landscape scale aspects in capercaillie–habitat relationships in the Alps.

The Alps contain the largest distribution range of the capercaillie south of the boreal forest (see Figure 12.2). The species occurs throughout alpine forests, but population density and trends vary locally. In the Bavarian Alps the landscape is characterized by mountain ranges reaching altitudes of 1000 to 3000 m with treeline at 1300 to 1800 m and separated by valleys up to a few kilometers wide. The valleys at 600 to 900 m altitude are mostly dairy farmland, and many are densely populated. Intensive forestry started as early as the seventeenth century and has created a mosaic of stands of varying successional stages. Thus, there is a hierarchical ordering of forest fragmentation: at the mountain range scale and at the forest stand scale (Figure 12.2). Among and within forest stands, the composition of the ground vegetation is highly variable (Figure 12.2). Similar landscape mosaics are found in most central European forests where capercaillie occur.

Teisenberg, a 50-km$^2$ mountain range in the foothills of the Bavarian Alps, Germany, (47°48'N, 12°47'E) was chosen because it had a stable capercaillie population estimated at 100 to 200 birds; forests were fragmented into stands of various age classes; and topography, vegetation, and forest management were typical of the region. Elevations ranged from 700 m to 1800 m, and treeline was around 1400 m in elevation. The climate was moist and temperate with mean annual temperatures of 5°C, precipitation of 1800 mm annually, and snow cover from December through April. Forests dominated by Norway Spruce (*Picea abies*) mixed with beech (*Fagus sylvatica*) and fir (*Abies alba*) covered 95% of the area, and were fragmented into a patchwork of stands of varying size (1 to 100 ha, median 7 ha) and successional stage. Capercaillie mostly used the old (>90 years) and middle-aged (50 to 90 years) forest stands, which covered one-fifth and one-third of the area, respectively. The remainder of the forest habitat comprised pole-stage (20 to 50 years old) and younger stands.

With a rather elusive, forest-dwelling, and mobile species such as the capercaillie, radio telemetry was the best choice to answer questions of habitat use and spacing patterns. Between 1988 and 1992, about 10,000 radio locations were obtained from 16 female and 24 male capercaillie that were radio-tracked over periods from 1 to 56 months. A variety of habitat variables were recorded on Teisenberg and were compiled into a digital habitat map based on a geographical information system. To analyze habitat use, radio locations of capercaillie were related to habitat composition of the study area and of individual home ranges. For details on the study area and methods, see Storch (1993a, 1993b, 1995c).

## 12.2.2 Habitat Selection at the Forest Stand Level

Forestry practices have major influences on the structure and dynamics of the habitats of forest-dwelling species. In the Bavarian Alps and most other central European regions, forests are managed for timber production. Forestry practices have resulted in a mosaic of even-aged stands, from clearcuts to stands of >100 years of age. Forest stands are the basic spatial unit of forest management. In order to incorporate capercaillie conservation measures into forest management, it is important to understand which forest stand habitat features are most attractive to the capercaillie.

Forest stands on Teisenberg differed in a variety of factors that might influence habitat suitability for the capercaillie: canopy height and cover, species composition and diversity of the tree layer, height, cover, species composition and diversity of the ground vegetation, number of vegetation layers, patch size, and amount of edge. Furthermore, they differed in topographical features, including elevation, steepness of slope, and exposure. The composition of the study area regarding these variables was extrapolated from the habitat map based on a set of 2000 random points. A comparison between the distribution of random points and locations of radio-marked birds revealed that only a few habitat features made a significant difference to the capercaillie. The key variables were percentage of canopy cover, percentage of ground cover, composition of the ground vegetation, and patch size.

In the boreal forest (Gjerde 1991b) as well as in the Alps (Storch 1993a, b), capercaillie preferred forest stands with moderate canopy cover of 50 to 60% (Figure 12.3). These stands are open enough for a big bird such as the capercaillie to fly in and permit enough sunlight to develop a rich ground vegetation (Storch 1994), a habitat feature particulary important to the capercaillie for food and cover in summer and autumn (Storch 1993b, 1994). In all seasons, capercaillie locations were in significantly more open cover than were random points (Mann-Whitney $U$ tests, $P < 0.001$) (Storch 1993a, b, c). In winter, when the birds spend most of their time in trees feeding on conifer needles, denser stands were selected, most probably for microclimatical reasons (Storch 1993b), than in summer ($P < 0.001$ for both sexes, Mann-Whitney $U$ tests). In managed central European forests, moderate canopy cover and a well-developed ground vegetation were best represented in the latest successional stage, here termed old forest. Correspondingly, alpine capercaillie preferred old forest throughout the year; e.g., on Teisenberg, both sexes used old forest significantly more often than expected from availability, leks were located in old forest (Storch 1993a, 1993b), and old forest was preferred by nesting females and by broods (Storch 1994). Also, on an annual basis, predation risk was lower in old forest than in other successional stages (Storch 1993b). Preference for the latest successional forest stage also has been shown in studies from other parts of the capercaillie range (see Klaus et al. 1989 for a review). In most areas, old natural or seminatural forests are the capercaillie's stronghold, because they best meet its structural habitat needs (Rolstad and Wegge 1987, Picozzi et al. 1992). However, capercaillie forests need to be neither old nor natural. As long as there are trees with branches strong enough to hold the birds, the actual age of a stand is of little significance. It is structure that is important. A similar dependence on structure has been reported for American marten by Bissonette et al. (this volume).

In old forest, capercaillie selected ground vegetation. They preferred stands with high proportions of bilberry (Figure 12.4), an important feature of capercaillie habitats (Storch 1995a). Bilberry is a major food plant of capercaillie in the snow-free seasons (Jacob 1987, Storch et al. 1991). It is rich with insects for chicks, and it provides optimal hiding and thermal cover for both adults and broods (Storch 1995b). In spring, summer, and autumn, locations of females and males were significantly greater in bil-

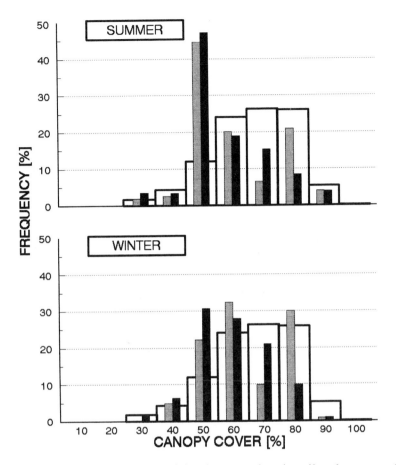

FIGURE 12.3. Distribution of capercaillie telemetry relocations (females grey, males black bars) from summer and winter in pole-stage and older forest stands on Teisenberg, Bavarian Alps, in relation to percentage of ranges of canopy cover. The open bars indicate availability. Canopy cover in pole-stage stands averaged 75 ± 11%, in middle-aged forest 70 ± 10%, and in old forest 56 ± 11% ($\bar{x}$ ± 1SD). In summer and in winter, female ($n$ = 437 and 344) and male ($n$ = 1352, 1417) locations were in significantly more open cover than a set of 2000 random points distributed over the area (Mann-Whitney $U$ tests, $P$ < 0.001) (recalculated based on Storch 1993a, 1993b).

berry cover than expected from a distribution of random points (Mann-Whitney $U$ tests, $P$ < 0.001).

Habitat use by capercaillie was affected by forest fragmentation, as indicated by the preference of both females and males for large stands. Within the preferred habitat of old forest (Figure 12.5), the birds selected the largest stands (Storch 1993a, 1993b, 1993c) throughout the year (Mann-Whitney $U$ tests, $P$ < 0.001 for spring, summer, autumn, winter, and for both

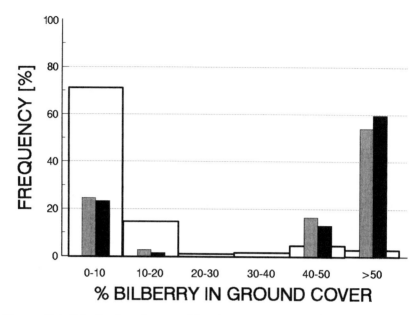

FIGURE 12.4. Distribution of capercaillie telemetry relocations (females grey, males black bars) from summer and autumn in old forest stands on Teisenberg, Bavarian Alps, in relation to the proportion of bilberry cover in the ground vegetation. The open bars indicate availability. Bilberry cover averaged 11 ± 18% in old forest ($\bar{x} \pm$ 1SD). Habitat use by capercaillie is based on 572 female and 1332 male radio locations (recalculated based on Storch 1993a, 1993b). Male and female locations were significantly greater in bilberry cover than a set of 2000 random points distributed over old forest stands (Mann-Whitney $U$ tests, $P < 0.001$).

sexes), although smaller stands did not differ from larger stands in canopy cover, ground vegetation, bilberry, or any other habitat feature. In the boreal forest, predation on capercaillie increased with increasing old forest fragmentation (Gjerde and Wegge 1989). Although the Teisenberg data did not allow me to address this hypothesis, it seemed likely that the selection of large forest stands is causally related to predator avoidance behavior.

## 12.2.3 Selection of Home Ranges

Preferences for particular habitat features may be apparent in the selection of home ranges as well as in the selection of locations within home ranges. Capercaillie on Teisenberg selected spring (Storch 1993c), summer (Storch 1993a, 1994), winter (Storch 1993b), and annual (Storch 1995c) home ranges with amounts of old forest greater than proportionally availably in the study area (Figure 12.6). Also within their home ranges, the birds used habitats nonrandomly: males in winter and both sexes in summer were located significantly more often in old forest than expected (Figure 12.6).

Middle-aged forest stands covered almost one-third of Teisenberg and of individual home ranges and were frequently used by capercaillie. Females in winter and males throughout the year neither preferred nor avoided this forest stage (Figure 12.6). Females, however, avoided middle-aged stands in summer that provide only sparse ground cover, a feature even more important for females and broods, who hide from predators, as contrasted with males, who stand watch and may eventually fight.

Besides old forest, bilberry was a habitat feature that played a role in the selection and use of home ranges by Teisenberg capercaillie. Annual home ranges of females and males had significantly more bilberry than expected from availability (Storch 1995b). The significance of bilberry in the selection of home ranges and in the use of locations within home ranges was most evident for broods (Storch 1994). Although bilberry was best developed in old forest stands, the spatial distribution of bilberry did not parallel the distribution of old forest on Teisenberg; the amounts of old forest and of bilberry cover within home ranges were not related.

## 12.2.4 Home Ranges and Habitat Fragmentation

In the past, capercaillie in central Europe have been regarded as sedentary birds, in part because movement data were lacking. An attempt to recon-

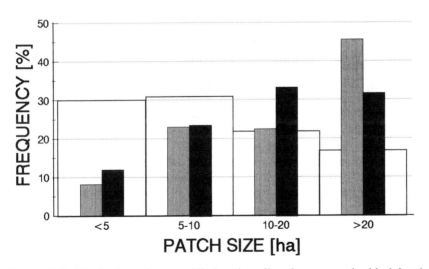

FIGURE 12.5. Distribution of capercaillie locations (females grey, males black bars) in old forest on Teisenberg, Bavarian Alps, in relation to availability (open bars) of old forest stands of various size. Habitat use by capercaillie is presented as the mean of seasonal (spring, summer, autumn, winter) distributions based on a total of 10,000 radio locations. In all seasons and for both sexes, bird locations were in larger stands than a set of 2000 random points distributed over old forest stands (Mann-Whitney $U$ tests, $P < 0.001$). Recalculated based on Storch (1993a, 1993b, 1993c).

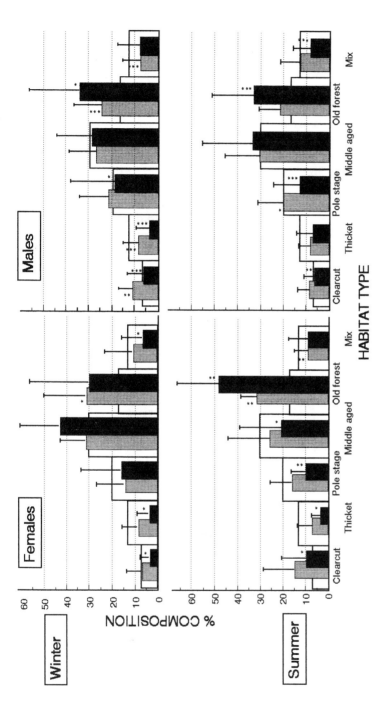

FIGURE 12.6. Percentage composition of habitat types on Teisenberg (open bars), in individual home ranges (grey, $\bar{x} \pm 1SD$) of female and male capercaillie and at locations within home ranges (black, $\bar{x} \pm 1SD$) in winter ($n = 8$ and 35 home ranges) and summer ($n = 13$ and 38 home ranges). Asterisks indicate differences between use and availability; above grey columns: home ranges vs. study area; above black columns: locations versus home ranges (Wilcoxon test: $*P < 0.05$, $**P < 0.01$, $***P < 0.001$). Home ranges were estimated as 100% convex polygons. Habitat types were defined as clearcuts (covered by natural rejuvenation), thickets (young forest before thinning), pole stage (after first thinning, approximately 20 to 50 yrs old), middle-aged forest (after second thinning, sparse ground cover, approximately 50 to 90 yrs), and old forest (final felling stage, rich ground cover, >90 yrs). Stands with <1 ha variation in composition of succession stages were described as mixed when no stage constituted >75% of the area (based on Storch 1993a, 1993b).

struct annual home ranges on the basis of sightings of individual birds distinguished by plumage characteristics in an area in Germany resulted in size estimates of 50 ha (Müller 1974). These figures may not be comparable with telemetry data, however, because it is likely that only the most sedentary birds could be followed, and movements may have remained unnoticed.

Telemetry data from Teisenberg and other parts of the distribution range (Rolstad and Wegge 1989, Menoni 1991) resulted in annual home ranges of a few hundred hectares in size. Estimated as 100% convex polygons, but excluding excursions to neighboring mountains (see below), the annual home ranges of capercaillie on Teisenberg varied from 132 to 1207 ha ($N =$ 26) and averaged 550 ha for both sexes; the mean distance between the two outermost locations within a home range was 3.9 km (Storch 1995c). Because the birds used seasonally different parts of their home ranges, seasonal ranges were smaller, e.g., winter ranges measured about 150 ha for females and adult males and 280 ha for subadult males (Storch 1993b); summer ranges averaged about 160 ha for females and broods and 250 ha for males (Storch 1993a, 1994).

Both habitat features, old forest and bilberry cover, that played a role in home range selection were patchily distributed on Teisenberg. Old forest stands had a median size of 7 ha and occurred throughout the area with a mean nearest neighbor distance measured from edge to edge of 150 m; maximum distance was 450 m. Bilberry occurred throughout the area. It developed best in old forest, but due to soil characteristics, was most abundant in the central part of Teisenberg. Compared to the size of capercaillie home ranges, the distribution of old forest and of bilberry was fine grained (Wiens 1976): preferred habitat patches were much smaller than home ranges, and each bird included several patches in its seasonal and annual ranges.

How, then, did the distribution of preferred habitats affect home range size? As previously shown in Norway (Wegge and Rolstad 1986, Gjerde and Wegge 1989), capercaillie home ranges on Teisenberg increased in size with the degree of habitat fragmentation. The greater the proportion of preferred habitat within a home range, the smaller was its size. Winter home range size was negatively related to the proportion of old forest (Storch 1993b), and summer home range size to bilberry cover (Storch 1993a). Eighty-three percent of the variation in annual home range size of females was explained by the amount of old forest, while late forest stages, old and middle-aged forest, and bilberry cover contributed 58% to the variation in annual ranges of males (Storch 1995c).

So far, I can only speculate on the consequences of enlarged home ranges in fragmented habitats. Increasing fragmentation results in larger home ranges, requiring longer movements to satisfy foraging and other life history needs. If the preferred old forest habitat also is the habitat where the birds are least exposed to predation risk, as indicated from preliminary data from Teisenberg (Storch 1993c), enlarged home ranges may result in reduced survival. Data from Norway strongly suggest reduced survival of capercaillie in fragmented habitats (Gjerde and Wegge 1989) that has been attributed to an increase in small- and medium-sized predators as a result of large-scale clearcutting and a subsequent increase in small mammals and other herbivores (Rolstad and Wegge 1989). In the Alps, however, the present degree of forest fragmentation has existed for centuries, and habitat-specific mortality in capercaillie may result from differential exposure to predators in various forest stages.

## 12.2.5  Spacing Patterns and Leks

The capercaillie is a lekking species. During April and May, males occupy permanent territories aggregated around lek centers where display activity is most intense (Wegge and Larsen 1987, Storch 1993c). Female spring ranges are spaced independently of leks (Wegge 1985), and on Teisenberg, most females visit one or several leks before mating (Storch 1993c). Leks may exist for many decades: there are old place names in the Alps such as "dancing-ridge" or "cocks hill"; quite a few of these places still hold capercaillie leks today. Due to the conspicuous display behavior of males, leks were well known in the Alps. For the rest of the year, the birds are rather elusive. Before telemetry data became available, capercaillie were believed to stay close to their leks after the mating season. A major aim of the Teisenberg study was to describe the spacing pattern of capercaillie during the remainder of the year after the breeding season. After marking the first seven males on a Teisenberg lek in spring 1988 and locating them in close vicinity of the lek center throughout May, we first believed the transmitters failed when suddenly in early June, signal reception from the same birds became difficult. Further work revealed that the birds had left the area, each moving up to a few km away from the lek locations.

In the course of the study it became clear that there were pronounced sexual and seasonal differences in the spacing behavior of capercaillie (Storch 1995c). Figure 12.7 shows the mean distances of capercaillie from their home lek in each month of the year. In March, April, and May, males stayed closest to leks, mean distance to the lek was about 0.5 km and most locations ($\bar{x} \pm 1\text{SD}$) were within a 1 km radius around the lek center. At the beginning of June, males moved to individual summer ranges. They spread out in all directions, and, until November, most of their locations were distributed within a 4-km radius from their home lek. In 16 of 20 males, summer ranges did not overlap display territories. The longest distance between lek and summer range recorded on Teisenberg was 7.3 km. In autumn, males moved closer to their lek again, and from December onwards most locations ($\bar{x} \pm 1\text{SD}$) were within 2 km of the lek center. Winter ranges included display territories, and most males stayed close to the lek throughout winter. Throughout the year, mean female locations were about

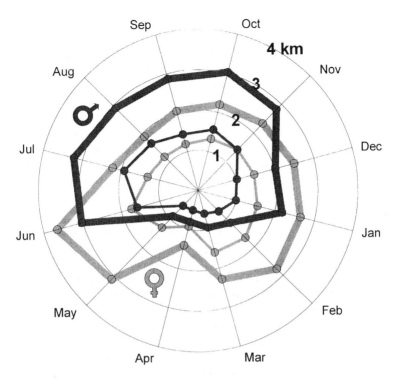

FIGURE 12.7. Mean distances of female (grey lines; $n = 5$ to 12 per month) and male (black lines; $n = 18$ to 29 per month) capercaillie from their most frequently used lek center in the Teisenberg area, Bavarian Alps, Germany, 1988–1992. For each bird, monthly home range centers were calculated as arithmetic means of >10 radio locations. The inner polygons give the mean distances over all individual home range centers, the outer polygons $\bar{x} + 1SD$ (recalculated based on Storch 1995c).

1.3 km from the lek they most frequently attended; most locations and nest sites were within a 3-km radius of the lek. Some females moved long distances to visit leks, and the longest recorded distance was 6.2 km on Teisenberg. There were no significant seasonal differences in the distances of females from leks.

Several radio-marked birds moved between Teisenberg and Sulzberg, the neighboring mountain range. They crossed a 2- to 3-km wide farmland valley with a busy tourist resort; the distance overflown was 5 to 7 km. We documented no movement between the mountain ranges on a daily basis within seasonal home ranges. The birds either moved on excursions lasting one to three weeks or on a seasonal basis. At least one male regularly moved between his winter range on Teisenberg and his summer range on Sulzberg; one female mated on Teisenberg and nested on Sulzberg (Storch 1993c). Although these data remained anecdotal, they indicate that nonforested areas a few-km wide can be overcome by capercaillie, although not necessarily on a daily basis. Flight energetics and exposure to raptors might preclude daily movements across long distances. Thus, a few kilometers of open landscape separating alpine forests need not mean an absolute barrier between local capercaillie populations, but may result in a metapopulation system with partially separated mountain range populations.

Seasonal movements may occur as a consequence of spatial separation of seasonal resources (Rolstad and Wegge 1989). However, this explanation was unlikely for Teisenberg capercaillie, because the preferred habitat of old forest was available throughout the area, and bilberry, the staple summer food, was most abundant in the central part of the study area where most winter ranges were located. Even though both sexes used the same food sources (Storch et al. 1991) and had similar seasonal habitat preferences (Storch 1993a, 1993b), seasonal movements occurred with males but not females.

Because they were independent of the distribution of resources, movements of males from leks to distinct summer ranges need to be explained in other terms; they may be related to predator avoidance. Mortality of male capercaillie peaks in winter and spring (Klaus 1985, Wegge et al. 1990) when males, but not females, live in territories aggregated around leks. This clumped distribution may attract predators, and thus it may be advantageous to leave the lek during the summer molt when the birds are restricted in their ability to fly and consequently are particularly susceptible to predation. Dispersion may reduce predation risk. Thus, male movements may occur independently of habitat features and may be a consequence of the lek mating system. This is supported by observations in the boreal forest, where males also move in unfragmented old-growth habitats (Wegge and Rolstad 1986). Because females are randomly dispersed throughout the year, seasonal movements related to predator avoidance may not be required.

## 12.3 Capercaillie Conservation in Central Europe

In central Europe, wildlife ecology and conservation have traditionally focused on local habitat structure and have almost completely neglected landscape scale issues of habitat suitability. At the landscape scale the distribution of forest versus agricultural land, and within forests the mosaic of age classes have existed for centuries and have changed only gradually; they were not perceived as habitat variables that affected viability of populations. The most obvious changes in central European wildlife habitats have occurred in vegetation structure. This is the feature a local land owner or forest manager can influence, whereas no mechanism exists to integrate management of landscape scale features. That local population density within a small area, e.g., a hunting district of a few hundred hectares size, may be affected by habitat quality within a radius of many km is not readily understood by capercaillie conservation practitioners. This is partly because larger scale ecological ideas are relatively new in conservation, and new concepts take time until they become understood and accepted by the semiprofessional or lay public. In Germany, the absence of a professional wildlife conservation agency and the minor role wildlife conservation plays in forest management policy may explain why landscape scale issues have been neglected in habitat conservation of forest dwelling species. In the Bavarian Alps, where capercaillie locally still occur in good numbers, capercaillie conservation is limited to the initiatives of a few wildlife-minded foresters who try to include habitat preservation in forest management. In areas where populations are close to extinction, conservation initiatives are directed mostly by local activist groups. In most regions, there is little coordination of local efforts.

Disadvantageous wildlife management structures may be an explanation, but not an excuse, for asking the wrong questions. In the case of the capercaillie, people debated about the optimal tree species composition and the minimum number of anthills per hectare required for chick food, details about which the birds are rather flexible (Storch 1995a), rather than estimating spatial requirements of individuals and populations. Even in attempts to reintroduce capercaillie to areas where they were extinct (e.g., Spittler 1994) habitat suitability has been evaluated without any consideration of spatial requirements.

## 12.4 Implications for Capercaillie Habitat Preservation

Studies throughout the distribution range confirm that capercaillie select habitats that resemble its primary habitat, the boreal forest. This has important implications for conservation. The features of the boreal forest may be used as a guideline for habitat management. Accordingly, at the forest stand level, forests managed for capercaillie should be dominated by coniferous trees and have moderate canopy cover and a well-developed ground vegetation rich in bilberry shrubs. More details for capercaillie habitat management have been listed elsewhere (Storch 1995a). Recommendations at the forest stand level are partly in agreement with conventional measures that have largely resulted from the perception by hunters.

Leks have received special emphasis in capercaillie conservation in central Europe. Because leks were known to disappear due to clearcutting or other marked habitat changes, habitat preservation at leks was considered a key to capercaillie conservation (Popp 1974, Stein 1974, Zeimentz 1974, Weiss 1988). The results from Teisenberg show that leks certainly play an important role in the spacing behavior of capercaillie. The birds return to the same lek year after year, and up to 20 or more males may use an area of not much more than $1 \, km^2$ around their lek jointly for several months in winter and spring. Therefore, habitat preservation at leks is a recommended measure for capercaillie conservation. However, there is more to capercaillie habitat than leks. The Teisenberg study and results from Scandinavia (Wegge 1985, Rolstad 1989, Gjerde 1991a) and the Pyrenees (Menoni 1991) showed that females throughout the year and males in summer and autumn are dispersed over an area of 3 to 4 km from the center of their home lek. Hence, a lek population, i.e., the birds that visit a particular lek, may use an area of 30 to 50 km$^2$ in the course of the year. The spatial requirements of a lek population may vary with the number of birds and with the degree of habitat fragmentation. It is evident that preservation of leks alone will not maintain a capercaillie population. Suitable habitats need to be available within a radius of several km from the lek as well.

Capercaillie habitats need not to be continuous, however. The birds can tolerate some degree of forest fragmentation. In terms of conservation, it is evident that fragmentation of suitable habitats should be kept at a minimum. According to a predictive model developed for capercaillie in the boreal forest (Rolstad and Wegge 1987), the density of lekking males will depend on the grain size of the forest mosaic and the percentage of old forest habitat. Telemetry studies showed that fragmentation of old forest may lead to enlarged home ranges and reduced survival (Rolstad 1989, Gjerde and Wegge 1989, Wegge et al. 1990, Storch 1995c). The number of cocks at a lek is limited by the area of old forest surrounding the lek. Patches smaller than 50 ha rarely contain leks (Rolstad and Wegge 1987, Picozzi et al. 1992), and on Teisenberg, capercaillie avoided small forest stands even if their vegetation structure was optimal (Storch 1993a, 1993b).

At least for the Alps, it is not as important for capercaillie conservationists to know critical fragmentation levels: the greater the percentage of old forest and the larger the old forest patches, the better. It also should be noted that habitat fragmentation may not result only from changes in the physical structure of the environment. Human disturbance may have similar effects: heavily frequented ski slopes and hiking trails may exclude capercaillie from otherwise suitable habitat (Meile 1982, Miquet 1986, Menoni et al. 1989, Zeitler 1995). Thus, besides physical habitat fragmentation, human disturbance may lead to functional fragmentation of the habitat (Storch 1995a).

Many capercaillie populations in central Europe are small and isolated. Many have become extinct during the last few decades, and in several areas, reintroduction attempts have been made without much success. Because reproductive success largely depends on the stochasticity of the weather (e.g., Kastdalen and Wegge 1985, Moss 1985), small capercaillie populations are highly vulnerable and may not respond to habitat improvement. Therefore, habitat conservation needs to be initiated long before a population is seriously threatened with extinction. In this context, estimates of the minimum viable population (MVP) size and its area requirements are needed to plan conservation measures and to assess their chances of success. Sound calculations of an MVP of capercaillie are still lacking. As a first operational guideline for conservation, an estimate of 500 birds for an isolated viable population has been proposed (Storch 1995c). Based on population densities in various parts of the range, this corresponds to area requirements between 25 and 250 km². As a conservation guideline, at least 100 km² of forest with a high amount of good habitat has been suggested for an isolated viable capercaillie population (Storch 1995a). These large spatial requirements and their implications for conservation is the point least accepted among capercaillie conservation practitioners in central Europe.

## 12.5 Capercaillie and Forest Policy

Capercaillie in central Europe have long been influenced by human land use practices. Compared to the primary boreal forest habitat of capercaillie, central European coniferous forests are dense and dark by nature and thus are suboptimal habitat. In historic times, e.g., during the Seventeenth and Eighteenth century, litter collection by peasants and cattle grazing kept the forests open and created optimal habitats for capercaillie (Klaus et al. 1989). During this century, standing timber volume increased and capercaillie habitats deteriorated. In the Bavarian Alps and other capercaillie areas in Germany, the goal of present-day forest policy is to bring back the "naturalness" of the forests. Longer rotation periods, selective cutting, and natural regeneration increasingly are being adopted. Additionally, location-specific, natural tree species composition, and multilayered stands with trees of various ages are being promoted. As a consequence, capercaillie will disappear from a few secondary habitats where man-made conifer forests will change to deciduous forests. In the Alps, there probably will be negative effects at the forest stand level as the oldest stands become multilayered and thus, at least at lower elevations, too dense to be optimal capercaillie habitat. At the same time, longer rotation periods and single-stem cutting will reduce the amount of unsuitable habitat types, e.g., thickets and pole-stage stands. Thus, the present forest policy may lead to reduced quality and increasing quantity of suitable stands and also to a reduced degree of fragmentation.

## 12.6 Conclusions

Based on the Teisenberg study, it is now possible to modify forest management in a way to favor capercaillie habitats both at the local and at the landscape scale. Understanding habitat relationships at all relevant levels of scale is a vital prerequisite for successful habitat conservation for endangered species. Structural habitat needs in relation to vegetational characteristics at the forest-stand scale (see Figure 12.2), spacing behavior of individuals in relation to old-forest fragmentation at the local scale (see Figure 12.2), and spatial requirements of populations and metapopulations at the regional scale (see Figure 12.2) need to be considered. Microhabitat needs at the forest-stand scale, as well as macrohabitat requirements at the landscape scale must be met in order to maintain viable populations. This may seem trivial. The case of the capercaillie in central Europe shows that it is not. The appropriate and relevant answers come only if the right questions are asked.

*12.7 Acknowledgments.* The Teisenberg study was a joint project of the Institute of Wildlife Research and Management at the Faculty of Forestry of the University of Munich and of the Munich Wildlife Society. The study initially received financial support from IBM Germany and in the later years from the Bavarian State Ministry of Agriculture.

## 12.8 References

Gjerde, I. 1991a. Winter ecology of a dimorphic herbivore: temporal and spatial relationships and habitat selection of male and female capercaillie Ph.D [Dissertation]. University of Bergen, Norway.

Gjerde, I. 1991b. Cues in winter habitat selection by capercaillie. I. Habitat characteristics. Ornis Scandinavica 22:197–204.

Gjerde, I., and P. Wegge. 1989. Spacing pattern, habitat use, and survival of capercaillie in a fragmented winter habitat. Ornis Scandinavica 20:219–225.

Jacob, L. 1987. Le régime alimentaire du grand tétras: synthèse bibliographique. Gibier Faune Sauvage 4:429–448

Johnsgard, P.A. 1983. The grouse of the world. University of Nebraska Press, Lincoln, Nebraska, USA.

Kastdalen, L., and P. Wegge. 1985. Animal food in capercaillie and black grouse chicks in south east Norway—a preliminary report. Proceedings of the International Symposium on Grouse 3:499–509.

Klaus, S. 1985. Predation among capercaillie in a reserve in Thuringia. Proceedings of the International Symposium on Grouse 3:334–346.

Klaus, S., A.V. Andreev, H.-H. Bergmann, F. Müller, J. Porkert, and J. Wiesner. 1989. Die Auerhühner. Die Neue Brehm Bücherei, Band 86. Westarp Wissenschaften, Magdeburg, Germany.

Meile, P. 1982. Skiing facilities in alpine habitat of black grouse and capercaillie. Proceedings of the International Symposium on Grouse 2:87–92.

Menoni, E. 1991. Écologie et dynamique des population du Grand Tétras dans les Pyrénées, avec des reférences spéciales à la biologie de la réproduction chez les poules—quelques applications a sa conservation Ph.D. [Dissertation]. University of Toulouse, France.

Menoni, E., C. Novoa, and E. Hansen. 1989. Impact de station de ski alpin sur des populations de Grand Tétras dans les Pyrénées. Actes du Colloque National de l'Association Française des Ingénieures Écologues en collaboration avec la Societé d'Écologie 5:427–449.

Miquet, A. 1986. Contribution a l'étude des relation entre Tétras Lyre et tourisme hivernal en Haute-Tarentaise. Acta Œcologica/Œcologia applicata 7:325–335.

Moss, R. 1985. Rain, breeding success and distribution of capercaillie and black grouse in Scotland. Ibis 128:65–72.

Müller, F. 1974. Territorialverhalten und Siedlungsstruktur einer mitteleuropäischen Population des Auerhuhns. Ph.D. [Dissertation]. University of Marburg, Germany.

Müller, F. 1979. A 15-year study of a capercaillie lek in the western Rhönmountains. Proceedings of the International Symposium on Grouse 1:120–129.

Picozzi, N., D.C. Catt, and R. Moss. 1992. Evaluation of capercaillie habitat. Journal of Applied Ecology 29:751–762.

Popp, D. 1974. Langfristige Schutzkonzeption für eine hessische Auerhuhnpopulation im Spessart. Allgemeine Forst Zeitschrift 39:839–840

Rolstad, J. 1989. Habitat and range use of capercaillie in southcentral Scandinavian boreal forests Ph.D. [Dissertation]. Agricultural University, Ås, Norway.

Rolstad, J., and P. Wegge. 1987. Distribution and size of capercaillie leks in relation to old forest fragmentation. Oecologia 72:389–394.

Rolstad, J., and P. Wegge. 1989. Capercaillie populations and modern forestry—a case for landscape ecological studies. Finnish Game Research 46:43–52.

Scherzinger, W. 1988. Fünf nach Zwölf für das Auerhuhn im Bayerischen Wald. Nationalpark 58:8–12.

Spittler, H. 1994. Wiedereinbürgerungsversuch mit Auerwild im Hochsauerland. Zeitschrift für Jagdwissenschaft 40:185–199.

Stein, J. 1974. Die qualitative Beurteilung westdeutscher Auerhuhnbiotope unter besonderer Berücksichtigung der Grenzlinienwirkung. Allgemeine Forst Zeitschrift 39:837–839.

Storch, I. 1993a. Habitat selection of capercaillie in summer and autumn: Is bilberry important? Oecologia 95:257–265.

Storch, I. 1993b. Patterns and strategies of winter habitat selection in alpine capercaillie. Ecography 16:351–359.

Storch, I. 1993c. Habitat use and spacing of capercaillie in relation to forest fragmentation patterns. Ph.D. [Dissertation]. University of Munich, Germany.

Storch, I. 1994. Habitat and survival of capercaillie nests and broods in the Bavarian Alps. Biological Conservation 70:237–243.

Storch, I. 1995a. Habitat requirements of capercaillie. Proceedings of the International Symposium on Grouse 6:151–154.

Storch, I. 1995b. The role of bilberry in central European capercaillie habitats. Proceedings of the International Symposium on Grouse 6:116–120.

Storch, I. 1995c. Annual home ranges and spacing patterns of capercaillie in central Europe. Journal of Wildlife Management 59:392–400.

Storch, I., C. Schwarzmüller, and D. von den Stemmen. 1991. The diet of capercaillie in the Alps: a comparison of hens and cocks. Pp. 630–635 in S. Csànyi and J. Ernhaft, editors. Transactions of the International Union of Game Biologists XX Congress, Gödöllö, Hungary: 630–635.

Wegge, P. 1985. Spacing patterns and habitat use of capercaillie hens in spring. Proceedings of the International Symposium on Grouse 3:261–274.

Wegge, P., and J. Rolstad. 1986. Size and spacing of capercaillie leks in relation to social behavior and habitat. Behavioral Ecology and Sociobiology 19:401–408.

Wegge, P., and B.B. Larsen. 1987. Spacing of adult and subadult male common capercaillie during the breeding season. Auk 104:481–490

Wegge, P., I. Gjerde, L. Kastdalen, J. Rolstad, and T. Storaas. 1990. Does forest fragmentation increase the mortality rate of capercaillie? Pp. 448–453 in S. Myrberget, editor. Transactions of the 19th International Union of Game Biologists Congress, Trondheim, Norway: 448–453.

Weiss, H. 1988. Auerwild im Nordschwarzwald. Die Pirsch 9:22–24.

Wiens, J.A. 1976. Population responses to patchy environments. Annual Review of Ecology and Systematics 7:81–120.

Zeimentz, K. 1974. Lebensraum und Bestandestendenz des Auerwildes in den Bayerischen Alpen. Allgemeine Forst Zeitschrift 39:824–825.

Zeitler, A. 1995. Skilauf und Rauhfusshühner. Der Ornithologische Beobachter 92:227–230.

# 13
# Landscape Heterogeneity and Ungulate Dynamics: What Spatial Scales Are Important?

Monica G. Turner, Scott M. Pearson, William H. Romme, and Linda L. Wallace

## 13.1 Introduction

Ungulates make foraging choices at a variety of spatial scales, but the environmental parameters that are most important at various scales are not well known. Clearly, the spatial arrangement and density of vegetation influences the success of herbivores in finding food (Kareiva 1983, Risch et al. 1983, Stanton 1983, Cain 1985, Bell 1991). Theoretical studies suggest that organisms must operate at larger spatial scales (i.e., search a larger area) as resources become scarce and clumped across a landscape (O'Neill et al. 1988, Turner et al. 1993). In addition, the effectiveness of different foraging tactics may vary with the spatial distribution of resources (e.g., Cain 1985, Roese et al. 1991). However, understanding the responses of animals to spatial pattern at multiple scales is in its infancy (Kotliar and Wiens 1990, Kareiva 1990, Hyman et al. 1991, Ward and Saltz 1994) and remains a high priority for ecology (Lubchenco et al. 1991, Levin 1992). In this chapter, we synthesize results from three studies of winter foraging by

Monica G. Turner is Associate Professor of Terrestrial Ecology in the Department of Zoology, University of Wisconsin. She coteaches a graduate class in Landscape Ecology, and her research interests focus on the causes and consequences of landscape heterogeneity for ecological processes.

Scott M. Pearson is Assistant Professor in the Biology Department, Mars Hill College. He teaches a variety of courses in ecology, evolution, and population dynamics. His research emphasizes the response of organisms, especially birds and plants, to landscape pattern.

William H. Romme is Professor in the Biology Department, Fort Lewis College. He teaches courses in ecology as well as interdisciplinary courses in the sciences, and his research focuses on the role of fire in the Rocky Mountains.

Linda L. Wallace is Professor in the Department of Botany and Microbiology, University of Oklahoma. She teaches courses in ecology and plant physiology, and her research has emphasized the complex interactions between plants and the animals that consume them.

elk (*Cervus elaphus*) and bison (*Bison bison*) in northern Yellowstone National Park in which we explored scale-dependent foraging patterns.

During winter, foraging ungulates seek resources that are highly variable in space and time and which vary across multiple scales. Spatially, forage may vary in abundance and in quality across a landscape as a function of plant species composition, moisture, soil fertility, and topography. This spatial variability ranges from meter-by-meter variation in forage quality or quantity to between-habitat variation over a scale of kilometers (O'Neill et al. 1989, 1991a, 1991b, S. Turner et al. 1991). The spatial pattern of winter forage also is modified by the distribution of snow, which may reduce or even eliminate potential foraging sites. Temporally, forage resources change daily as the resource is depleted by grazing and as snow conditions change. While depletion by grazing is gradual and patchy, a major snow event can rapidly change the distribution of forage availability across an entire landscape. Ungulates respond to this variability in space and time by making foraging decisions hierarchically (Senft et al. 1987, Kotliar and Wiens 1990, Danell et al. 1991, Schmidt 1993), but the scales at which decisions are made and the cues at each scale are not well understood.

Highly mobile animals such as ungulates are capable of first choosing a large area that is likely to contain suitable foraging conditions and then making additional decisions at finer and finer scales eventually to position individual foraging efforts. The scale of these choices is bounded at the broadest scale by the movement capability of the organism and at the finest scale by the spatial and temporal extent of an individual foraging event. The resulting pattern of foraging depends on the organism's foraging choices and the spatial heterogeneity of forage (Morse and Fritz 1982). It has been suggested that large ungulates make decisions at scales ranging from choosing a position in the landscape (e.g., a geographic region or watershed) for feeding down to selecting a specific portion of a plant to be eaten (Senft et al. 1987). Indeed, a variety of authors (e.g., Addicott et al. 1987, Wiens 1976, 1989, Kotliar and Wiens 1990, Orians and Wittenberger 1991, Bergin 1992, Russell et al. 1992, Pearson 1993, Pearson et al. 1995) have suggested that animals make scale-dependent choices in habitat use and/or foraging.

By creating heterogeneity, natural events (e.g., fires, storms, and droughts) provide valuable opportunities to study ungulate responses to landscape heterogeneity. The 1988 fires in Yellowstone National Park (YNP), unusual in both size and heterogeneity (Christensen et al. 1989, Knight and Wallace 1989, Turner et al. 1994a) created such an opportunity. From 1989 to 1991, we studied the effects of fire size and pattern on the foraging patterns and survival of wintering elk and bison in YNP. As part of this work, we examined the spatial scales at which ungulates appeared to respond to heterogeneity in the Yellowstone landscape. In this chapter, we synthesize our results to suggest the spatial scales that might be most important for wintering ungulates. We organize the chapter by the scale of heterogeneity and response explored: (1) fine-scale (within $900\,m^2$) foraging patterns, (2) broad-scale (150 to 500 ha) foraging patterns, and (3) foraging patterns and ungulate survival with alternative scales of heterogeneity across the entire winter range landscape (78,000 ha).

## 13.2 Study Area

Yellowstone National Park (YNP) was established in 1872 and encompasses $9000\,km^2$ in the northwest corner of Wyoming and adjacent parts of Montana and Idaho. Elevations in the park range from 1500 to more than 3000 m, and much of the area is covered by Quaternary volcanic deposits that underwent at least three extensive glaciations (Houston 1982). General descriptions can be found for park geology (Keefer 1972); and for physiography, soils, and vegetation (Meagher 1973, Barmore 1980, Houston 1982, Despain 1991). The climate of YNP is characterized by long, cold winters and short, cool summers (Diaz 1979, Dirks and Martner 1982). Our study focused on the northern 20% of the park, which is primarily a lower elevation grassland or sagebrush steppe (Figure 13.1). The northern Yellowstone elk and bison migrate seasonally between a high-elevation summer range and this lower elevation winter range (Craighead et al. 1972, Barmore 1980, Houston 1982).

The northern range extends about 80 km down the Lamar, Yellowstone, and Gardner River drainages (Houston 1982). Approximately 83% of the winter range for elk is included within YNP (Houston 1982), comprising nearly 80,000 ha; it has a warmer, drier climate than the rest of the park (Diaz 1979, Dirks and Martner 1982). We focused solely on the winter range within the park boundaries. Most of the soils derive from glacial till deposited during the Pinedale glaciation (Houston 1982) and tend to be higher in silt, clay, and organic matter than soils derived from rhyolite, the parent rock of most YNP (Despain 1991). The drier grasslands are dominated by big sagebrush (*Artemisia tridentata*), bluebunch wheatgrass (*Agropyron spicatum*), and Idaho fescue (*Festuca idahoensis*). Wet sites are dominated by bearded wheatgrass (*Agropyron caninum*), sedges (*Carex* spp.), and introduced graminoids such as Kentucky bluegrass (*Poa pratensis*). More continuous forest occurs at high elevations and on north slopes (see detailed descriptions in Houston 1982 and Despain 1991). At lower elevations, the sagebrush grasslands are interspersed with coniferous forest, primarily Douglas fir (*Pseudotsuga menziesii*), and lodgepole pine (*Pinus contorta* var. *latifolia*); aspen groves (*Populus tremuloides*); and riparian willow communities.

Natural fires have influenced plant succession on the winter range for a long time (Houston 1973). Tree-ring evidence suggests that 8 to 10 extensive fires occurred in the area during the last 300 to 400 years, but no large

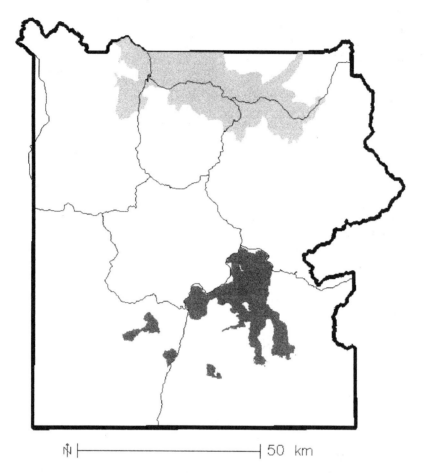

FIGURE 13.1. Map of Yellowstone National Park (YNP) showing the 78,000-ha northern winter range study area in light gray and large lakes in dark gray. Solid lines indicate major park roads.

fires have occurred previously in this century (Houston 1973). Approximately 34% of the winter range burned during the 1988 fires, including 22% of the grasslands (Despain et al. 1989). The vegetation has been dynamic during the past century (Houston 1982). Forest cover has increased, mostly on north slopes, and forests have recolonized burned areas, although aspen appears to have declined from 4–6% to 2–3% of the winter range. Sagebrush increased in extent on many slopes and exposures. Willows and associated riparian vegetation have shown a net decrease. The production of herbaceous vegetation on the northern range shows substantial annual fluctuations resulting from varied growing conditions, primarily precipitation (Houston 1982, Merrill et al. 1993).

## 13.3 Fine-Scale Feeding Patterns

Ungulate feeding patches (i.e., a bounded area where animals invested time and energy in finding food) are easily detectable in winter. Elk dig craters to access snow-covered forage, and bison "till" a foraging area by using head swings. Feeding craters are particularly useful indicators of foraging activity. Studies of cratering by reindeer have shown that crater area is linearly related to duration of cratering activity and the number of digging strokes of the leg (Collins and Smith 1991). Thus, crater area offers a rough index of the amount of time and energy invested in obtaining forage in a patch. Craters are visited only once by a forager simply because disturbance of the snow causes a manyfold increase in its hardness, which thereby precludes animals from reusing previously excavated areas (Pruitt 1959, Collins and Smith 1991). Because crater area is roughly proportional to foraging effort, feeding patches can be functionally identified independent of any a priori definition of patches on the landscape.

To explore fine-scale winter foraging patterns, we examined the distribution of individual feeding stations (i.e., craters) within sites of high feeding activity on the northern winter range of YNP during the 1989 and 1990 winters to address two questions (Wallace et al. 1995): (1) are the locations of ungulate feeding stations within a site predictable based upon an underlying fine-scale pattern in forage vegetation at that site? and (2) does the location of ungulate foraging sites across the landscape reflect broad-scale patterns of winter forage quantity and quality? The 30-m × 30-m study sites each encompassed a feeding patch. Feeding stations were mapped by recording $x/y$ coordinates of each feeding crater within eight sites during the winter of 1988–1989. Forage abundance was mapped by imposing a regular 15 × 15 grid of sampling points spaced at 2-m intervals ($n$ = 225 point samples per site) at six sites during October 1990. Sites mapped in 1990 were nearby and visually similar to sites in which feeding stations were mapped during 1988–1989. Although an a priori measurement of biomass distributions would have been optimal, it is not possible to predict exactly where ungulates will forage during any given winter.

Results demonstrated that both forage biomass and the distribution of feeding craters were spatially random within the feeding patches, except when a relatively large area was devoid of vegetation, e.g., with a large rock covering a substantial portion of the feeding patch. Thus, ungulates appeared to select feeding stations at random within a feeding patch as long as biomass was present. Ungulates apparently did not differentiate between areas with low and high biomass at the scale of selection of individual feeding stations. By simulating the use of a "smart" feeding strategy, in which an ungulate was assumed to have knowledge of the underlying distribution of biomass and could select feeding stations that would yield the most biomass, we observed that ungulates could potentially double their forage intake compared to use of a random feeding strategy (Wallace et al. 1995).

Mean forage biomass differed more than twofold among the sampling grids. To further examine the variability in forage biomass abundance across the northern-range landscape, forage biomass abundance was determined at 38 locations within nine general community types (Wallace et al. 1995). There were large differences in forage abundance between habitats. Mesic grasslands that had burned in 1988 had ~3 times the biomass observed in comparable unburned areas (170.5 and 63.3 g m$^{-2}$, respectively). Biomass also increased along a moisture gradient, with dry areas having low biomass (e.g., 34 g m$^{-2}$) and wet grasslands having high biomass (>200 g m$^{-2}$). This landscape-scale variability in forage abundance suggests that ungulates could respond to this broader-scaled spatial pattern by selecting feeding patches that would generally yield more resources, even though they might crater at random within the patches. These broad-scale patterns are addressed next.

## 13.4 Broad-Scale Foraging Patterns

The distribution of feeding patches across the landscape provides a snapshot of the patterns and intensity of ungulate foraging activity. Broad-scale estimates of foraging dynamics can be inferred by monitoring these patterns through time. Spatial patterns of winter foraging by elk and bison in northern YNP were studied in 15 study areas encompassing 7500 ha of winter range during 1991 and 1992 (Pearson et al. 1995). This study was designed to determine whether ungulates preferentially used burned areas of the landscape in which forage quantity and/or quality might be enhanced during the first few years following fire. The study also examined whether ungulates preferentially used particular habitats or landscape positions for feeding, whether the geographic pattern of habitat use changed through the winter, and whether the use of burned areas was influenced by the spatial pattern of burning. Finally, we analyzed the environmental data at four spatial scales to explore the scale(s) at which most of the variation in cumulative grazing intensity during each winter was explained; these results are summarized here.

Field studies were established in January 1991 by using a vegetation classification generated with data from YNP stored in a geographic information system (GIS) to assure that all habitats were represented in our sampling (Pearson et al. 1995). Each of the 15 study areas provided a view of several hundred hectares of the northern range from a fixed position of relatively high elevation. These 15 viewing areas encompassed approximately 10% of the northern range. The boundaries of each area were demarcated on topographic maps and digitized in a GIS. Each viewing area was visited at two-week intervals from mid-January through late March during 1991 and 1992. At each visit, the location of feeding patches (i.e., contiguous area within which animals have grazed with the same level of intensity as indicated by the abundance of feeding craters) was mapped, and grazing intensity was measured as the percentage in 20% increments of the area cratered or tilled. The minimum mapping unit for feeding patches was 1 ha. Cumulative grazing intensity for each 1-ha grid cell over each winter was determined by summing the biweekly data in the GIS.

To test whether ungulate grazing intensity measured at the scale of 1 ha might be influenced by environmental variation at several scales, multiple regression was used to explore the relationship between cumulative grazing intensity and environmental characteristics at scale ≥1 ha (Pearson et al. 1995). Grazing data were first subsampled to remove the effect of spatial autocorrelation in cumulative grazing intensity. Based upon semivariance analysis, grid cells used in the regression were separated by a distance of 600 m. Factor analysis was used to identify the covariance structure of the environmental data and to reduce the data to a smaller number of "factors." Six environmental variables were reduced to three factors: (1) grassland habitat, precipitation, and elevation; (2) slope and aspect; and (3) burn status. To quantify the characteristics of the 1-ha cells and the surrounding landscape, environmental factor scores were measured at four spatial scales: 1, 9, 81, and 225 ha for each cell. At the 1-ha scale, environmental factor scores were used directly from the appropriate 1-ha map cell. To measure habitat at scales >1 ha, a square was centered on the sampling point (e.g., a 3 × 3 cell square for 9 ha) and the mean score of all cells in the square was calculated to represent habitat at a particular scale. Thus, a total of 12 variables was used in the regression analysis describing habitat variability due to three factors at four spatial scales. In addition, distance to cover, which may influence ungulate feeding-site selection (Grover and Thompson 1986) was used in the regression.

Results indicated that broad-scale measurements of environmental variables were more useful for statistically predicting cumulative grazing intensity on a per-grid-cell basis than fine-scale measurements (Table 13.1). Burn status (Factor 3) and slope and aspect (Factor 2) were important at the broadest scale (225 ha) for both years during mid- and late winter. Burn status at the scale of 1 ha was significant for mid- and late winter in 1991, but its explanatory power was low. Elevation, precipitation, and grassland habitat type (Factor 1) were significant at finer scales (1 to 9 ha) during both years. Interestingly, environmental characteristics were not useful for predicting ungulate foraging during early winter, a time when resources are less limited. Overall, ungulates appeared to respond strongly in mid- to late winter to broad-scale variation (225 ha) in topography and the presence of burning and to fine-scale variation (1 ha) in grassland habitat type.

The importance of broad-scale environmental heterogeneity in predicting ungulate grazing intensity might result for several reasons. Ungulates

TABLE 13.1. Results from stepwise regression of cumulative grazing intensity. Independent variables included distance to cover and factor scores for environmental parameters at four spatial scales: 1, 9, 81, and 225 ha. Factor 1 represents elevation, precipitation, and grassland habitat type. Factor 2 includes slope and aspect. Factor 3 is burn status. Analyses were conducted for three observation periods in 1991 (early, mid- and late winter) and two periods (early and late winter) in 1992. Only the factors that were significant in the model are listed. From Pearson et al. (1995)

| Variable | Partial $r^2$ | P | Coefficient | N |
|---|---|---|---|---|
| | | *1991* | | |
| Early winter | | | | |
|   Factor 2—1 ha | 0.020 | 0.055 | 2.50 | 174 |
|   Model total | 0.020 | | | |
| Midwinter | | | | |
|   Factor 3—225 ha | 0.191 | 0.0001 | 3.51 | 140 |
|   Factor 2—225 ha | 0.109 | 0.001 | 4.35 | |
|   Distance to cover | 0.028 | 0.018 | −15.50 | |
|   Factor 1—9 ha | 0.016 | 0.071 | 4.10 | |
|   Model total | 0.344 | | | |
| Late winter | | | | |
|   Factor 3—225 ha | 0.135 | 0.0001 | 5.08 | 135 |
|   Factor 1—1 ha | 0.038 | 0.026 | −2.07 | |
|   Factor 2—225 ha | 0.050 | 0.004 | 1.99 | |
|   Factor 3—1 ha | 0.020 | 0.069 | −1.24 | |
|   Model total | 0.243 | | | |
| | | *1992* | | |
| Early winter | | | | |
|   Factor 3—225 ha | 0.050 | 0.038 | 5.23 | 86 |
|   Model total | 0.050 | 0.038 | | |
| Late winter | | | | |
|   Factor 3—225 ha | 0.153 | 0.0001 | 10.97 | 101 |
|   Factor 2—225 ha | 0.027 | 0.074 | 5.07 | |
|   Factor 1—1 ha | 0.034 | 0.042 | −4.62 | |
|   Model total | 0.214 | | | |

may actually be making foraging choices based on landscape cues that are apparent at broad spatial scales. Snow may obscure fine-scale cues that might be used to select the most profitable foraging sites during other seasons, and features like topography might be good indicators of areas most suitable for foraging. Ungulates may also increase the duration of time and number of foraging bouts in areas, such as burned sites, that contain greater biomass. Finally, because our study was of relatively coarse temporal grain (two-week intervals rather than observations over minutes to hours), and broader scale environmental patterns could become more influential as the temporal scale of observation increases. Clearly, however, our results suggest that interpreting or predicting ungulate grazing at any given location requires an understanding of the environmental heterogeneity of the surrounding landscape, not simply a description of local site attributes.

# 13.5 Landscape Heterogeneity Within and Between Habitats

Experimentation, or even field observation, at the scale of an entire landscape is often logistically infeasible. For example, in the northern range of YNP, researchers could not impose alternative scales of heterogeneity across the landscape, then observe ungulate foraging responses. Simulation models, when combined with empirical studies, are valuable tools for exploring the implications of alternative scenarios. An individual-based, spatially explicit simulation model (NOYELP) of winter foraging by ungulates in northern YNP (Turner et al. 1994b) was used to examine the effects of heterogeneity at various spatial scales on winter habitat use and survival of elk and bison. Spatial heterogeneity in the distribution of biomass and snow across the landscape were varied in the simulations (Turner and O'Neill 1995).

The Northern Yellowstone Park (NOYELP) model simulates the search, movement, and foraging activities of individuals or small groups of elk and bison (Turner et al. 1994b). The model was developed to explore the effects of fire scale and pattern on the winter foraging dynamics and survival of these free-ranging ungulates in northern YNP. The 77,020-ha landscape is represented as a gridded irregular polygon with a spatial resolution of 1 ha. The model simulates daily forage intake as a function of an animal's initial body weight, the absolute amount of forage available on a site, and the depth and density of snow. Energy balances are computed daily, with energy gain a function of forage intake and energy cost a function of baseline metabolic costs and travel costs. When the energy expenditures of an animal exceed the energy gained during a day, the animal's endogenous reserves are reduced to offset the deficits. Simulations are conducted with a one-day time step for a duration of 180 days, approximately November 1 through April 30. The model has been used to examine the effects on ungulate survival and habitat use of fire size, fire pattern, winter weather conditions, and initial ungulate numbers (Turner et al. 1994b).

Simulation experiments were designed to assess effects of forage heterogeneity at multiple scales (Turner and O'Neill 1995). The model includes six different vegetation classes, each of which can be either burned or unburned. The initial biomass present on a grid cell at the beginning of winter is determined by the vegetation class and burn status by using the mean and 95% confidence interval obtained from field data (Wallace et al. 1995); mean forage abundance within each habitat type matches the mean obtained from the field data, but any individual grid cell within a habitat type has a forage value assigned from the distribution. We refer to this as within-habitat variability. For the simulations reported here, we used the winter of 1988–1989. Because the fires occurred in the summer of 1988, areas of the northern range that were burned had no forage biomass present during the first winter following the fires.

To explore the effects of aggregating across this spatial heterogeneity, a series of three successive aggregations was conducted (Figure 13.2). In all cases, the total forage biomass available across the landscape was held constant. First, the within-habitat heterogeneity was removed by assigning to each grid cell the mean forage biomass for its habitat type. Second, the between-habitat heterogeneity was removed by assigning to each vegetated

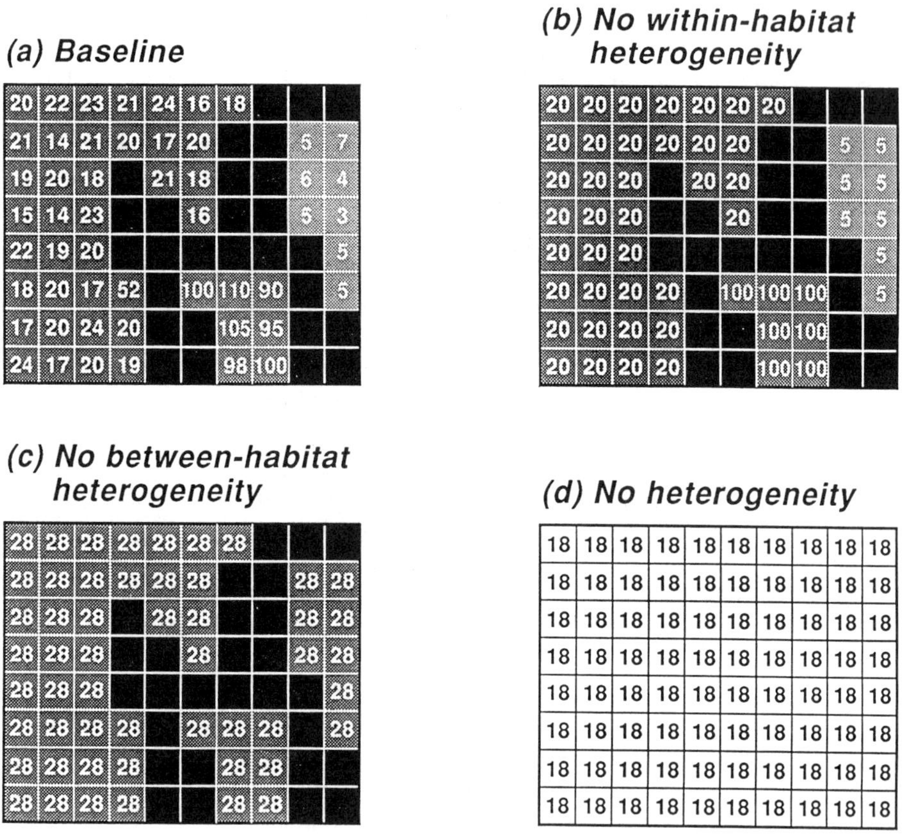

### (a) Baseline

### (b) No within-habitat heterogeneity

### (c) No between-habitat heterogeneity

### (d) No heterogeneity

FIGURE 13.2. Schematic representation of aggregation across three scales of forage heterogeneity simulated in a spatially explicit individual-based model. The initial map with three habitat types is shown in (a). Burned areas, which have no forage, are in solid black, and forage abundance is indicated with numbers for the three habitats. When within-habitat heterogeneity is removed (b), each grid cell in a given habitat class has the same forage. When between-habitat heterogeneity is removed (c), each grid cell in one of the three habitats has the same forage. Finally, when no heterogeneity is present, all grid cells have equal forage. In all cases, total forage across the map remains constant.

grid cell the mean biomass present across the landscape, obtained by dividing the total landscape biomass by the number of unburned grid cells; burned areas continued to have zero biomass. Finally, all spatial heterogeneity was eliminated, and each grid cell was assigned an average value for the entire landscape.

The removal of within-habitat heterogeneity had little effect on elk survival, but aggregating habitats such that between-habitat heterogeneity was eliminated reduced winter ungulate survival by 30%. Removal of all spatial heterogeneity in forage abundance across the landscape reduced survival by another 20%. These results illustrate the importance for wintering ungulates of maintaining coarse-scale variability in forage biomass. The distribution across the landscape of areas of high and low biomass substantially enhances ungulate survival. Similar results were obtained when the heterogeneity of snow depth and density across the landscape was varied (Turner and O'Neill 1995); maintenance of spatial variability in snow conditions, especially the presence of areas with reduced snow mass, was also important for winter ungulate survival.

## 13.6 Discussion

Winter foraging patterns of elk and bison in northern YNP clearly differ with spatial scale. Ungulates respond to relatively coarse variability in the landscape and appear to select general locations for feeding (i.e., feeding patches) based on environmental heterogeneity at broad scales (81 to 255 ha). Cues appear to be related to topography, habitat type, and, for the winters we studied, whether a site was burned. Once positioned within snow-covered feeding patches, however, selection of feeding stations appears to occur at random as long as some forage biomass is present. This random pattern is typically expected when forage quality is generally low, as for most herbaceous winter forage (Wickstrom et al. 1984, Klein and Bay 1990).

At a fine spatial scale ($900\,\mathrm{m}^2$), we did not observe the overmatching aggregate response pattern predicted by Senft et al. (1987) in which forage or nutrient intake would be maximized. Matching occurs when the foraging effort at a site is proportional to resource profitability. Undermatching occurs when the foraging effort is less than proportional to the resource, and overmatching would occur if foraging effort was greater than proportional to resource profitability. Two factors may be of particular importance. First, Abrahams (1986) suggests that when an organism is unable to discriminate between resource profitabilities (e.g., an elk unable to predict forage biomass or quality under snow), it will allocate itself randomly to a site, leading to an undermatching relationship between organisms and resources. Furthermore, an increase in the number of organisms will increase the degree of undermatching. Second, prewinter forage abundance within our study area spans an order of magnitude ($5 \times 10^2$ to $2 \times 10^3$ kg/ha), but the costs associated with obtaining forage in snow can potentially range much more broadly. Parker et al. (1984) found that the energy costs of traveling in snow increased exponentially as a function of snow density and relative sinking depth as compared to the no-snow condition. Foraging theory suggests that as the next patch becomes more difficult to reach, the forager should tolerate a smaller likelihood that the present patch is good before leaving (Stephens and Krebs 1986, p. 93). Therefore, as foraging conditions deteriorate due to accumulated snow mass, the energetic costs associated with moving to another feeding site increase, and animals should spend a greater amount of time at a given site before moving elsewhere.

Interpreting ungulate grazing at a specific location requires understanding heterogeneity in the surrounding landscape. Our work supports the notion that the landscape matrix must be considered when the conservation and management of populations are considered (e.g., Franklin 1993, Andren 1994, Wiens 1997). Other studies have also demonstrated an important influence of landscape context on populations at a given site. Pearson (1993) found that the structure of the surrounding landscape explained as much as 74% of the variance in habitat occupancy of some wintering bird species in the Georgia Piedmont. Lindenmayer and Nix (1993) found that the occupancy of corridors by arboreal marsupials in Australia could not be predicted by habitat features within the corridor; information on the composition of the surrounding landscape was required. However, landscape controls over foraging patterns and movements of ungulates have not received much study.

Foraging theory has made a large set of predictions about when animals should abandon one resource patch for another (e.g., Charnov 1976, Belovsky 1984, Real and Caraco 1986, Stephens and Krebs 1986, Mangel and Clark 1988, Belovsky et al. 1989, Dannell et al. 1991, McNamara and Houston 1992, Laca et al. 1994, and many others). Most of these predictions consider the resource availability within the landscape to be fixed. However, resources across the winter landscape change dramatically as snow and previous grazing modify resource availability. Predicting ungulate foraging patterns in a large, spatially heterogeneous landscape in which resource availability (abundance and spatial pattern) can change dramatically in a matter of days remains a challenge that requires a linkage between population and landscape ecology. Furthermore, the linkage between the decisions of individual animals and the landscape-scale dynamics of an entire population are not well known. Development of models and empirical studies that consider these complex spatio-temporal dynamics would greatly enhance our understanding of ungulate dynamics.

Revealing the causes and consequences of spatial heterogeneity in natural systems has emerged as a fundamental challenge for contemporary ecologists (Turner 1989, Kareiva 1990, Kolasa and Pickett 1991, Levin 1992, Kareiva 1994). An important aspect of this challenge is understanding the ways that consumers create and respond to heterogeneity in the resources they use. Large, mobile herbivores discriminate among spatially variable food resources, thereby altering the structure of plant communities and the rates of ecosystem processes. Improving our knowledge of the responses of large herbivores to spatial heterogeneity can contribute to understanding the workings of many other ecological processes.

Ecologists have come to recognize that many of the patterns and processes they study are scale dependent, that what happens at one scale of observation does not necessarily translate into the same thing at other scales (Wiens 1996). We make no contentions that the observations reported here for particular scales (both grain and extent) will hold for other scales not yet examined. Rather, we hope that these results will stimulate the development of studies that explicitly link the scales of heterogeneity in the environment with the decisions of individuals and the dynamics of populations in dynamic landscapes.

## 13.7 Summary

Ungulates make foraging choices at a variety of spatial scales, but the environmental parameters that are most important at various scales are not well known. In this chapter, results from three studies of winter foraging by elk and bison in northern Yellowstone National Park are synthesized to explore scale-dependent foraging patterns. Measurements of forage abundance and feeding station distribution within 30-m × 30-m plots suggested that the spatial pattern of foraging by ungulates was random within feeding patches. Ungulates appeared to select feeding stations at random within a feeding patch as long as biomass was present. However, large differences in forage abundance across the landscape suggested that ungulates might make choices at broader spatial scales. Analyses of grazing intensity in a set of 15 study areas encompassing 7500 ha of the northern range revealed that grazing intensity on a per-hectare basis could best be predicted by environmental variability at broad scales (100 to 500 ha) rather than by per-hectare environmental variability. A spatially explicit, individual-based simulation model of elk and bison foraging on the northern range was also used to explore scale-dependent foraging. Simulation experiments designed to assess the effects of forage heterogeneity at multiple spatial scales revealed strong responses in ungulate survival to removal of between-habitat variation in forage abundance but not within-habitat variation, and also indicated that snow heterogeneity enhanced ungulate survival. Our field and modeling results suggest strong effects of resource heterogeneity and environmental cues at broad scales on ungulate habitat use and survival. Interpreting or predicting ungulate grazing at any given location requires an understanding of the environmental heterogeneity of the surrounding landscape, not simply a description of local site attributes. Future studies that explicitly link the scales of heterogeneity in the environment with the decisions of individuals and the dynamics of populations in dynamic landscapes represent an important research need.

*13.8 Acknowledgments.* This paper was based on a presentation by MGT in a symposium, Managing Ungulates as Components of Ecosystems, at the annual meeting of The Wildlife Society in Portland, Oregon, 16 September 1995. John Bissonette's invitation to contribute to this volume is greatly appreciated. This work was supported by the U.S. National Park Service and U.S. Forest Service through a grant from the University of Wyoming–National Park Service Research Center; by Oak Ridge Associated Universities through Faculty Participation Awards to LLW and WHR and an Alexander Hollaender Fellowship to SMP; and by the Ecological Research Division, Office of Health and Environmental Research, U.S. Department of Energy, under Contract No. DE-AC05-84OR21400 with Martin Marietta Energy Systems, Inc.

## 13.9 References

Abrahams, M.V. 1986. Patch choice under perceptual constraints: a cause for departures from an ideal free distribution. Behavioral Ecology and Sociobiology 19:409–415.

Addicott, J.F., J.M. Aho, M.F. Antolin, J.S. Richardson, and D.A. Soluk. 1987. Ecological neighborhoods: scaling environmental patterns. Oikos 49:340–346.

Andren, H. 1994. Effects of habitat fragmentation on birds and mammals in landscapes with different proportions of suitable habitat: a review. Oikos 71:355–366.

Barmore, W.J. 1980. Population characteristics, distribution and habitat relationships of six ungulates in northern Yellowstone National Park. Final Report. Research Division, Yellowstone National Park, Wyoming, USA.

Bell, W.J. 1991. Searching behavior. Chapman and Hall, London, UK.

Belovsky, G.E. 1984. Herbivore optimal foraging: a comparative test of three models. American Naturalist 124:97–115.

Belovsky, G.E., M.E. Ritchie, and J. Moorehead. 1989. Foraging in complex environments: when prey availability varies over time and space. Theoretical Population Biology 36:144–160.

Bergin, T.M. 1992. Habitat selection by the western kingbird in western Nebraska: a hierarchical analysis. Condor 94:903–911.

Cain, M.L. 1985. Random search by herbivorous insects: a simulation model. Ecology 66:876–888.

Charnov, E.L. 1976. Optimal foraging: The marginal value theorem. Theoretical Population Biology 9:129–136.

Christensen, N.L., J.K. Agee, P.F. Brussard, J. Hughes, D.H. Knight, G.W. Minshall, J.M. Peek, S.J. Pyne, F.J. Swanson, J.W. Thomas, S. Wells, S.E. Williams, and H.A. Wright. 1989. Interpreting the Yellowstone fires of 1988. BioScience 39:678–685.

Collins, W.B., and T.S. Smith. 1991. Effects of wind-hardened snow on foraging by reindeer (*Rangifer tarandus*). Arctic 44:217–222.

Craighead, J.J., H. Atwell, and B.W. O'Gara. 1972. Elk migrations in and near Yellowstone National Park. Wildlife Monographs 29:6–48.

Dannell, K., L. Edenius, and P. Lundberg. 1991. Herbivory and tree stand composition: moose patch use in winter. Ecology 72:1350–1357.

Despain, D.G. 1991. Yellowstone vegetation: consequences of environment and history. Roberts Rinehart Publishing Co.

Despain, D., A. Rodman, P. Schullery, and H. Shovic. 1989. Burned area survey of Yellowstone National Park: the fires of 1988. Unpublished report, Division of Research and Geographic Information Systems Laboratory, Yellowstone National Park, Wyoming, USA.

Diaz, H.F. 1979. Ninety-one years of weather records at Yellowstone National Park, Wyoming, 1887–1977. National Oceanic and Atmospheric Administration, Environmental Data and Information Service, National Climatic Center, Asheville, North Carolina, USA.

Dirks, R.A., and B.E. Martner. 1982. The climate of Yellowstone and Grand Teton National Parks. Occasional Paper No. 6, U.S. National Park Service, Washington, DC, USA.

Franklin, J.F. 1993. Preserving biodiversity: species, ecosystems, or landscapes? Ecological Applications 3:202–205.

Grover, K.E., and M.J. Thompson. 1986. Factors influencing spring feeding site selection by elk in the Elkhorn Mountains, Montana. Journal of Wildlife Management 50:466–470.

Houston, D.B. 1973. Wildfires in northern Yellowstone National Park. Ecology 54:1111–1117.

Houston, D.B. 1982. The northern Yellowstone elk: ecology and management. Macillan Publishing Co., New York.

Hyman, J.B., J.B. McAninch, and D.L. DeAngelis. 1991. An individual-based simulation model of herbivory in a heterogeneous landscape. Pp. 443–475 in M.G. Turner and R.H. Gardner, editors. Quantitative methods in landscape ecology. Springer-Verlag, New York, USA.

Kareiva, P.M. 1983. Influence of vegetation texture on herbivore populations: resource concentration and herbivore movements. Pp. 259–289 in R.F. Denno and M. McClure, editors. Variable plants and herbivores in natural and managed systems. Academic Press, New York, USA.

Kareiva, P.M. 1990. Population dynamics in spatially complex environments: theory and data. Philosophical Transactions of the Royal Society of London B 330:175–190.

Kareiva, P. 1994. Space: the final frontier for ecological theory. Ecology 75:1.

Keefer, W.R. 1972. Geologic story of Yellowstone National Park. Geological Survey Bulletin 1347. U.S. Government Printing Office, Washington, DC, USA.

Klein, D.R., and C. Bay. 1990. Foraging ecology of bison in aspen boreal habitats. Holarctic Ecology 13:269–280.

Knight, D.H., and L.L. Wallace. 1989. The Yellowstone fires: issues in landscape ecology. BioScience 39:700–706.

Kolasa, J., and S.T.A. Pickett, editors. 1991. Ecological heterogeneity. Springer-Verlag, New York, USA.

Kotliar, N.B., and J.A. Wiens. 1990. Multiple scales of patchiness and patch structure: a hierarchical framework for the study of heterogeneity. Oikos 59:253–260.

Laca, E.L., R.A. Distel, T.C. Griggs, and M.W. Demment. 1994. Effects of canopy structure on patch depletion by grazers. Ecology 75:701–716.

Levin, S.A. 1992. The problem of pattern and scale in ecology. Ecology 73:1943–1983.

Lindenmayer, D.B., and H.A. Nix. 1993. Ecological principles for the design of wildlife corridors. Conservation Biology 7:627–630.

Lubchenco, J., A.M. Olson, L.B. Brubaker, S.R. Carpenter, M.M. Holland, S.P. Hubbell, S.A. Levin, J.A. McMahon, P.A. Matson, J.M. Mellillo, H.A. Mooney, C.H. Peterson, H.R. Pulliam, L.A. Real, P.J. Regal, P.G. Risser. 1991. The Sustainable Biosphere Initiative: an ecological research agenda. Ecology 72:371–412.

Mangel, M., and C.W. Clark. 1988. Dynamic modeling in behavioral ecology. Princeton University Press, Princeton, New Jersey, USA.

McNamara, J.M., and A.I. Houston. 1992. Risk-sensitive foraging: a review of the theory. Bulletin of Mathematical Biology 54:355–378.

Meagher, M.M. 1973. The bison of Yellowstone National Park. National Park Service Monograph Series, Number One.

Merrill, E.H., M.K. Bramble-Brodahl, R.W. Marrs, and M.S. Boyce. 1993. Estimation of green herbaceous phytomass from Landsat MSS data in Yellowstone National Park. Journal of Range Management 46:151–157.

Morse, D.H., and R.S. Fritz. 1982. Experimental and observation studies of patch choice at different scales by the crab spider *Misumena vatia*. Ecology 63:172–182.

O'Neill, R.V., B.T. Milne, M.G. Turner, and R.H. Gardner. 1988. Resource utilization scales and landscape pattern. Landscape Ecology 2:63–69.

O'Neill, R.V., A.R. Johnson, and A.W. King. 1989. A hierarchical framework for the analysis of scale. Landscape Ecology 3:193–205.

O'Neill, R.V., R.H. Gardner, B.T. Milne, M.G. Turner, and B. Jackson. 1991a. Heterogeneity and spatial hierarchies. Pp. 85–96 in J. Kolasa and S.T.A. Pickett, editors. Ecological Heterogeneity. Springer-Verlag, New York, USA.

O'Neill, R.V., S.J. Turner, V.I. Cullinen, D.P. Coffin, T. Cook, W. Conley, J. Brunt, J.M. Thomas, M.R. Conley, and J. Gosz. 1991b. Multiple landscape scales: an intersite comparison. Landscape Ecology 5:137–144.

Orians, G.H., and J.F. Wittenberger. 1991. Spatial and temporal scales in habitat selection. American Naturalist 137:S29–S49.

Parker, K.L., C.T. Robbins, and T.A. Hanley. 1984. Energy expenditures for locomotion by mule deer and elk. Journal of Wildlife Management 48:474–488.

Pearson, S.M. 1993. The spatial extent and relative influence of landscape-level factors on wintering bird populations. Landscape Ecology 8:3–18.

Pearson, S.M., M.G. Turner, L.L. Wallace, and W.H. Romme. 1995. Scale and pattern of winter habitat use by ungulates following fires in northern Yellowstone National Park. Ecological Applications 5:744–755.

Pruitt, W.O. 1959. Snow as a factor in the winter ecology of the barren ground caribou. Arctic 12:159–179.

Real, L., and T. Caraco. 1986. Risk and foraging in stochastic environments. Annual Review of Ecology and Systematics 17:371–390.

Risch, S.J., D. Andow, and M.A. Altieri. 1983. Agroecosystem diversity and pest control: data, tentative conclusions and new research directions. Environmental Entomology 12:625–629.

Roese, J.H., K.L. Risenhoover, and L.J. Folse. 1991. Habitat heterogeneity and foraging efficiency: an individual-based model. Ecological Modelling 57:133–143.

Russell, R.W., G.L. Hunt, Jr., K.O. Coyle, and R.T. Cooney. 1992. Foraging in a fractal environment: spatial patterns in a marine predator–prey system. Landscape Ecology 7:195–209.

Schmidt, K. 1993. Winter ecology of nonmigratory Alpine red deer. Oecologia 95:226–233.

Senft, R.L., M.B. Coughenour, D.W. Bailey, L.R. Rittenhouse, O.E. Sala, and D.M. Swift. 1987. Large herbivore foraging and ecological hierarchies. BioScience 37:789–799.

Stanton, M.L. 1983. Spatial patterns in the plant community and their effects upon insect search. Pp. 125–157 in S. Ahmad, editor. Herbivore insects: host-seeking behavior and mechanisms. Academic Press, New York, USA.

Stephens, D.W., and J.R. Krebs. 1986. Foraging theory. Princeton University Press, Princeton, New Jersey, USA.

Turner, M.G. 1989. Landscape ecology: the effect of pattern on process. Annual Review of Ecology and Systematics 20:171–197.

Turner, S.J., R.V. O'Neill, W. Conley, M.R. Conley, and H.C. Humphries. 1991. Pattern and scale: statistics for landscape ecology. Pp. 17–1 in M.G. Turner and R.H. Gardner, editors. Quantitative methods in landscape ecology. Springer-Verlag, New York, USA.

Turner, M.G., Y. Wu, W.H. Romme, and L.L. Wallace. 1993. A landscape simulation model of winter foraging by large ungulates. Ecological Modelling 69:163–185.

Turner, M.G., W.W. Hargrove, R.H. Gardner, and W.H. Romme. 1994a. Effects of fire on landscape heterogeneity in Yellowstone National Park, Wyoming. Journal of Vegetation Science 5:731–742.

Turner, M.G., Y. Wu, W.H. Romme, L.L. Wallace, and A. Brenkert. 1994b. Simulating interactions between ungulates, vegetation and fire in northern Yellowstone National Park during winter. Ecological Applications 4:472–496.

Turner, M.G., and R.V. O'Neill. 1995. Exploring aggregation in space and time. Pp. 194–208 in C.G. Jones and J.H. Lawton, editors. Linking species and ecosystems. Chapman and Hall, New York, USA.

Wallace, L.L., M.G. Turner, W.H. Romme, R.V. O'Neill, and Y. Wu. 1995. Scale of heterogeneity of forage production and winter foraging by elk and bison. Landscape Ecology 10:75–83.

Ward, D., and D. Saltz. 1994. Foraging at different spatial scales: dorcas gazelles foraging for lilies in the Negev desert. Ecology 75:45–58.

Wickstrom, M.L., C.T. Robbins, T.A. Hanley, D.E. Spalinger, and S.M. Parish. 1984. Food intake and foraging energetics of elk and mule deer. Journal of Wildlife Management 48:1285–1301.

Wiens, J.A. 1976. Population responses to patchy environments. Annual Review of Ecology and Systematics 7:81–120.

Wiens, J.A. 1989. Spatial scaling in ecology. Functional Ecology 3:385–397.

Wiens, J.A. 1996. Wildlife in patchy environments: metapopulations, mosaics, and management. Pp. 53–84 in D. McCullough, editor. Metapopulations and wildlife conservation. Island Press, Washington, DC, USA.

Wiens, J.A. 1997. Metapopulation dynamics and landscape ecology. Pp. 43–62 in I. Hanski and M. Gilpin, editors. Metapopulation Dynamics: Ecology, Genetics and Evolution. Academic Press, London.

# 14
# The Influence of Landscape Scale on the Management of Desert Bighorn Sheep

PAUL R. KRAUSMAN

## 14.1 Introduction

Historically, there were more desert bighorn sheep (*Ovis canadensis californiana, O.c. mexicana, O.c. nelsoni, O.c. cremnobates*) inhabiting desert mountain ranges of the southwestern United States than there are today. Seton (1929) estimated there were approximately 2,000,000 bighorn sheep in the contiguous United States in pristine times; about half were desert bighorn sheep (Buechner 1960, Cooperrider 1985). However, Valdez (1988) doubted that wild sheep numbers ever exceeded 500,000 in North America. Bighorn sheep are highly selective in their habitat preference and were not distributed uniformly throughout the mountains of western North America. However, in recent times, numbers are down; drastic habitat alteration and destruction by humans eliminated or reduced desert bighorn herds, and in 1991 approximately 25,000 desert bighorn sheep existed in isolated populations scattered throughout their former range (Valdez and Krausman, in press). Desert bighorn sheep populations of the western United States and Mexico have suffered severe declines due to human impacts. Desert bighorns have become one of the rarest ungulates in North America and have been eliminated in Texas and the Mexican states of Chihuahua and Coahuila. Transplant programs have been successful in reestablishing populations in localized areas in many American states

Paul R. Krausman is Professor of Wildlife and Fisheries Science in the School of Renewable Natural Resources, The University of Arizona, Tucson. His research interests have centered around the influences of human disturbances (i.e., canals, recreation, military overflights, housing developments) on large mammals and their habitats and the relationship between large mammals and their habitats. Dr. Krausman teaches classes in wildlife management, big game ecology, and research methodology at the graduate and undergraduate level. He is also the Associate Director of the Arizona Agricultural Experiment Station and assists with the research administered by the College of Agriculture. In addition to research in the southwestern United States Dr. Krausman is involved in a long-term project in India that is examining the influence of ungulates on two national parks.

where they were extirpated (Buechner 1960, Trefethen 1975). Desert big-horns have increased significantly in several areas within the contiguous United States; increasing from approximately 20,000 in 1978 (Monson 1980) to approximately 25,000 in 1991 (Valdez and Krausman, in press). Low population numbers for desert bighorn are improving, but numbers in individual areas present more of a problem. For example, Weaver (1975) identified 77 populations of mountain sheep in California; only 11 con-tained more than 100 individuals. In Arizona, there are at least 59 populations, but only 7 have populations in excess of 100 animals (D.E. Brown, personal communication). The pattern is similar in Utah, New Mexico, and Texas; small isolated populations of fewer than 100 individuals constitute a significant proportion of the remaining desert bighorn sheep.

These small populations are threatened by continued habitat destruction and disturbance. The maintenance and management of each small popula-tion presents a challenge to managers, especially with the high demand for trophy hunting. Geist (1975, p. 93) maintained that "... relict, natural sheep populations of less than a hundred individuals should not be subjected to hunting, until research has clarified what kind of hunting is still compatible with the maintenance of the populations." Wilson (1979) argued that 125 animals was the minimum number for maintaining a viable population of North American wild sheep. The minimum population size needed for desert races to survival is arguable, but many researches accept as a rule of thumb a minimum population size of 50 to preserve fitness and 500 to maintain genetic variance or genetic adaptation in mammals (Frankel 1983). Regardless, it appears that most populations of desert bighorn sheep are at, below, or are approaching what many researchers would consider minimum viable population levels.

When examining minimum population size on a landscape level, Berger (1990) calculated a minimum viable population size (MVP) using empirical data. He presented data on localized extinctions of mountain sheep in five weatern states and concluded that these extirpations were not caused by food shortages, severe weather, predation, or interspecific competition. Rather, he suggested that extinction times were related to initial population size. Native populations with less than 50 sheep were subject to rapid extinction; only populations with more than 100 animals were found to exist for more than 70 years. The implication of this finding to management is clear; unless small (less than 50) populations are enhanced somehow (e.g., transplants or habitat improvement), they will become extinct. Berger (1990) did not include populations from Arizona in his study, thus, desert bighorn sheep in Arizona provide a data set with which to examine Berger's (1990) predictions about persistence times on a smaller scale (i.e., five western states vs. Arizona). Krausman et al. (1993) discussed some of the differences found between their data and Berger's, but Berger (1993) ar-gued that Krausman et al. (1993) did not provide sufficient information to support the Arizona data. Those data were supplied in a later paper by

Krausman et al. (1996). Krausman et al. (1996) followed Berger's (1990) methods as closely as possible and reviewed Arizona Game and Fish Department census information from 38 desert populations of bighorn sheep in Arizona. They examined populations with and without reintroduced sheep but did not include populations that had been augmented in the analysis. They excluded populations with less than two size estimates or less than 10 years of data. Populations were classified into five size groups to examine population persistence over time. When no sheep were observed during two or more consecutive censuses the population was considered extinct. Krausman et al. (1996) and Berger (1990) assumed that census data were unbiased over time despite different survey techniques.

The results of Berger (1990) were on a large landscape scale (sensu King, this volume) including parts of five states, >5,000 km$^2$ and did not agree with those of Krausman et al. (1996), who examined the problem from a more regional level (parts of one state, <1000 km$^2$). Berger's (1990) prediction was that native populations of bighorn sheep of less than 50 individuals were unable to resist extinction. Krausman et al. (1996) found that 6 of 20 populations became extinct during the 35-year survey in Arizona, but they did not find a relationship between the probability of extinction and initial population size. Extinction occurred in 4 of 12 populations with less than 50 individuals, compared with 2 of 8 populations with more than 50 individuals, and at least one extinction occurred in each population size class (i.e., 1 to 15, 16 to 30, 31 to 50, 51 to 100, more than 100).

Several explanations have been offered to explain the differing results. Krausman et al. (1996) argued that several assumptions made in both studies were questionable; e.g., all races of mountain sheep in all parts of their range have identical habitat requirements, separate mountain ranges constitute discrete populations, and that the census data are accurate. Goodson (1994) examined the work of Berger (1990) and Krausman et al. (1993) and concluded that it was not "... small population size per se but continuation of management practices that resulted in declines that caused extinctions. With improved management many small populations of mountain sheep can persist and increase to healthy population levels."

Population dynamics, environmental fluctuations, and management strategies govern persistence times. When subspecies are combined so that Rocky Mountain bighorn sheep and desert bighorn sheep are considered as having similar dynamics and populations on the fringe of their range are grouped with other populations, as Berger (1990) did in his analysis, many important effects may become lost in arriving at answers due to the loss of resolution. A loss of resolution leads to grosser, larger scale interpretations, where important population-specific influences may be lost. The broad landscape scale is certainly an important way to examine dynamics of systems, but individual (i.e., local) situations need to be considered on a case-by-case basis because of the influence of different management scenarios.

Much of the management of bighorn sheep populations has been at the local (i.e., mountain range) scale and not at the landscape (i.e., meta-population) scale. Habitat within mountain ranges has been enhanced, but the projects should be conducted ". . . with the awareness that all areas used by mountain sheep may be essential for their long-term survival. For viable populations of mountain sheep to persist, more than 'mountain islands within desert seas' must be protected" (Bleich et al. 1990). Without some form of protection the consequences of fragmentation as outlined by Wilcox and Murphy (1985) will become evident: populations may be destroyed, reduced, or subdivided, resulting in the conversion of natural habitat and hindering of immigration, thus losing potential sources of emigrants. These concerns are important to the conservation of desert bighorn sheep (Bleich et al. 1990). The contemporary conservation of desert bighorn sheep in a wild state demands that managers consider all scales as important in influencing sheep demographics.

## 14.2 Metapopulations and Desert Bighorn Sheep

Levins (1970) introduced the metapopulation concept as ". . . a population of populations which go extinct locally and recolonize. A region is suitable for a metapopulation if the mean extinction rate is less than the migration rate. Thus a species can survive even if it does not form a part of any stable local community." Metapopulations are not isolated and need to be considered with local population dynamics through an array of other ecological factors including behavior, migration, genetic exchange, and community structure (Figure 14.1). The metapopulation concept has been used as a popular framework to understand threats to species in fragmented habitats (Harrison 1994). For example, Bleich et al. (1990) reviewed the management of desert bighorn sheep on a population level and argued for the ". . . long-term maintenance of genetically viable populations of desert-dwelling mountain sheep (*Ovis canadensis* ssp.) . . ." via connectivity. Here, *connectivity* is defined as a management strategy that conserves numerous habitat patches and the potential for dispersal between them (Harrison 1994). Connectivity, or the lack thereof (i.e., isolation), are central themes in conservation biology (Soule and Wilcox 1980, Bleich et al. 1990, 1996, Schoener 1991, Harrison 1994). Two approaches have been made to increase connectivity, or reduce framgentation, in bighorn sheep populations (Bleich et al. 1990): supplemental transplants to reestablish populations on ranges that will decrease interdeme distances, facilitating gene migration (Frankel 1983, Bleich et al. 1990), and maintenance of corridors that connect fragments (Noss 1987, Harrison 1994). Bleich et al. (1990) reviewed the importance of connectivity to desert-dwelling mountain sheep that I review below. The latter part of this chapter will use a case history to demonstrate how the lack of connectivity can lead to the extinction of populations.

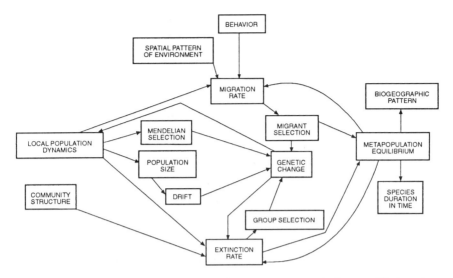

FIGURE 14.1. Ecological factors associated with metapopulation equilibrium (redrawn from Levins 1970).

## 14.3  Importance of Connectivity

### 14.3.1  Overlooked Movements

Habitat requirements of bighorn sheep were discussed earlier, but two components need to be emphasized: bighorn sheep are usually found in proximity to escape cover and rely on keen vision. Both are adaptations to avoid predation (Geist 1971). In general, desert bighorn sheep do not remain in areas where they cannot see long distances or escape predators. Geist (1971) indicated that bighorn sheep were poor dispersers in northern climates. However, desert bighorn sheep exist in mountain ranges separated by relatively flat deserts with limited vegetation and perhaps fewer predators than northern races are exposed to (Bleich et al. 1990). Hence, one may expect increased dispersal between mountain ranges. This, indeed, has been the case throughout the Southwest (Witham and Smith 1979, Elenowitz 1982, Cochran and Smith 1983, King and Workman 1983, Ough and deVos 1984, Krausman and Leopold 1986, Warrick and Krausman 1987, Alderman et al. 1989, Krausman et al. 1989, Bleich et al. in press). Historically, desert bighorn sheep probably used or moved to every mountain range they could see (Krausman 1993). The importance of the intermountain habitats separating desert mountain ranges has been elaborated in numerous studies (Ough and deVos 1984, Krausman and Leopold 1986, Schwartz et al. 1986, Krausman et al. 1989). Bleich et al. (1990) summarized the importance of the areas between habitat patches for genetic and ecological considerations.

The genetics of populations in separate desert mountain habitats was probably maintained due to dispersal that minimized inbreeding (Simberloff and Cox 1987). Small, isolated populations of bighorn sheep are not necessarily doomed to extinction from various forces (e.g., inbreeding depression or demographic stochasticity) if their habitats are protected from humans (Simberloff et al. 1992). However, as barriers to intermountain movement increase, negative effects on mountain sheep may be enhanced (Geist 1975, Noss 1987, Simberloff and Cox 1987).

Ecological considerations (e.g., disease transmission) may strongly influence desert bighorn sheep. Desert sheep traditionally have been managed on a case-by-case basis, usually within a mountain range. However, the flats between mountains may act as corridors and are important for sheep to gain access to other mountain ranges for lambing and foraging. However, when flat intermountain areas are grazed by livestock, diseases may be spread into the bighorn population (Clark et al. 1985, Wehausen 1987, Krausman 1996). "The potential ecological importance of these and similar areas should not be underestimated" (Bleich et al. 1990).

Bleich et al. (1990) proposed a model for the conservation of desert bighorn sheep (Figure 14.2), and Schwartz et al. (1986) were among the first to suggest a management strategy on a landscape basis for desert bighorn sheep. Overall, they suggested that management needs to "... seriously consider intermountain travel corridors for sheep, taking steps to minimize potential barriers such as range fences and motorized recreational activities." To maintain connectivity, use of the corridors by domestic stock (especially sheep) should be minimized to avoid transmission of disease (Technical Staff of the Desert Bighorn Council 1990, Bureau of Land Management 1995, Krausman 1996). Schwartz et al. (1986) concluded: "We still have the raw materials; what is needed is a commitment to protect and manage them properly. Only with the recognition that stewardship responsibilities extend beyond areas of 'traditional' habitat and what are perceived to be 'variable' populations will we be assured the long-term stability of desert-dwelling mountain sheep and other fragile species that similarly inhabit naturally fragmented habitat." This is overt recognition of scale effects. Managing bighorn sheep at a geographical scale (i.e., the scale of the entire geographical range of a species; individuals have typically no possibility of moving to most parts of the range, Hanski and Gilpin 1991) is impractical. Although most data on bighorn sheep have been gathered at the local scale (i.e., the scale at which individuals move and interact with each other in the course of their routine foraging and breeding activities, Hanski and Gilpin 1991), managers need to consider management at the metapopulation scale (i.e., the scale at which individuals frequently move from one place to another, typically across areas that are not suitable for feeding and breeding activities and often with substantial risk of failing to locate another suitable habitat patch in which to settle, Hanski and Gilpin 1991).

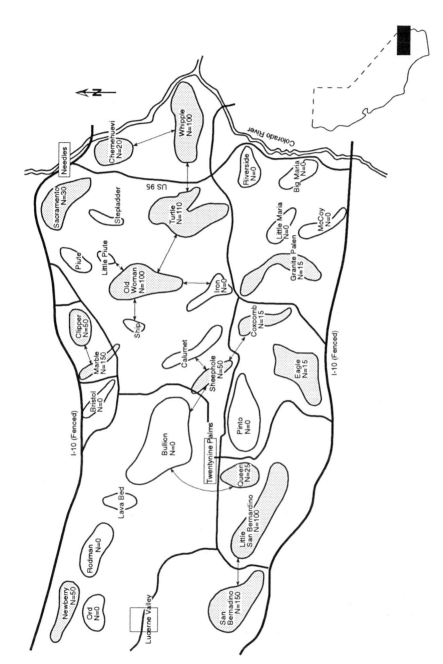

FIGURE 14.2.  Map of a metapopulation of bighorn sheep in southeastern California. Stippled mountain ranges have resident populations of the approximate size indicated. Mountain ranges with $N = 0$ are extirpated populations; ranges with no $N$ value listed are not known ever to have had resident populations. Arrows indicate documented inter-mountain movements by sheep. (redrawn from Bleich et al. 1990).

## 14.3.2  The Metapopulation Controversy

The concept of metapopulations is not embraced as a solid conservation measure by all. Harrison (1994), for example, argues that the concept does not account for (i.e., measure) variability in the size or longevity of local populations, rates of colonization that vary with interpatch distance, behavioral responses to conspecifics, or temporal correlations in extinction probabilities between nearby populations. She further argues that strictly defined metapopulations are scarce because ". . . their essence, persistence in a balance between the extinction and recolonization of populations, is an improbable condition. To attain it, local populations must be roughly equal in size (or longevity); isolated enough to constitute separate populations; and yet interconnected enough to permit recolonization to balance extinction." Failing any of these criteria, the system is more aptly described as one of three quasi-(or less strictly defined) metapopulations: (1) "mainland-island metapopulation," where one or more populations persist indefinitely; (2) "patchy populations" that are found on sets of patches fragmented at a finer scale than that of the population; and (3) "non-equilibrium metapopulations" where conspecific populations are virtually or completely isolated from one another (Harrison 1994). In each of these situations, population persistence is more dependent on within-population dynamics than on metapopulation dynamics, and Harrison (1994) argues that is the case for many ". . . superficially metapopulation-like systems." She presents three reasons why metapopulation considerations are not used as critical conservation strategies.

1. At least one population in the "metapopulation" is more persistent, or can be made so through conservation efforts, thus reducing the importance of the other populations.
2. Populations are so isolated from one another that they are managed individually.
3. Neighboring populations may not be independent enough of one another for a metapopulation strategy to be effective. If there is constant interaction between "populations," the dynamics are those of a single entity.

However, on the one hand, Harrison (1994, pp. 117–118) uses as an example desert bighorn sheep where metapopulations aspects of viability are of critical importance because they exist as networks of small populations, ". . . avoiding both local and regional extinction through infrequent dispersal." On the other hand, she claims (p. 119) the evidence of bighorns traversing "non-habitat areas" is circumstantial. The citations (Witham and Smith 1979, Cochran and Smith 1983, Ough and deVos 1984, Krausman and Leopold 1986, Krausman et al. 1989, Bleich et al. 1990) used to document movement are not circumstantial. Further, recent studies by

Bleich et al. (1996) and Boyce et al. (in review) provide strong evidence that bighorn sheep constitute valid metapopulations. The debate of whether desert bighorn sheep exist in a metapopulation structure will continue, but the work mentioned above provides solid conservation models for populations that are naturally fragmented. How do populations impacted by humans respond, and what are the available conservation strategies? Harrison (1994) argues that species in human-fragmented habitats respond differently from those in naturally fragmented habitats as described by Bleich et al. (1990). In the former, dispersal between fragments occurs so seldom that natural recolonization is scarcely possible. With low or no recolonization the fate of species will probably be decided by factors other than metapopulation dynamics (Harrison 1994). Such is the case with desert bighorn sheep in the Pusch Ridge Wilderness, Santa Catalina Mountains, near Tucson, Arizona (Figure 14.3).

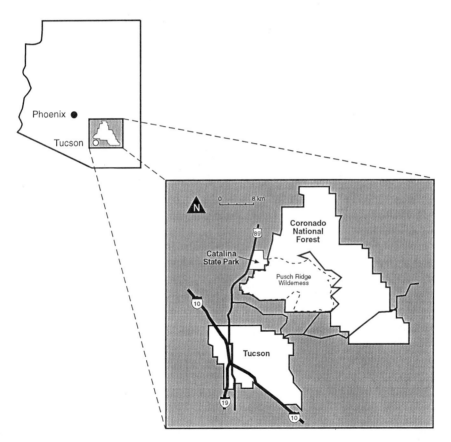

FIGURE 14.3. Desert bighorn sheep habitat in the Santa Catalina Mountains, Arizona.

## 14.4 Desert Bighorn Sheep in a Human-Fragmented Habitat: The Decline of the Tucson Basin Metapopulation

Desert bighorn sheep are rapidly declining in the mountain ranges around Tucson, Arizona (Krausman et al. 1994) (Figure 14.4). Accurate estimates of the populations in the Rincon, Santa Catalina, Tortilla, Tucson, Silverbell, and Santa Rita mountains and adjacent ranges were never made. However, estimates in the late 1800s were first reported for this complex near Tucson. By 1995 sheep were eliminated from all ranges except the Pusch Ridge Wilderness in the Santa Catalina ($N < 20$) and Silverbell Mountain ($N = 30$ to $50$) because of human activity (Krausman et al. 1979). Since the 1920s desert bighorn sheep in the Pusch Ridge Wilderness have declined from more than 200 to less than 20 (Figure 14.5). The limited management of sheep in the Tucson complex was orientated towards sheep in the Catalinas and even those efforts were limited in the face of extensive

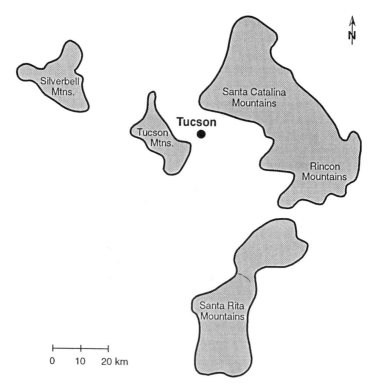

FIGURE 14.4. Mountain ranges surrounding Tucson, Arizona that provided habitat for desert bighorn sheep.

human pressure that fragmented habitats as outlined by Krausman et al. (1994).

## 14.4.1 The Pusch Ridge Wilderness

The Pusch Ridge Wilderness (Figure 14.3) was established 24 February 1978 through the Endangered American Wilderness Act (PL 95-237 [H.R. 3454], 16 USC 1132). One of the major goals of the 22,837-ha wilderness was to protect habitat for desert bighorn sheep (Anonymous 1978), but only within the widerness. The Pusch Ridge Wilderness forms the southwest portion of the Santa Catalina Mountains located in the Coronado National Forest, Arizona. The Santa Catalina Mountains are roughly triangular in shape with an east-west base of about 32 km and the apex 32 km north of the base (Krausman et al. 1979). Elevations ranged from >2745 m at Mount Lemon to 854 m at the southwestern base of the range (Whittaker and Niering 1965).

The Santa Catalina Mountains are unique among Arizona and New Mexico mountain ranges because they possess a full sequence of plant communities from subalpine fir (*Abies lasiocarpa*) forests to Sonoran desert (Blumer 1909, Martin and Fletcher 1943, Nichol 1952, Wallmo 1955, Lowe 1961, Whittaker and Niering 1964).

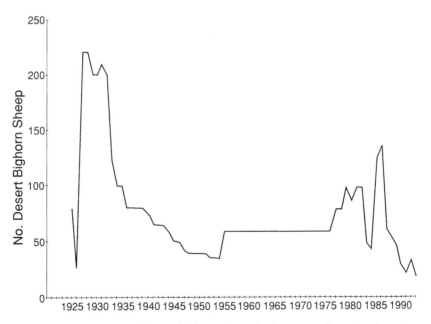

FIGURE 14.5. Estimates of desert bighorn sheep in the Santa Catalina Mountains, Arizona.

The Pusch Ridge Wilderness consists of steep, highly erosive areas with large, deep canyons that support riparian vegetation. Ridges rise from the desert floor to higher elevations forming vertical rock faces and various geologic formations. Vegetation varies from desert grassland at the lower elevations to ponderosa pine (*Pinus ponderosa*) and mixed conifers at higher elevations (Krausman et al. 1979). Whittaker and Niering (1964, 1965) provide a physical and vegetation analysis of the Santa Catalina Mountains. The Pusch Ridge Wilderness is further described by deVos (1983), Gionfriddo and Krausman (1986), Etchberger et al. (1989, 1990), and Mazaika et al. (1992).

Historically, desert bighorn sheep used other (Figure 14.4) mountains surrounding the Pusch Ridge Wilderness, but in the development of the wilderness no consideration was given to the importance of those ranges. As a result, the management and habitat improvement for the sheep was on a local scale and concentrated in a <20 km$^2$ area. During the development of management plans, little or no thought was given to corridors linking other mountains; bighorns in the wilderness areas were essentially considered an isolated population.

## 14.4.2 Previous Studies

Krausman et al. (1979) recommended that the ". . . future well-being of this population will require management and planning based on a sound understanding of basic biological parameters of the herd and of human intrusions into its habitat." Krausman et al. (1979) recommended that information on density, distribution, lambing, habitat, fire, recreation, and human impacts as they relate to desert bighorn sheep were needed for informed management. Each of these arenas has been addressed to a *limited* degree at the local scale and research has been conducted in two major areas: habitat and biology, and human influences.

Krausman et al. (1979) reviewed the literature and status of bighorn sheep in Pusch Ridge Wilderness and recommended that more information was needed to understand the ecological relationships between sheep and their habitat and between humans and bighorn sheep in Pusch Ridge Wilderness. Since then studies have been conducted to collect data to be used as a basis for management decisions (de Vos 1983) and to learn important habitat characteristics (Gionfriddo and Krausman 1986, Etchberger et al. 1989, 1990, Mazaika et al. 1992, Krausman et al. in press). Throughout these studies at least three consistent patterns were recognized in relation to bighorn sheep management in Pusch Ridge Wilderness.

1. Habitat features for bighorn sheep in Pusch Ridge Wilderness are similar to habitat features for other bighorn sheep in the Southwest.
2. Fire suppression is reducing visibility for bighorn sheep and effectively reducing Pusch Ridge Wilderness as a bighorn sheep habitat.
3. Human disturbance and activities, including housing developments and recreation on forest lands, are eliminating habitat available for sheep.

The last issue was supported by sociological studies by Purdy and Shaw (1981), King et al. (1986, 1988), Harris and Shaw (1993), and Harris et al. (1995).

## 14.5 Local Scale Versus Metapopulation Scale Management

If desert bighorn sheep in the mountains surrounding Tucson had been managed with a metapopulation approach (i.e., conserving numerous habitat patches [mountain ranges] and the potential for dispersal between them), would their numbers be as low as they are today? Harrison (1994) argues against the metapopulation paradigm because it stresses the maintenance of an extinction–colonization balance. Because the dynamics of the mountain sheep populations in the Tucson area (or most other areas) are not well known, metapopulation theory has not been widely applied. However, the sheep in the Tucson area could be placed in parts of at least two of Harrison's (1994) less strictly defined metapopulations. Because the Santa Catalina Mountain's Pusch Ridge Wilderness has historically provided habitat for the most sheep and has persisted longer than the surrounding ranges, the argument can be made for the existence of a mainland—island metapopulation where "extinction and colonization occur among some populations, but are nearly irrelevant to metapopulation persistence, since this is assumed by the mainland." Mainland–island dynamics are the rule in many natural metapopulations of long-lived species (Schoener 1991).

There is less support for the patchy population classification as applied to bighorn sheep in the Tucson area. Available data indicate that habitat patches have decreased due to human activity (i.e., fire suppression, recreation, housing), and that these activities have influenced the entire population. In many cases humans have precluded the existence of sheep in habitat that sheep once used.

The nonequilibrium metapopulation classification where ". . . conspecific populations are virtually or completely isolated from one another" would be difficult to defend. Desert bighorn sheep have recently been observed in the Rincon Mountains adjacent to the Santa Catalina Mountains; and when sheep were introduced into a mountain range approximately 60 km from the Pusch Ridge Wilderness, at least one bighorn moved to the Wilderness. As late as 1996 one sheep was observed halfway between these two ranges. Managers were surprised at these movements in areas not normally used by sheep to reach more traditional habitat.

Most extinctions in the Tucson area have been anthropogenic and probably precluded any equilibrium that may have existed before human interference. Clearly, this is an academic question, but had human developments not occurred and habitats not been altered, the corridors connecting the ranges in the Tucson Basin might have ensured a higher population than exists today.

Had metapopulation theory been developed and incorporated into the management of sheep, their survival may have been enhanced. However, the establishment of the Pusch Ridge Wilderness concentrated on a relatively small area (22,837 ha), and enhancing of corridors between other mountain ranges was not considered. Given the rapid urban development in the Tucson area, management options are now limited. The time when managers might have considered all patches (mountain ranges) in developing management plans is past. Bighorn sheep were eliminated from the mountains in the Tucson area (except the Pusch Ridge Wilderness and Silverbell Mountains) more than 50 years ago, and the remaining populations are declining rapidly. In the Pusch Ridge Wilderness no sheep were observed in the 1994 annual surveys, two and only two in 1995. During field studies in 1995 my graduate students and I have observed only two sheep.

## 14.5.1 Metapopulation Extinction

Hanski (1991) discussed four causes of metapopulation extinction, all of which can be applied to the bighorn sheep near Tucson, Arizona.

1. The most fundamental cause of metapopulation extinction operates when the rate of establishment of new local populations is lower than the rate of local extinctions. As local populations became extinct around Tucson the opportunities for reestablishment were precluded by human interference.

2. Without alternative equilibria, a small metapopulation becomes deterministically extinct. Even a large population may become extinct if it happens to become smaller than a certain threshold size. Human activities have precluded the maintenance of corridors between local populations around Tucson.

3. Immigration—extinction stochasticity (i.e., randomness in migration numbers) can cause extinction when the number of local populations is small. For example, the number of habitat patches is small around Tucson, for bighorn sheep. Movement possibilities to suitable habitat are limited.

4. Regional stochasticity is probably a threat, but its extent is debatable. How scale operates in synchronized population dynamics is largely unexplored.

Bleich et al. (1996) examined metapopulations of desert in sheep in Southern California and concluded that "The ramifications of a metapopulation structure for mountain sheep conservation are clear: managers must act to ensure that anthropogenic extirpations are minimized, and that opportunities for natural recolonizations by females and the migration of nuclear genes via males are not impeded."

Unfortunately the biological data and sociological data required to understand population dynamics of bighorn sheep were insufficient for managing the local populations that existed around Tucson. Humans have created a barrier around Pusch Ridge Wilderness, effectively fencing bighorns in (Krausman 1993) while at the same time reducing their habitat. Now that the population has declined (Figure 14.5) managers will need to decide what needs to be done to make Pusch Ridge Wilderness suitable habitat for bighorn sheep. Unfortunately, the modifications made by humans have precluded the continued existence of bighorn sheep in the Pusch Ridge Wilderness. The most likely decision will be to do nothing and accept the decline of the herd as inevitable. Metapopulation management has not occurred for bighorn sheep in Pusch Ridge Wilderness and the indigenous herd is nearly gone.

## 14.5.2 The Upshot

Bighorn sheep in Pusch Ridge Wilderness of the Santa Catalina Mountains may be the next indigenous herd to be eliminated and perhaps replaced with transplants. If <50 sheep exist, a transplant may be warranted, *but* only after the human disturbance, including fire suppression that has been instrumental in eliminating sheep habitat, has been minimized. Also, metapopulation theory and scale issues have to be considered in the management of desert bighorn sheep on Pusch Ridge Wilderness. Operating on a local scale, when a metapopulation scale was called for, contributed to the decline of sheep in this area. However, the reality of management precludes this possibility, and corridors to other mountains have been eliminated, with one exception. Desert bighorn sheep in the Galiuro Mountains moved across the San Pedro Valley to the Pusch Ridge Wilderness in the 1980s, and in February 1996 a young ram was seen in Oracle, Arizona between these two areas. This region on the north end of the Santa Catalina Mountains is the only remaining corridor to other ranges. Any future management without considering means to keep that corridor open would probably be doomed to failure. Hanski et al. (1996) demonstrated that not all of the patches of a particular habitat can be conserved and managers need to set priorities; unfortunately bighorn sheep were a priority only at the local scale in the Tucson Basin.

Could this situation have been avoided? It is unlikely. As humans encroached into bighorn sheep habitat with houses, roads, malls, mines, livestock, and other aspects of human society, the priority to conserve sheep was rarely, if ever, seriously considered. The Pusch Ridge Wilderness is a clear example of managing on a small scale when management at the metapopulation scale was needed. Leopold (1933) could have predicted the outcome; he classified mountain sheep a wilderness species because they fail to thrive in contact with human development. Neglecting to consider influence of scale on the management of desert bighorn sheep has been one of the shortcomings of wildlife management. That shortcoming has recently been recognized, and efforts are underway to include scale issues in bighorn sheep management.

## 14.6 References

Alderman, J.A., P.R. Krausman, and B.D. Leopold. 1989. Diel activity of female desert bighorn sheep in western Arizona. Journal of Wildlife Management 53:264–271.

Anonymous. 1978. The 95th Congress: big gains for the wilderness system. Wilderness Report 15:3.

Berger, J. 1990. Persistence of different sized populations: an empirical assessment of rapid extinctions in bighorn sheep. Conservation Biology 4:91–98.

Berger, J. 1993. Persistence of mountain sheep: methods and statistics. Conservation Biology 7:219–220.

Bleich, V.C., J.D. Wehausen, and S.A. Holl. 1990. Desert-dwelling mountain sheep: conservation implications of a naturally fragmented distribution. Conservation Biology 4:383–390.

Bleich, V.C., J.D. Wehausen, R.R. Ramey II, and J.L. Rechel. 1996. Pp. 353–373 in D.R. McCullough, editor. Metapopulations and wildlife conservation. Island Press, Covelo, California, USA.

Blumer, J.C. 1909. On the plant geography of the Chiricura Mountains. Science 32:72–74.

Boyce, W.M., P.W. Hedrick, N.E. Muggli-Cockett, S. Kalinowski, M.C.T. Penedo, and R.R. Ramey II. (In review). Genetic variation of major histocompatibility complex and microsatellite loci; a comparison in bighorn sheep.

Buechner, H.K. 1960. The bighorn sheep in the United States, its past, present and future. Wildlife Monograph 4:1–174.

Bureau of Land Management. 1995. Mountain sheep ecosystem management strategy in the 11 western states and Alaska. BLM/SC/PL-95/001 + 6600. Colorado: Bureau of Land Management.

Clark, R.K., D.A. Jessup, M.D. Kock, and R.A. Weaver. 1985. Survey of desert bighorn sheep in California for exposure to selected infectious disease. Journal of the American Veterinarian Medical Association 187:1175–1179.

Cochran, M.H., and E.L. Smith. 1983. Intermountain movements by a desert bighorn ram in western Arizona. Transactions of the Desert Bighorn Council 27:1–2.

Cooperrider, A.Y. 1985. The desert bighorn. Pp. 473–485 in T.L. Di Silvestro, editor. Audubon Wildlife Report 1985. National Audubon Society, New York, New York, USA.

deVos, J.C., Jr. 1983. Desert bighorn sheep in the Pusch Ridge Wilderness area. Final Report R381151. U.S. Forest Service, Coronado National Forest, Tucson.

Elenowitz, A. 1982. Preliminary results of a desert bighorn transplant in the Peloncillo Mountains, New Mexico. Transactions of the Desert Bighorn Council. 26:8–11.

Etchberger, R.C., P.R. Krausman, and R. Mazika. 1989. Mountain sheep habitat characteristics in the Pusch Ridge Wilderness, Arizona. Journal of Wildlife Management 53:902–907.

Etchberger, R.C., P.R. Krausman, and R. Mazaika. 1990. Pp. 53–57 in P.R. Krausman and N.S. Smith, editors. Effects of fire on desert bighorn sheep habitat: Managing Wildlife in the Southwest symposium. Arizona Chapter of The Wildlife Society, Phoenix, Arizona, USA.

Frankel, O.H. 1983. The place of management in conservation. Pp. 1–14 in C.M. Schonewald-Cox, editor. Genetics and conservation. Benjamin/Cummings Publishing Company, Menlo Park, California, USA.

Geist, V. 1971. Mountain sheep, a study in behavior and evolution. University of Chicago Press, Chicago, Illinois, USA.

Geist, V. 1975. On the management of mountain sheep: theoretical considerations. Pp. 77–98 in J.B. Trefethen, editor. The wild sheep in modern North America. Winchester Press, New York, New York, USA.

Gionfriddo, J.P., and P.R. Krausman. 1986. Summer habitat use by bighorn sheep. Journal of Wildlife Management 50:331–336.

Goodson, N.J. 1994. Persistence and population size in mountain sheep: why different interpretations? Conservation Biology 8:617–618.

Hanski, I. 1991a. Single-species metapopulation dynamics: concepts, models and observations. Biological Journal of the Linnean Society 42:17–38.

Hanski, I., and M. Gilpin. 1991. Metapopulation dynamics: Brief history and conceptual domain. Biological Journal of the Linnean Society 42:3–16.

Hanski, I., A. Moilanen, T. Pakkala, and M. Kuussaari. 1996. The quantitative incidence function model and persistence of an endangered butterfly metapopulation. Conservation Biology 10:578–590.

Harris, L.K., and W.W. Shaw. 1993. Conserving mountain sheep habitat near an urban environment. Transactions of the Desert Bighorn Council 37:16–19.

Harris, L.K., P.R. Krausman, and W.W. Shaw. 1995. Human attitudes and mountain sheep in a wilderness setting. Wildlife Society Bulletin 23:66–72.

Harrison, S. 1994. Metapopulations and conservation. Pp. 111–128 in P.J. Edwards, R.M. May, and N.R. Webb, editors. Large-scale ecology and conservation biology. Blackwell Scientific Publication, London, UK.

King, M.M., and G.W. Workman. 1983. Preliminary report on desert bighorn movements on public lands in southeastern Utah. Transactions of the Desert Bighorn Council 27:4–6.

King, D.A., D.J. Bugarsky, and W.W. Shaw. 1986. Contingent valuation: an application to wildlife. International Union of Foresty Research Organizations World Congress 18:1–11.

King, D.A., D.J. Flynn, and W.W. Shaw. 1988. Pp. 243–264 in J.B. Loomis, compiler. Total and existence values of a herd of desert bighorn sheep. Western Regional Research Publication W-133. University of California, Davis, California, USA.

Krausman, P.R. 1993. The exit of the last wild mountain sheep. Pp. 241–250 in G.P. Nabhan, editor. Counting sheep. University of Arizona Press, Tucson, Arizona, USA.

Krausman, P.R. 1996. Problems facing bighorn sheep in and near domestic sheep allotments. Pp. 59–64 in W.D. Edge and S.L. Olson-Edge, editors. Sustaining rangeland ecosystems. Special Report 953. Oregon State University Extension Service, Corvallis, Oregon, USA.

Krausman, P.R., W.W. Shaw, and J.L. Stair. 1979. Bighorn sheep in the Pusch Ridge Wilderness area, Arizona. Transactions of the Desert Bighorn Council 23:40–46.

Krausman, P.R., and B.D. Leopold. 1986. The importance of small populations of desert bighorn sheep. Transactions of the North American Wildlife and Natural Resource Conference 51:52–61.

Krausman, P.R., B.D. Leopold, R.F. Seegmiller, and S.G. Torres. 1989. Relationships between desert bighorn sheep and habitat in western Arizona. Wildlife Monograph 102:1–66.

Krausman, P.R., R.C. Etchberger, and R.M. Lee. 1993. Persistence of mountain sheep. Conservation Biology 7:219.

Krausman, P.R., W.W. Shaw, R.C. Etchberger, and L.K. Harris. 1994. The decline of bighorn sheep in the Santa Catalina Mountains, Arizona. Pp. 245–250 in P.F. Ffolliott and A. Ortega-Rubio, editors. Biodiversity and management of the Madrean Archipelago: the sky islands of southwestern United States and northwestern Mexico. USDA Forest Service Report Number RM-GTR-264. Rocky Mountain Forest and Range Experiment Station, Fort Collins, Colorado, USA.

Krausman, P.R., R. Valdez, and J.A. Bissonette. 1996. Bighorn sheep and livestock. Pp. 237–243 in P.R. Krausman, editor. Rangeland wildlife. Society for Range Management, Denver, Colorado, USA.

Krausman, P.R., R.C. Etchberger, and R.M. Lee. 1996. Persistence of mountain sheep populations in Arizona. Southwestern Naturalist 41: In Press.

Krausman, P.R., G. Long, and L. Tarango. 1996. Desert bighorn sheep and fire, Santa Catalina Mountains, Arizona. Pp. 162–168 in P.F. Ffolliott, L. De Bono, M.B. Baker, Jr., G.J. Gottfried, G. Solis-Garza, C.B. Edminster, D.G. Neary, L.S. Allen, and R.H. Hamre, Technical Coordinators. Effects of fire on Madrean Province ecosystems. USDA Forest Service General Technical Report RM-GTR-289, Rocky Mountain Forest and Range Experiment Station, Fort Collins, Colorado, USA.

Leopold, A. 1933. Game management. Charles Scribner's Sons, New York, New York, USA.

Levins, R. 1970. Extinction. Pp. 75–107 in Some mathemathical qustions in Biology, American Mathemathical Society, Volume 2, Providence, Rhode Island. Lecture on mathematics in the life sciences.

Lowe, C.H., Jr. 1961. Biotic communities in the Sub-Mogollon region of the island. Southwestern Arizona Academy of Science Journal 2:40–49.

Martin, W.P., and J.E. Fletcher. 1943. Vertical zonation of great soil groups on Mt. Graham, Arizona, as correlated with climate, vegetation, and profile characteristics. University of Arizona Agricultural Experimental Station Technical Bulletin 99:89–153.

Mazaika, R., P.R. Krausman, and R.C. Etchberger. 1992. Forage availability for mountain sheep in Pusch Ridge Wilderness, Arizona. The Southwestern Naturalist 37:372–378.

Monson, G. 1980. Distribution and abundance. Pp. 40–51 in G. Monson, and L. Sumner, editors. The desert bighorn. University of Arizona Press, Tucson, Arizona, USA

Nichol, A.A. 1952. The natural vegetation of Arizona. The University of Arizona Agricultural Experimental Station Technical Bulletin 1271:189–230.

Noss, R.F. 1987. Corridors in real landscapes: a reply to Simberloff and Cox. Conservation Biology 1:159–164.

Ough, W.D., and J.C. deVos, Jr. 1984. Intermountain travel corridors and their management implications for bighorn sheep. Transactions of the Desert Bighorn Council 28:32–36.

Purdy, K.G., and W.W. Shaw. 1981. An analysis of recreational use patterns in desert bighorn habitat: the Pusch Ridge Wilderness case. Transactions of the Desert Bighorn Council 25:1–5.

Schoener, T.W. 1991. Extinction and the nature of the metapopulation. Acta Oecologica 12:53–75.

Schwartz, O.A., V.C. Bleich, and S.A. Holl. 1986. Genetics and the conservation of mountain sheep *Ovis canadensis nelsoni*. Conservation Biology 37:179–190.

Seton, E.T. 1926. Lives of game animals. Vol 3, Part 2. Doubleday Page and Company. New York, New York, USA.

Simberloff, D., and J. Cox. 1987. Consequences and costs of conservation corridors. Conservation Biology 1:63–71.

Simberloff, D., J.A. Farr, J. Cox, and D.W. Mehlman. 1992. Movement corridors: conservation bargains or poor investments? Conservation Biology 6:493–504.

Soulé, M.E., and B.A. Wilcox. editors. 1980. Conservation Biology. Sinauer Associates. Sunderland, Massachusetts, USA.

Technical Staff of the Desert Bighorn Council. 1990. Guidelines for management of domestic sheep in the vicinity of desert bighorn habitat. Transactions of the Desert Bighorn Council 34:33–35.

Trefethen, J.B. 1975. The wild sheep in modern North America. Boone and Crockett Club and Winchester Press. New York, New York, USA.

Valdez, R. 1988. Wild sheep and wild sheep hunters of the New World. Wild Sheep and Goat International. Mesilla, New Mexico, USA.

Valdez R., and P.R. Krausman. Mountain sheep in North America. University of Arizona Press, Turrson, Arizona. In Press.

Wallmo, O.C. 1955. Vegetation of the Huachuca Mountains, Arizona. American Midland Naturalist. 54:466–480.

Warrick, G.D., and P.R. Krausman. Barrel cacti consumption by desert bighorn sheep. The Southwestern Naturalist. 34:483–486; 19.

Weaver, R.A. 1975. Status of the bighorn sheep in California. Pp. 58–64 in J.B. Trefethen, editor. The wild sheep in modern North America. The Boone and Crockett Club and Winchester Press. New York, New York, USA.

Wehausen, J.D. 1987. Some probabilities associated with sampling for diseases in bighorn sheep. Transactions of the Desert Bighorn Council 31:8–10.

Whittaker, R.H., and W.A. Niering. 1964. Vegetation of the Santa Catalina Mountains, Arizona. 1. Ecological classification of distribution of species. Journal of the Arizona Academy of Science 8:9–34.

Whittaker, R.H., and W.A. Niering. 1965. Vegetation of the Santa Catalina Mountains, Arizona: a gradient analysis of the south slope. Ecology 46:429–452.

Wilcox, B.A., and D.D. Murphy. 1985. Conservation strategy: the effects of fragmentation on extinction. American Naturalist 125:879–887.

Wilson, L.O. 1979. North American wild sheep: some management choices. Wild Sheep International 1:8–15.

Witham, J.H., and E.L. Smith. 1979. Desert bighorn movements in a southwestern Arizona mountain complex. Transactions of the Desert Bighorn Council 23:20–23.

# 15
# The Influence of Spatial Scale and Scale-Sensitive Properties on Habitat Selection by American Marten

John A. Bissonette, Daniel J. Harrison, Christina D. Hargis, and Theodore G. Chapin

## 15.1 Introduction

Ecological explanation exists at all hierarchical levels and spatial scales (Pickett et al. 1994). Thus, relying upon only a single level or scale of investigation limits ecological understanding. One reason for this is that a limited observation set underlies every data gathering process. An *observation set is* "the phenomena of interest, the specific measurements taken, and

John A. Bissonette is Leader of the Utah Cooperative Fish and Wildlife Research Unit and Professor in the Department of Fisheries and Wildlife at Utah State University. His research and teaching emphasize landscape ecology and the influence of large-scale processes on wildlife populations and habitat.

Daniel Harrison is Associate Professor in the Department of Wildlife Ecology at the University of Maine. He teaches a field course in research methods, a course in wildlife-habitat relationships to seniors and graduate students, and a graduate course in predator ecology. His research addresses the influence of forest mangement practices on area-sensitive wildlife species, as well as habitat relationships of mammalian predators and their prey. Current research topics address the influence of spatial scale on habitat selection by generalist versus specialized carnivores, as well as an 11-year study of the influence of forest harvesting on population performance and habitat selection by American marten, evaluating marten responses to habitat features at multiple spatial scales.

Christina D. Hargis has recently completed her doctorate in wildlife ecology at Utah State University, Logan, Utah, where she investigated the effects of forest fragmentation and landscape pattern on the American marten. Prior to this work, she was employed by the Forest Service and examined home range use by goshawks at two spatial scales. She currently works for the US Forest Service in Flag Staff, Arizona.

Theodore Chapin is a recent graduate of the University of Maine with a Master of Science in wildlife ecology. His research interests include the role of landscape pattern and spatial scale in wildlife-habitat relationships, with a focus on GIS and statistical techniques for analyzing habitat selection. Mr. Chapin currently is a GIS specialist with Ecology and Environment, Inc. in Lancaster, New York, where he continues to apply GIS to natural resource issues.

the techniques used to analyze the data" (O'Neill et al. 1986, Bissonette this volume). Every study is limited by inherent restrictions in the phenomena of interest (e.g., competition, predation), limitations in the kinds of measurements (e.g., population density, number surviving, longevity, fitness), and the specific techniques used to analyze the data (e.g., by hypothetico-deductive science). For example, at the population level, the processes of competition and predation can provide compelling mechanistic explanations for population persistence. At the same time, larger scale effects, involving differences in the relative proportions of available habitat, the frequency distribution of habitat patch sizes, and the degree of isolation of remnant patches from unfragmented habitat influence species' responses and provide contextual explanation and understanding of the effects of larger scale landscape changes on those same species (Andrén 1994).

A second shortcoming of the single-scale approach is that it does not reflect the hierarchical way in which organisms perceive habitat. An individual organism responds to its environment over several spatial scales and hierarchical levels, with the smallest scale corresponding to the grain of the organism, and the largest scale being at least its home range (Kotliar and Wiens 1990). A study conducted at one hierarchical level may provide a parsimonious, mechanistic explanation for habitat selection at the next level, i.e., a triadic approach (O'Neill et al. 1986); however, this explanation may not hold when the study is scaled several levels upward. Different aspects of an organism's life history may motivate habitat selection at each scale, e.g., foraging decisions may occur at a fine scale, while finding a mate may occur at a coarse scale. Moreover, there are inherent differences in biotic and abiotic processes operating at each scale (Urban et al. 1987) that may influence habitat selection. The question being asked should determine the appropriate scale of habitat studies, recognizing that animals may exhibit different, and perhaps bipolar, responses at different scales of investigation. Studies conducted over several scales more readily integrate scale-dependent relationships into a meaningful whole to allow greater understanding of how organisms assimilate information (Ritchie, this volume, 1997) and make decisions that influence habitat choice and ultimately, fitness.

The range of spatial scales over which an organism can respond is determined largely by its mobility. Also, response is based on decision hierarchies (Holling 1992), which appear to be adjudicated through sensory perception. For terrestrial species there exists an approximate allometric relationship between body size and spatial scale of habitat selection (Milne this volume, Ritchie in press). Fine-scale decisions are made with every stride, which may be a millimeter for arthropods and over a meter for large carnivores, whereas large-scale decisions are made over the maximum area traversed by an individual. Individuals generally do not respond to scales larger than their home range, except in the cases of dispersal and migration. However, there is value in examining habitat selection at scales larger than the response range of an organism because scale-sensitive processes and patterns may emerge that directly or indirectly influence habitat selection. These scale-sensitive properties may operate by downward causation (Kawata 1995) or by imposing constraints on the range of habitat choices available to an individual responding at smaller scales.

The concept of upper level constraints is a basic principle of rate-structured hierarchy theory (O'Neill et al. 1986). Ecological processes occur at different rates. Process rates and spatial scales are linked, with faster rates often occurring at a finer level of resolution (O'Neill et al. 1986, Urban et al. 1987). Fine-scale decisions, e.g., changes in the direction of travel, are made frequently, whereas large-scale decisions such as selection of home range boundaries occur at slower [temporal] rates. Generally, large-scale, slower processes tend to place constraints on finer scale processes operating at faster frequencies. For example, long-term, large-scale weather patterns constrain the maximum annual biomass accumulation of plants, and patterns of vegetation succession occurring over several decades influence habitat selection and diet choice in mammals. When multiscale studies are restricted to the scales at which individuals can respond, these upper level, large-scale constraints cannot be observed.

Dominant patterns in landscapes are often determined by geological, biological, and human-induced influences that operate at scales much larger than the response range of most organisms and which become gradually evident at a slow rate relative to the mean generation time of most organisms. When a critical threshold is reached, landscape pattern can act as a constraint on animal movement (Bissonette et al. 1989, Chapin 1995, Hargis 1996) and demography (Fredrickson 1990, Hargis 1996). For example, landscape pattern can mediate organism response, influencing fitness and survival. Such properties may be considered emergent if they cannot be fully explained mechanistically at lower levels (Bissonette this volume).

## 15.2 Emergent Landscape Properties: A Working Model

We propose that disturbance-created habitat fragmentation and the resulting changes in landscape pattern may be an example of an emergent property (see Bissonette this volume for definition) that acts as a slow-rate, large-scale constraint on habitat selection. Andrén (1994) suggested that as a landscape is fragmented, some species may decrease in greater proportion than expected solely by loss of suitable habitat. He suggested that the combination of increasing isolation and decreasing patch size of suitable habitat may accelerate the negative effects of simple habitat loss. There is a characteristic quality associated with, for example, degree of isolation and patch size, that occurs at a larger scale than the response range of individuals, yet influences species response. How, specifically, a species may respond to habitat structure, composition, and pattern at several spatial scales can be illustrated by our collective studies on the American marten (*Martes americana*). Martens are an appropriate species to study because they are closely associated with mature forests (e.g., Steventon and Major 1982, Hargis and McCullough 1984, Thompson 1988, Bissonette et al. 1989, Brainerd 1990, Katnik 1992, Buskirk and Powell 1994), are thought to be susceptible to forest fragmentation (Buskirk and Powell 1994, Chapin 1995), and were studied by us in several landscapes where clear-cut logging was a predominant silvicultural practice. Our approach was to compare and contrast results collected at different spatial scales among study sites in Utah, Wyoming, Maine (United States), and Newfoundland (Canada).

As a framework for discussion, we propose the following model of marten response to habitat loss, both with and without the assumption of emergent landscape properties. Under both assumptions, we anticipate a decline in marten numbers with increasing loss of forest habitat, because martens are associated closely with mature forests (Bissonette et al. 1989, Brainerd 1990, Hargis and McCullough 1994). If martens respond solely to loss of habitat, the expected relationship would be linear, with marten numbers proportional to habitat availability (Figure 15.1, curve A), and

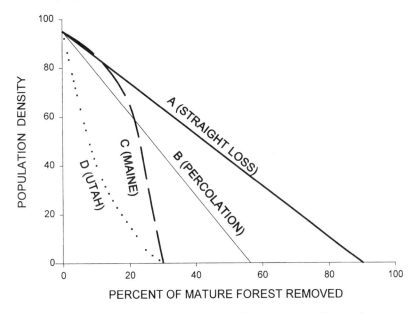

FIGURE 15.1. Responses by marten to habitat fragmentation. Curve A represents the expected response by marten to increasing fragmentation if they are influenced only by habitat loss. Curve B represents the expected relationship if marten are responding to loss of habitat connectivity. Curves C and D represent actual marten responses to increasing habitat fragmentation levels in Maine and Utah, respectively, suggesting that habitat pattern is imposing constraints on marten demography.

approaching zero when all forest habitat is lost. Alternatively, marten populations may reach zero when the forest area is reduced to about 60% of the landscape, because at this point, the forest approaches discontinuity, and martens would need to cross open areas to access isolated forest patches (Figure 15.1, curve B). Andrén (1994) proposed that effects of landscape pattern may not occur until approximately 70–90% of the habitat is lost, and accelerated declines in population densities would occur thereafter.

However, if emergent properties influence marten habitat selection, we would anticipate a nonlinear relationship between marten numbers and loss of habitat (Figure 15.1, curve C). Alternatively, response to landscape pattern could produce an exponential loss in animal numbers at low levels of fragmentation, with the magnitude of effects increasing at lower levels of fragmentation (Figure 15.1, curve D). We discuss the biological reasons for the different nonlinear response curves in section 15.7.

We investigated habitat selection of the American marten at four spatial scales: (1) the microhabitat or substand scale; (2) the stand scale (i.e., selection among available forest stands occurring within a marten's home range; (3) the home range scale (i.e., composition of habitats within the home range relative to availability within a marten's dispersal radius), and (4) the landscape scale (i.e., equivalent to at least the dispersal distance to individual marten, on the order of tens of km$^2$). Individual martens respond directly to habitat at the first three scales. We will suggest why we observed differences in marten response at the stand scale between our Maine and Newfoundland study sites, why we observed different stand- and home range-scale responses by martens to habitat in Maine, and, contrarily, why we observed striking similarities in marten response to fragmentation at the landscape scale in Maine and Utah.

## 15.3  Relevant Scales

Each scale represents an increase in both grain and extent. At the microhabitat scale, the grain is the smallest unit of resolution at which martens perceive heterogeneity in habitat. Although martens may detect patterns on the scale of millimeters, we will assume that habitat selection decisions are potentially made with every stride, and therefore, landscape grain for a marten is perhaps the length of stride, approximately 0.5 m or some small multiple. Correspondingly, microhabitat extent encompasses a few tens of meters, or in silvicultural terms, a substand.

At the stand scale, the grain is roughly equivalent to the resolution at which the quality of attributes within a forest stand can be assessed. We presume that the suitability of a forest stand is determined by its size, structural characteristics, and prey availability.

At the home range scale, the forest stand serves as the grain at which habitat quality assessment is made, and *extent* is defined as the area of the home range, which can be one to several kilometers (Powell 1994) During dispersal (Phillips 1994), martens presumably exhibit some choice when deciding where to position their home range (Katnik 1992), although intraspecific competition and access to other breeding adults also play major roles in home range location (Katnik et al. 1994). Home range location incorporates an individual marten's perception of features such as habitat heterogeneity, stand interspersion, and juxtaposition.

At the landscape scale, forest stands or vegetation associations are aggregated into two classes, mature forest habitat and open areas, and landscape extent encompasses the home ranges of many martens over tens to hundreds of kilometers. At this scale, the distinguishing variable in marten habitat selection is the presence of overhead canopy cover. Cover in itself is not a mechanistic explanation for habitat selection, but serves as a surrogate for a suite of mechanisms that are known to be important to martens: thermal advantages (Buskirk 1984, Buskirk et al. 1988, 1989, Taylor and Buskirk 1994), predator avoidance (Drew 1995), subnivean access (Bissonette and Sherburne 1993, Sherburne and Bissonette 1993, 1994), and prey availability (Bissonette and Sherburne 1993, Clevenger 1993, Buskirk and Powell 1994). In our analyses, regenerating forest stands and brushy vegetation are classified as open areas because they do not provide the structural attributes associated with older forest stands.

## 15.4 Mechanisms of Marten Habitat Selection

Throughout North America, martens generally are associated with mature forests and have demonstrated an avoidance of large, unforested areas (see review by Buskirk and Powell 1994). Habitat selection is based on the functional values provided by structure. For example, coarse woody debris (CWD) associated with mature forests intercepts snow during the winter, creating subnivean spaces that provide thermal shelter (Buskirk 1984, Buskirk et al. 1989). Because of their elongate morphology, coupled with their northern and boreal distribution, martens face homeothermic problems during the winter (Buskirk et al. 1988, Taylor 1993, Harlow 1994). Indeed, the energetic implications of individual body size, shape, age, sex, activity level, pregnancy, and lactation status have been studied in weasels (Iverson 1972, Powell et al. 1985, King 1989, Sandell 1989), and in fishers (Davison et al. 1978, Powell 1979, Powell and Leonard 1983), but only limited work has been done for martens (Buskirk et al. 1988, Adair and Bissonette, in progress).

Coarse woody debris also provides access to subnivean prey. In Wyoming, Sherburne and Bissonette (1993, 1994) and Bissonette and Sherburne (1993) showed that martens used subnivean access holes differentially, based upon prey availability. For every 50 g increase in prey biomass, martens were 1.37 times more likely to use that access hole.

Additionally, tree boles provide vertical escape routes for martens. Drew (1995), in a series of bait-box field experiments, demonstrated that martens in Newfoundland did not venture more than 25 to 50 m into clearcut areas, even when food was provisioned, yet they foraged through coniferous forest completely defoliated by spruce budworm (*Choristoneura fumiferana*) and hemlock looper (*Lambidina fiscellaria*). Furthermore, Chapin et al. (in press) observed selection for budworm-killed conifer stands over residual conifer stands in Maine, and Sherburne (1992) showed that martens in Yellowstone National Park, Wyoming used lodgepole pine (*Pinus contorta*) stands that had been defoliated by the fires of 1988 similarly to unburned stands. Given the high prevalence of predation by mammalian and avian predators on the untrapped marten population studied by Hodgman et al. (in press), vertical escape routes may be important habitat features influencing marten survival.

Martens across North America depend primarily upon forest-dwelling microtines, e.g., red-backed voles (*Clethrionomys* sp.) (Zielinski et al. 1983, Martin 1994). In western North America, red-backed voles are associated with mature forest stands (Ramirez and Hornocker 1981, Raphael 1988, Tallmon and Mills 1994, Mills 1995), although populations also have been reported in clearcuts with high amounts of lichen-covered woody debris (Gunther et al. 1983). As a consequence, foraging by martens is concentrated in mature forest stands and other vegetation types where vole habitat is located.

## 15.5  Habitat Selection at the Microhabitat or Sub-Stand Scale

In this section we draw upon data from our study sites in Newfoundland, Maine, Utah, and Wyoming to make several comparisons; however, not all comparisons involve all sites. Although many studies of marten habitat selection have indicated a close association between marten habitat and mature forest stands with abundant coarse woody debris, we have noted regional variations from this pattern at the microhabitat scale. In Maine, microhabitat variables associated with understory characteristics and structure near the ground had little influence on the selection of resting sites during the summer or winter, based on multivariate logistic regression (Chapin, Phillips, Harrison et al., in press), whereas several microhabitat features appear to be associated with use of habitat for foraging (D. Payer and D. Harrison, in progress). Apparently, the structural attributes required by martens are provided in a wide variety of forest types in Maine; in all forests types >12 m in height, structure appears to occur above thresholds at which habitat features are limiting. Thus, Maine's forests differ greatly from other areas throughout the range of the American marten where structural features associated with foraging and energetic needs are insufficient to maintain martens, except in mature, conifer-dominated forests (Clark et al. 1987, Thompson and Harestad 1994, Buskirk and Powell 1994). Apparently, the high diversity of broad-leafed trees in Maine, coupled with their structural complexity, enhances the functional value of deciduous and mixed forests as habitats for martens, as compared to many areas of the intermountain west and boreal regions of Canada and Alaska, where even-aged, clonal stands of aspen (*Populus* spp.) represent the single dominant deciduous tree.

In Newfoundland, martens are more closely tied to mature forests than in some other parts of North America, primarily due to limited availability of prey. Red-backed voles are absent, and only seven small mammal species are available on the island. Of these, only meadow voles (*Microtus pennsylvanicus*) and masked shrews (*Sorex cinereum*) occur in any abundance, while in Labrador, across the Strait of Belle Isle, less than 10 miles from Newfoundland, 17 species are available (Table 15.1). Meadow voles in Newfoundland are found throughout the forest landscape, but are concentrated in openings and gaps. Tucker (1988) documented that voles were very scarce in second growth forest or in clearcuts, 1, 13, or 23 years after logging (Sturtevant and Bissonette 1996). Hence, the depauperate food base has had a defining influence on marten habitat selection and appears to account for the differences in habitat selection in Newfoundland compared with Maine.

TABLE 15.1. Small mammal species found in Labrador and Newfoundland (from Tucker 1988)

| Labrador | Newfoundland |
|---|---|
| Meadow vole (*Microtus pennsylvanicus*) | yes |
| Red-backed vole (*Clethrionomys gapperi*) | no |
| Heather vole (*Phenacomys intermedius*) | no |
| Rock vole (*Microtus chrotorrhinus*) | no |
| Deer mouse (*Peromyscus maniculatus*) | yes |
| Meadow jumping mouse (*Zapus hudsonius*) | no |
| Woodland jumping mouse (*Napaeozapus insignis*) | no |
| Snowshoe hare (*Lepus americanus*) | yes |
| Arctic hare (*Lepus arcticus*) | yes |
| Red squirrel (*Tamiasciurus hudsonicus*) | yes |
| Northern flying squirrel (*Glaucomys sabrinus*) | no |
| Starnosed mole (*Condylura cristata*) | no |
| Northern bog lemming (*Synaptomys borealis*) | no |
| Collared lemming (*Dicrostonyx hudsonius*) | no |
| Masked shrew (*Sorex cinereus*) | yes |
| Northern water shrew (*Sorex palustris*) | no |
| Pigmy shrew (*Microsorex hoyi*) | no |
| No | eastern chipmunk (*Tamias striatus*) |

In Maine, martens prey primarily on red-backed voles, but also show substantial use of snowshoe hares (*Lepus americana*), red squirrels (*Tamisciurus hudsonicus*) (Soutiere 1979), deer mice (*Peromyscus maniculatus*), and a variety of seasonal (June through December) fruits (H. Lachowski and D. Harrison, in progress). Red-backed voles are distributed in a wide variety of forest types, but are associated with structural variables that are positively correlated with degree of stand maturity (H. Lachowski and D. Harrison, in progress). Capture rates of red squirrels do not differ significantly in deciduous, coniferous, and mixed stand types, but squirrels are effectively absent from young regenerating clearcuts. Hares in Maine previously have been shown to be associated with stands characterized by high stem densities of regenerating coniferous and deciduous tree species (Litvaitis et al. 1985), and plants that produce the fruits consumed by martens in Maine are most common in regenerating cuts. Furthermore, other species occasionally consumed by martens (e.g., *Peromyscus maniculatus, Sorex* spp., *Napaeozapus insignis, Zapus husonicus*) are found in varying densities in a variety of forested and regenerating habitats in Maine. Thus, the relative differences in habitat quality among stand types are probably less in Maine than in Newfoundland, and in the more xeric or less structurally diverse forests of the intermountain west and boreal Canada.

## 15.6 Habitat Selection at the Stand Scale

The home ranges of martens generally encompass several forest stands that differ in age, structure, species composition, and understory associations. Selection of home range locations is based on the presence of forest stands with sufficient attributes to meet all life-history requirements while avoiding competition from same-sex conspecifics (Katnik et al. 1994).

In Newfoundland, Snyder and Bissonette (1987) reported that martens used residual coniferous stands remaining after clearcutting, regardless of size, but used recent clearcuts up to 15 years of age much less than expected. For example, on their study site, clearcuts constituted 41% of the study area, but only 26% of marten activity was recorded in these areas and mostly by juvenile animals. Bissonette et al. (1989), working in the same general area, further documented that martens did not use second-growth forests or recent clearcuts, although other investigators had reported consistent activity by martens in burns in other areas (Buskirk 1984, Magoun and Vernam 1986, Vernam 1987, Johnson et al. 1995).

Information from ongoing studies indicate that the paradigm of the marten as a strict specialist for mature, closed-canopy, conifer-dominated forest does not apply in Maine. Katnik (1992), working in an industrial forest landscape where half of the forest had been clearcut during the preceding 20 years, reported that martens did not select forest stands within their home range based on height, canopy closure, and species composition. Conifer, deciduous, mixed, and regenerating forests all received use in proportion to availability within the home range, suggesting that once a territory was established, martens utilized all available forest stands within the territory. Fager (1991) and Kujala (1993) reported a similar pattern for Montana.

Similarly, Chapin, Berusan, and Phillips (in press) reported that, in a protected park dominated by mature forest >12 m in height, martens did not exhibit selection among the conifer, mixed, and deciduous-dominated stands occurring within their territories. However, selection indices were highest for spruce-budworm killed stands characterized by overstory canopy closure <30%, scattered snags, high amounts of coarse woody debris, and dense regeneration of conifer and deciduous trees <6 m in height. Although the percentage of territories of resident female martens consisting of mature conifer forest ranged from only 4% to 26%, nearly all (9 of 10) adult (≥2 years old) females captured during spring were successful reproducers, based on documented lactation (D. Phillips and D. Harrison, 1989–1994, unpublished data).

The apparent differences in habitat selection between Maine and other sites may be due to the habitat associations of potential prey, as well as differences in the structural attributes of early successional stands in Maine compared to other sites in North America. Presumably, all of the mature, immature, and regenerating forests in Maine maintained the complex structural features that are generally thought to influence habitat quality functionally for martens (Buskirk and Powell 1994).

## 15.7 Habitat Selection at the Home Range Scale

Despite the lack of selection by martens for specific cover types at the stand scale (Katnik 1992), highly significant differences in selection among forest age classes was observed at the scale of the home range. Occupied territories of resident martens on the industrial forest site contained proportionally less clearcut and regenerating forest (<6 m), and more mature forest (>6 m) than was available (Katnik 1992); the entire site should have been accessible to all martens based on observed dispersal distances of juveniles (Phillips 1994). Apparently, martens used all habitats, including clearcut stands within their territories, but chose a mosaic of stands that contained little clearcut habitat relative to other potential territories. For example, martens in a forest preserve occupied nearly all (>80%) of the available habitat; however, in the industrial forest, which consisted of a matrix of residual stands, regenerating clearcuts, and recent cuts, marten territories occupied less than one-third of the available landscape (Phillips 1994). This suggests that martens may be considering habitat heterogeneity, interspersion, and juxtaposition when establishing a territory. This is supported by results that suggested that martens did not exhibit home range-scale selection in an adjacent forest preserve where clearcuts and regenerating forest stands were not present (Chapin et al. in press). This process, whereby martens positioned their territories within predominately residual forest, allowed populations occupying the intensively clearcut, industrially managed landscape to maintain comparable survival (Hodgman et al., in press) and reproductive success (Phillips and Harrison, 1989–1994 unpublished data) to marten populations within the forest preserve.

## 15.8 Habitat Selection at the Landscape Scale

Despite very different patterns of habitat selection observed at smaller scales in various sites throughout North America, the patterns show strong convergence at the landscape scale. In both Maine and Utah, on sites where we have conducted large-scale studies, martens appear to avoid landscapes with more than 25% to 30% of the total area in vegetation types other than intact older forests. In industrial forests of Maine, no adult female marten territory occurred where >31% of the residual forest has been removed (Chapin and Harrison, unpublished data). In Utah, within a series of 18 landscapes having 2% to 42% open areas, marten densities declined to near zero as the representation of open areas reached 25% to 30%. At these fragmentation levels, the forest was still the matrix in the landscape, and it was possible for a marten to traverse the landscape and not cross an opening. Were there no scale-sensitive properties operating, then one would expect marten numbers to decline with decreasing older forest habitat until approximately 55% to 65% remained, which represents the proportion at which the matrix changes from forest to disturbed habitat. This was not the case. We hypothesize that martens respond not only to loss of habitat, but to scale-sensitive properties associated with landscape pattern.

Martens in Maine were sensitive to area and isolation effects (Chapin and Harrison, in review) associated with landscape fragmentation (Turner 1989). In the industrial forest, <10% of residual stands received use by resident martens. The median area of residual stands receiving use by martens was 18 times larger than stands that received no use, and residual stands <2.7 ha were not used (Chapin and Harrison, in review). In Newfoundland, isolated remnant spruce and fir stands <15 ha were used much less than larger remnant patches (Snyder and Bissonette 1987). No adult marten occupied territories where greater than 40% of the residual forest had been removed (median = 22% for adult males and 20% for adult females (Chapin and Harrison, in review). Furthermore, forest patches used by martens were closer to large patches and to the forest preserve than patches that did not receive use. Thus, landscape characteristics seemed to have an overriding influence on patterns of habitat occupancy by martens that were not detectable at finer scales of resolution. If based solely on results at the scale and substand scales, erroneous conclusions that martens in Maine were not sensitive to forest harvesting activities would likely have resulted.

The similarity in landscape proportions at which the geographically distant studies of Maine and Utah observed a response by martens suggests a point of upper level constraint on habitat selection at the landscape scale. The shapes of the response curves suggest that different properties may be operating at each site. Data from Maine indicate a curvilinear relationship between marten densities and percentage of habitat loss similar to the predicted response curve of Figure 15.1, line C). Here martens appear to compensate behaviorally and may actually benefit from enhanced access to alternate foods (e.g., fruits) at low levels of habitat fragmentation. However, once a threshold level of fragmentation is reached, the population declines exponentially. In Utah, however, the marten density curve suggests an exponential decline from the start, as predicted by Figure 15.1, line D. Nevertheless, in both Maine and Utah, landscape pattern appears to constrain marten habitat use when the proportion of forest cover habitat falls below approximately 70%.

## 15.9 Discussion

In our analysis of marten habitat selection at microhabitat, stand, and home range scales, we observed disparities in habitat use among the geographically distant sites of Newfoundland, Maine, Utah, and Wyoming that seem attributable to regional variation in vegetation types, stand structure, and prey base. These factors may differentially influence the nature of the scale-sensitive properties observed at the landscape scale in two areas where large-scale habitat studies were conducted: Maine and Utah.

In Maine, there are fewer differences in prey availability among seral stages than in Utah, and open areas thus provide potential prey as well as a variety of seasonal fruits. These openings also can contain sufficient structure for martens to avoid predation while foraging. Although some specific stand types may support higher prey densities, average foraging success in fragmented landscapes may be sufficient to maintain a marten if the home range is dominated by residual forest; thus, adverse effects from habitat fragmentation are not apparent until a threshold is reached, which appears to occur when older forest habitat constitutes <75% of the landscape. At this point, high levels of fragmentation may force a resident marten either to increase travel costs to forage selectively in residual stands or to reduce foraging success when traveling nonselectively throughout a fragmented landscape. This appears to result in an exponential decline in habitat quality at high levels of fragmentation, as depicted in Figure 15.1, curve C.

In contrast, clearcuts and natural openings in Utah generally function as nonhabitat for martens, because they do not provide habitat for red-backed voles or contain sufficient structural components for predator avoidance or thermal protection. Because of this, any additional openings within a landscape represent direct loss of habitat for martens. This in itself results in a loss of habitat at lower levels of fragmentation in Utah relative to Maine. However, the exponential relationship observed in Utah is caused, we surmise, not only by the quality of the open habitat, but also by changes in quality of the remaining forest habitat. As the number of open areas increase, a greater proportion of the remaining forest occurs as narrow strips between openings, and these strips could function as edge habitat rather than forest interior. There may be associated prey-base changes as well as different predator risk levels in narrow strips that influence marten use. For example, Mills (1995) found lower densities of California red-backed voles (*Clethrionomys californicus*) near edges than in forest interiors, and if red-backed voles are influenced by edge also, marten foraging success may be reduced in stands lacking interior habitat. Thermal advantages may also be lost along forest edges because higher wind speeds and lower temperatures have been documented along edges relative to forest interior (Waring and Schlesinger 1985). These relationships require additional study.

The scale-sensitive properties operating in Maine appear similar to those proposed by Andrén (1994), i.e., patch size and isolation, although the response curve for marten densities with increasing fragmentation occurs at lower fragmentation levels. Fragmentation effects are apparent as patch size is sufficiently reduced and isolation between patches is sufficiently increased to affect marten energetics. In Utah, loss of forest interior emerges as an additional constraint on marten habitat selection, due to the greater contrast in structure and food base between forests and open areas. Because of this added limitation, the response curve initially declines more sharply in Utah; however, the curves for Maine and Utah converge to indicate that habitat becomes unsuitable on both sites when the residual forest represents <70% of the landscape. Thus, martens in Utah may be sensitive to lower levels of fragmentation than martens in Maine, but in both areas, a similar upper threshold level to fragmentation is apparent.

Events that determine the size, isolation, and quality of forest patches (interior versus edge), are large-scale processes occurring at slower rates than marten foraging periods and reproductive cycles, and therefore appear to fit the definition of extrasystem or extrahierarchical boundary constraints in rate-structured hierarchy theory (O'Neill et al. 1986, King this volume). Although habitat fragmentation can occur quickly, causing local changes within one breeding season, the recovery time based on the growth rate of trees ultimately limits habitat availability. For example, regeneration of harvested conifers into mature and late-seral stages occurs within an 80- to 100-year cycle or longer, depending on tree species and site quality, whereas martens reproduce annually and forage daily. Decisions regarding any aspect of life history are therefore constrained by large-scale, slower process rates.

Large-scale studies provide greater understanding of marten habitat selection by revealing the nature of upper level constraints, but we reiterate that habitat selection decisions are not made at these scales. Martens are limited by their range of sensory reception and locomotory abilities and are not able to respond at larger scales, even if they are aware of large-scale patterns of habitat availability through daily travels and during juvenile dispersal. The mechanisms responsible for habitat selection are explained primarily at scales below or at the scale of the home range. We have discussed several possible mechanisms influencing habitat selection in the sections on microhabitat, stand, and home range scales. However, we have also shown that scale-sensitive properties at the landscape scale not only play a powerful role in habitat selection, but cannot be observed if studies are limited to the scales of organismal response.

*15.10 Acknowledgments.* Funding for projects in Maine were provided by the Maine Department of Inland Fisheries and Wildlife, Federal Aid in Wildlife Restoration Project W-82-R-II-368, Maine Forest Service, National Council of the Paper Industry for Air and Stream Improvement, Maine Agricultural and Forest Experiment Station, and Department of Wildlife Ecology, University of Maine. We gratefully acknowledge the field efforts of W. Giuliano, T. Hodgman, D. Katnik. H. J. Lachowski, D. Payer, D. Phillips, M. Saeki, and E. York. Funding for research conducted in Utah was provided by the Utah Division of Wildlife Resources, the Utah Wilderness Society, and the Wasatch-Cache and Ashley National Forests. We deeply appreciate field assistance in Utah from A. Krawarik, K. Lischak, P. Lortz, D. Masters, R. Rood, S. Kuerner, T. Sajwaj, H. Sloane, R. Vinkey, and C. Zank. Funding for research conducted in Wyoming was provided by the University of Wyoming National Park Service Research Center. We thank them for their support. We also thank B. K. Gilbert, J. A. Gessaman, S. L. Durham, and T. A. Crowl for their help. We are grateful to personnel at Yellowstone National Park for logistical assistance and support, especially J. Varley, and our field assistants in Yellowstone, K. Maloney M. Rowell, and L. Studley, and who all enthusiastically sported S. Sherburne's work. Funding for research carried out in Newfoundland was supported by the Newfoundland and Labrador Wildlife Division, Corner Brook Pulp and Paper Company, and the Western Newfoundland Model Forest, Inc. We are most grateful for the long-term support of J. Brazil, K. Curnew, D. Fillier, B. Greene, J. Hancock, K. Knox, L. Mayo, and I. Pitcher. Many technicians help in data collection over the years. They include M. Boyer, K. Boyes, D. Brink, K. Chaulk, R. Collins, B. Dennis, S. Green, T. Newbury, S. Perin, D. Perry, and S. Tsang.

## 15.11 References

Andrén, H. 1994. Effects of habitat fragmentation on birds and mammals in landscapes with different proportions of suitable habitat: a review. Oikos 75:355–366.

Bissonette, J.A., R.J. Fredrickson, and B.J. Tucker. 1989. American marten: a case for landscape-level management. Transactions of the North American Wildlife and Natural Resources Conference 54:89–101.

Bissonette, J.A., and S.S. Sherburne. 1993. Subnivean access: the prey connection. Transactions of the International Union of Game Biologists Congress 21:225–228.

Brainerd, S.M. 1990. The pine marten and forest fragmentation: a review and general hypothesis. Transactions of the International Union of Game Biologists Congress 19:421–434.

Buskirk, S.W. 1984. Seasonal use of resting sites by marten in south-central Alaska. Journal of Wildlife Management 48:950–953.

Buskirk, S.W., H.J. Harlow, and S.C. Forrest. 1988. Temperature regulation in American marten, (*Martes americana*) in winter. National Geographic Research 4:208–218.

Buskirk, S.W., H.J. Harlow, and S.C. Forest. 1989. Winter resting site ecology of marten in the central Rocky Mountains. Journal of Wildlife Management 53:191–196.

Buskirk, S.W., and R.A. Powell. 1994. Habitat ecology of fishers and American martens. Pp. 283–296 in S.W. Buskirk, A.S. Harestad, M.G. Raphael, and R.A. Powell, editors. Martens, sables, and fishers: biology and conservation. Cornell University Press, Ithaca, New York, USA.

Chapin, T.G. 1995. Influence of landscape pattern and forest type on use of habitat by marten in Maine [thesis]. University of Maine, Orono, Maine, USA.

Chapin, T.G., D.J. Harrison, and D.M. Phillips. 1997. Seasonal habitat selection by marten in an untrapped forest preserve. Journal of Wildlife Management 61:707–717.

Chapin, T.G., D.M. Phillips, D.J. Harrison, and E.C. York. 1997. Seasonal selection of habitats by resting marten in Maine. In G. Proulx, H.N. Bryant, and P.M. Woodard, editors. Martes: Taxonomy, ecology, techniques, and management. Provincial Museum of Alberta, Edmonton Alberta Canada.

Clark, T.W., E. Anderson, C. Douglas, and M. Strickland. 1987. *Martes americana*. Mammalian Species 298:1–8.

Clevenger, A.P. 1993. Spring and summer food habits and habitat use of the European pine marten (*Martes martes*) on the island of Minorca, Spain. Journal of Zoology, London 229:153–161.

Davison, R.P., W.W. Mautz, H.H. Hayes, and J.B. Holter. 1978. The efficiency of food utilization and energy requirements of captive female fishers. Journal of Wildlife Management 42:811–821.

Drew, G. 1995. Winter habitat selection by American marten (*Martes americana*) in Newfoundland: why old growth? [dissertation]. Utah State University, Logan, Utah, USA.

Fager, C.W. 1991. Harvest dynamics and winter habitat use of the pine marten in southwest Montana [thesis]. Montana State University, Bozeman, Montana, USA.

Fredrickson, R.J. 1990. The effects of disease, prey fluctuation, and clear-cutting on American marten in Newfoundland [thesis]. Utah State University, Logan, Utah, USA.

Gunther, P.M., B.S. Horn, and G.D. Babb. 1983. Small mammal populations and food selection in relation to timber harvest practices in the western Cascade Mountains. Northwest Science 57:32–44.

Hargis, C.D. 1996. The influence of habitat fragmentation and landscape pattern on American marten and their prey [dissertation]. Utah State University, Logan, Utah, USA.

Hargis, C.D., and D.R. McCullough. 1984. Winter diet and habitat selection of marten in Yosemite National Park. Journal of Wildlife Management 48:140–146.

Harlow, H.J. 1994. Trade-offs associated with the size and shape of American martens. Pp. 391–403 in S.W. Buskirk, A.S. Harestad, M.G. Raphael, and R.A. Powell, editors. Martens, sables, and fishers: biology and conservation. Cornell University Press, Ithaca, New York, USA.

Hodgman, T.P., D.J. Harrison, D.M. Phillips, and K.D. Elowe. 1997. Survival of American marten in an untrapped forest preserve in Maine. In G. Proulx, H.N. Bryant, and P.M. Woodard, editors. Martes: Taxonomy, ecology, techniques, and management. Provincial Museum of Alberta, Edmondton, Canada.

Holling, C.S. 1992. Cross-scale morphology, geometry, and dynamics of ecosystems. Ecological Monographs 62:447–502.

Iverson, S.A. 1972. Basal energy metabolism of mustelids. Journal of Comparative Physiology 81:341–344.

Johnson, W.N., T.F. Paragi, and D.D. Katnik. 1995. The relationship of wildland fire to lynx and marten populations and habitat in interior Alaska. Final Report. United States Department of Interior Fish and Wildlife Service, Galena, Alaska, USA.

Katnik, D.D. 1992. Spatial use, territoriality, and summer-autumn selection of habitat in an intersively harvested population of martens on commercial forestland in Maine [thesis]. University of Maine, Orono, Maine, USA.

Katnik, D.D., D.J. Harrison, and T.P. Hodgman. 1994. Spatial relations in a harvested population of martens in Maine. Journal of Wildlife Management 58:600–607.

Kawata, M. 1995. Emergent and effective properties in ecology and evolution. Researches in Population Ecology 37:93–96.

King, C.M. 1989. The advantages and disadvantages of small size to weasels, *Mustela* spp. Pp. 302–334 in J.L. Gittleman, editor. Carnivore behavior, ecology and evolution. Cornell University Press, Ithaca, New York, New York, USA.

Kotliar, N.B., and J.A. Wiens. 1990. Multiple scales of patchiness and patch structure: a hierarchical framework for the study of heterogeneity. Oikos 59:253–260.

Kujala, Q.J. 1993. Winter habitat selection and population status of pine marten in southwest Montana [thesis]. Montana State University, Bozeman, Montana, USA.

Litvaitis, J.A., J.A. Sherburne, and J.A. Bissonette. 1985. Influence of understory characteristics on snowshoe hare habitat use and density. Journal of Wildlife Management 49:866–873.

Magoun, A.J., and D.J. Vernam. 1986. An evaluation of the Bear Creek burn as marten (*Martes americana*) habitat in Interior Alaska. Bureau of Land Management and Alaska Department of Fish and Game Special Cooperative Project. AK-950-CAH-0, Final Report, Fairbanks, Alaska, USA.

Martin, S.K. 1994. Feeding ecology of American marten and fishers. Pp. 297–315 in S.W. Buskirk, A.S. Harestad, M.G. Rapheal, and R.A. Powell, editors. Martens, sables, and fishers: biology and conservation. Cornell University Press, Ithaca, New York, USA.

O'Neill, R.V., D.L. DeAngelis, J.B. Waide, and T.F.H. Allen. 1986. A hierarchical concept of ecosystems. Monographs in Population Biology No. 23. Princeton University Press, Princeton, New Jersey, USA.

Mills, L.S. 1995. Edge effects and isolation: red-backed voles on forest remnants. Conservation Biology 9:395–403.

Phillips, D.M. 1994. Social and spatial characteristics, and dispersal of marten in a forest preserve and industrial forest [thesis]. University of Maine, Orono, Maine, USA.

Pickett, S.T.A., J. Kolasa, and C.G. Jones. 1994. Ecological understanding: the nature of theory and the theory of nature. Academic Press, San Diego, California, USA.

Powell, R.A. 1979. Ecological energetics and foraging strategies of the fisher (Martes pennanti). Journal of Animal Ecology 48:195–212.

Powell, R.A. 1994. Structure and spacing of Martes populations. Pp. 101–121 in S.W. Buskirk, A.S. Harestad, M.G. Raphael, and R.A. Powell, editors. Martens, sables, and fishers: biology and conservation. Cornell University Press, Ithaca, New York, USA.

Powell, R.A., and R.D. Leonard. 1983. Sexual dimorphism and energy expenditure for reproduction in female fisher Martes pennanti. Oikos 40:166–174.

Powell, R.A., T.W. Clark, L. Richardson, and S.C. Forrest. 1985. Blackfooted ferret Mustela nigripes energy expenditure and prey requirements. Biological Conservation 34:1–15.

Ramirez, P. Jr., and M. Hornocker. 1981. Small mammal populations in different-aged clearcuts in northwestern Montana. Journal of Mammalogy 62:400–403.

Raphael, M.G. 1988. Habitat associations of small mammals in a subalpine forest, southeastern Wyoming. Pp. 359–366 in R.C. Szaro, K.E. Severson, and D.R. Patton, technical coordinators. Management of amphibians, reptiles, and small mammals in North America. United States Department of Agriculture Forest Service General Technical Report RM-GTR-166.

Ritchie, M. In press. Scale-dependent foraging and patch choice in fractal environments. Evolutionary Ecology.

Sandell, M. 1989. Ecological energetics, optimal body size, and sexual dimorphism: a model applied to the stoat, Mustela erminea L. Functional Ecology 3:315–324.

Sherburne, S.S. 1992. Marten use of subnivean access points in Yellowstone National Park, Wyoming [thesis]. Utah State University, Logan, Utah, USA.

Sherburne, S.S., and J.A. Bissonette. 1993. Squirrel middens influence marten (Martes americana) use of subnivean access points. American Midland Naturalist 129:204–207.

Sherburne, S.S., and J.A. Bissonette. 1994. Marten subnivean access point use: response to subnivean prey levels. Journal of Wildlife Management 58:400–405.

Snyder, J.E., and J.A. Bissonette. 1987. Marten use of clear-cuts and residual forest stands in western Newfoundland. Canadian Journal of Zoology 65:169–174.

Soutiere, E.C. 1979. Effects of timber harvesting on marten in Maine. Journal of Wildlife Management 43:850–860.

Steventon, J.D., and J.T. Major. 1982. Marten use of habitat in a commercially clear-cut forest. Journal of Wildlife Management 46:175–182.

Sturtevant, B.J., and J.A. Bissonette. 1996. Temporal and sporting dynamics in western Newfoundland: Silvicultural implications for marten habitat management. Forest Ecology and Management 87:13–25.

Tallmon, D., and L.S. Mills. 1994. Use of logs within ranges of California red-backed voles on a remnant of forest. Journal of Mammalogy 75:97–101.

Taylor, S.L. 1993. Thermodynamics and energetics of resting site use by the American marten (*Martes americana*) [thesis]. University of Wyoming, Laramie, Wyoming, USA.

Taylor, S.L., and S.W. Buskirk. 1994. Forest microenvironments and resting energetics of the American marten *Martes americana*. Ecography 17:249–256.

Thompson, I.D. 1988. Habitat needs for furbearers in relation to logging in boreal Ontario. Forestry Chronicle 65:251–261.

Thompson, I.D., and A.S. Harestad. 1994. Effects of logging on American martens, and models for habitat management. Pp. 355–367 in S.W. Buskirk, A.S. Harestad, M.G. Rapheal, and R.A. Powell, editors. Martens, sables, and fishers: biology and conservation. Cornell University Press, Ithaca, New York, USA.

Tucker, B.J. 1988. The effects of forest harvesting on small mammals in western Newfoundland and its significance to marten [thesis]. Utah State University, Logan, Utah, USA.

Turner, M.G. 1989. Landscape ecology: the effect of pattern and process. Annual Review of Ecology and Systematics 20:171–197.

Urban, D.L., R.V. O'Neill, and H.H. Shugart, Jr. 1987. Landscape ecology: a hierarchical perspective can help scientists understand spatial patterns. Bioscience 37:119–127.

Vernam, D.J. 1987. Marten habitat use in the Bear Creek burn, Alaska [thesis]. University of Alaska, Fairbanks, Alaska, USA.

Waring, R.H., and W.H. Schlesinger. 1985. Forest ecosystems: concepts and management. Academic Press, San Diego, California, USA.

Zielinski, W.J., W.D. Spencer, and R.H. Barrett. 1983. Relationship between food habits and activity patterns of pine martens. Journal of Mammalogy 64:387–396.

# 16
# Adaptive Policy Design:
# Thinking at Large Spatial Scales

CARL WALTERS

## 16.1 Introduction

Forest management in North America is undergoing major changes in response to initiatives related to environmental protection, sustainable harvesting, and maintenance of biodiversity. In this new policy environment, it is fair to say that there are no longer any reliable standard operating procedures and decision rules, so that every management decision and initiative should be viewed in some sense as an experiment with highly uncertain outcomes in terms of new performance measures (like biodiversity, forest health, and recreational potential). This situation has been widely acknowledged, and there is much interest in designing a so-called adaptive management approach to testing new policy initiatives. The essential idea of adaptive management is to recognize explicitly that management policies can be applied as experimental treatments,

Carl Walters is a Professor in the Fisheries Centre and Zoology Department, University of British Columbia. He has been on the UBC faculty since 1969, doing research and teaching in fisheries population dynamics and applied ecology. His university degrees are in fisheries; he did M.S. and Ph.D. degrees at Colorado State University on fish populations in high mountain lakes. Since coming to UBC, he has worked mainly on Pacific salmon and other marine fisheries, with emphasis on development of methods for managing sustainably in the face of high uncertainty and limited information. He has authored over a hundred scientific papers and two books on this subject. He has been active in public debates about management and conservation of Pacific salmon and has advocated major changes in management approaches to meet biological and institutional requirements for conservation. Besides fisheries research, he has been active in development of an interdisciplinary approach to dealing with natural resource and environmental management issues by using computer modelling and field experimentation to clarify and test policy options; this approach is now widely known as adaptive management. He is an avid sport fisherman.

without pretense that they are sure to work, so that management becomes an active process of learning what really works. On the large space and time scales over which forest management impacts occur, adaptive management may well mean treating the whole forest landscape as a set of experimental treatment units, to be monitored and evaluated over many decades.

The term *adaptive management* was first introduced to the natural resources literature in a fisheries paper (Walters and Hilborn 1976) that discussed how scientific research conducted separately from management was not producing useful predictions for fisheries managers about the consequences of management initiatives that would take fish populations into domains of abundance for which there were no historical data or experience to help guide the development of predictions. We advocated the idea of treating management decisions as experimental treatments, with dual effects of producing value and producing information to improve capability for future managers to produce value. Holling (1978) and I (Walters 1986) further expanded the idea of treating natural resource management as deliberate experimentation, and Lee (1994) has brought further broad attention to the concept. As these developments proceeded, the term *adaptive management* came into wide use by natural resource managers, but usually in reference to (and justification for) trial-and-error or monitor-and-correct management schemes that only represent new labels for traditional ways of doing management (and that we would not consider to be sound adaptive management; see Halbert 1993, Walters 1993).

The following paragraphs provide an example of how to design an adaptive management program, using forests in British Columbia, Canada as an example. This program is very different from a traditional trial-and-error approach. In particular, it begins with a careful and explicit analysis of policy options and admission of major uncertainties, and this analysis is used as a basis for restructuring management over the entire operable forest of British Columbia, in full recognition of how uncertain we are about the future of every bit of that forest.

We may be concerned about whether adaptive management is appropriate to forestry, where very long response times make it difficult to learn by doing, compared to other resources like fisheries that respond quickly to management. That concern is nonsense. Forest responses will occur on many time scales, permitting at least some types of corrective learning quite soon. And whatever the delay in learning, management must somehow go on. Eventually wise experimental decisions will help to guide long-term husbandry of the resource, and an experimental approach now will at least prevent broad application of any single policy that might not work and would preclude options for change in the future. If forests are forever as the Canadian forest industry and government assert, there should be no fear of planning for the long term.

## 16.2  Step 1: Start by Defining Policy Options and Policy Performance Measures

Adaptive management is about dealing with uncertainty. But if we begin a policy-design process by trying to identify scientific uncertainties about any managed ecosystem, that process will fail simply because there is literally an infinite number of such uncertainties. Adaptive policy design has to begin with the observation that we must somehow proceed with management in any case, so the issue is not uncertainty per se but rather what the management options are and how uncertain we are about the consequences of those specific options. So we begin an adaptive/experimental design process with at least an initial layout of the strategic options and explicit statements about how we will decide which is best in terms of specific policy performance measures.

Policy options vary widely in space/time scale of implementation and impact. To ensure that policy design does not become impossibly complex, it is wise to start the option identification process by defining a useful basic scale for treatment comparisons. In forest-management situations, this scale generally is a medium-sized (50,000 to 100,000 ha) watershed unit. Much smaller units are not managed for locally sustainable production, and much larger ones will have highly heterogeneous forest management practices and options within them. We recognize in identifying this nominal scale for initial discussions that it will be possible later to deal with other scales by using concepts of nested experimental design (testing smaller scale treatments within each larger scale experimental unit), and that later analysis of uncertainties may force us to revise the initial focus scale considerably.

As I understand forest policy problems, there are four main policy components that might form the basis for long term experimental comparisons:

1. Spatial harvest scheduling and pattern—cut block sizes, corridor patterns, buffer strips, access development management.
2. Harvesting methods and transport of product—selective versus clearcut, small-scale methods such as small tractor or horse logging versus large-scale industrial methods, etc.
3. Silvicultural treatments—site preparation methods, species/type diversity in restocking and brush control, pre- and commercial thinning policies.
4. Watershed restoration measures—slope and road management, restoration of stream channel and riparian zone integrity.

When we examine these in experimental policy planning, the first priority should be to determine just how much flexibility or how wide a range of options is really practical in each policy category. We look not for a best method or prescription, but rather for untested opportunity to evaluate new methods. It will likely be worthwhile and necessary to seek strong input from industry and environmental stakeholders in this step of the policy development (see last section, below).

Discussion of policy options will quickly reveal key indicators and performance measures by which the options can be compared or ranked. Such measures constitute the basic experimental response measurement set for an adaptive management program. Note that this set is likely to go well beyond standard biological and physical performance measures such as timber yield and biological diversity; it will very likely have economic-performance measures as well, such as cost and profitability of harvesting and total job creation. Indeed, some of the largest uncertainties may well be about the economic performance of alternative timber-harvesting systems.

## 16.3 Step 2: Identify Major Uncertainties by Trying to Predict the Comparative Outcomes of Policy Alternatives

After identifying a candidate set of policy options/prescriptions/strategies fo further evaluation, the next step in adaptive policy design is to test or challenge current understanding of the consequences that would follow from each option, by trying to predict what that option would do to the performance measures. This attempt to make predictions should be carried out in as thorough and careful a manner as possible, most likely using various formal models and simulations, so that in the end it is easy to pinpoint those predictions/outcomes that are truly uncertain (and would not be resolved simply by further simple research and/or modeling).

This step should be focused explicitly upon, and restricted to, prediction of *comparative* differences in performance measures between management treatments. It is not necessary to pretend that all scientifically interesting dynamics must be predicted or that absolute predictions of change over time can be made in the face of unpredictable climatic and economic changes. A focus on questions like "will policy A do better than policy B in terms of performance measure C" can go a long way toward avoiding spending a lot of time (and research) on uncertainties that either can never be resolved or are not directly relevant to the future of forest management.

Predictions and the uncertain limits of these predictions should be defined as explicit temporal projections (trajectories of response), not as simpler before-after or static comparisons. This demand for clarity in definition of response-time scales will be important later in identification and scheduling of priorities for experimental monitoring.

## 16.4 Step 3: Use Policy Screening Models to Define a Good Set of Policy Treatments

Many potential policies or management regimes have hidden pitfalls or deleterious cumulative impacts. For example, green-up restrictions on cutting around recent cut blocks can drive the harvest scheduling system to spread harvesting, road development, and a variety of other impacts far more widely over the landscape than would be ecologically desirable. In contrast, some relatively expensive environmental management policies that are often suggested, like putting logging roads off limits to travel, might have little real impact on important performance measures. Often such cumulative effects and ineffective policies can be made obvious by relatively simple modeling exercises and management gaming procedures.

The process of weeding out policies that are not worth further testing has come to be called policy screening (see, for example, Walters et al. 1992). Screening is particularly important in forest management situations where policy testing may commit relatively large parts of the landscape to particular regimes for very long periods of time; the cumulative cost of inappropriate commitments can be very large, justifying a substantial front-end investment in careful screening.

## 16.5 Step 4: Partition the Landscape into Experimental Units at Scales Appropriate to the Uncertainties

If we begin with the view that there is no standard best policy or operating procedure any more for North American forests, so that all management prescriptions are to be treated as experimental, then it is relatively easy to lay out experimental designs over the entire forest landscape such that every point on the landscape is assigned to one experimental regime or other. This is a very different view of adaptive management than the simplistic notion of pilot testing that has been applied to particular, short-term experimental questions about the efficacy of management measures such as site-preparation treatments in silviculture.

Treating the whole landscape as experimental units reduces the experimental-design problem to two tractable questions: (1) what proportion of the land should be devoted to each of the basic treatment regimes that have been identified as having major uncertainty and potential opportunity for long-term improvement? and (2) how large should each experimental unit be (an *experimental unit* is an area subject to one experimental policy option or regime)? Obviously there is a trade-off here. The smaller the experimental units, the more replication and variety of policy options that can be tested. But the smaller the unit, the higher is the risk that the results, at least in terms of animal ecology performance measures, will be dominated by edge effects and/or will not reveal effects of large-scale spatial processes. For example, simulations of proposed conservation areas for spotted owls in British Columbia indicate that fragmentation of the landscape into conservation areas with a few birds each could dramatically increase the risk of extinction for the species (by reducing large-scale dispersal success of juvenile birds). So to meet conservation objectives for the spotted owl, it might be better to have a single, very large conservation area near the United States border. This is a case where there is only one opportunity or experimental unit of the scale needed to test one of the policy options (single, large area near border), and to proceed with experimental comparison in such cases means accepting a substantial risk that the experimental results will not be interpretable.

A critical requirement for good experimental design and long-term evaluation is to insist on replication of treatments wherever physically possible. When we do an unreplicated experiment, for financial or monitoring convenience or because there is only one treatment opportunity, the outcome is almost always to leave a legacy of uncertainty almost as bad as what we started with. An unreplicated experiment demonstrates *only* that experimental units are different from one another (and any two pieces of our landscape are always going to be different from one another), not that the difference is due to any difference in the treatments applied. We can be sure that critics of a particular policy will make this basic scientific point some day, when we try to argue that the observed difference was due to policy. The bottom line of this argument is a working recommendation: experimental units should be made as small as possible, subject to constraints set by minimum scale needed to see effects of all critical processes that lead to policy uncertainty, so as to maximize the number of replicate applications of each treatment. A side benefit of this approach to setting unit size is to minimize the risk of being trapped over large areas by irreversible policy impacts (i.e., the risk of putting all the eggs in one basket).

It may not be possible a priori to define or agree upon an acceptable minimum size for experimental units aimed at testing certain policy options such as providing migration corridors for birds and mammals. This is because no matter how large an area is examined, there will be some ecological processes (like bird migration) that transcend this scale. Should unresolvable debates arise in the design process at this point, it may be necessary to make experimental unit size itself a design variable, i.e., test some policy prescriptions on a set of experimental units that range widely in size.

Several concepts from experimental design should enter the analysis at this point. One should make use whenever possible of split-plot and nested (hierarchical) experimental designs to increase the richness of policy combinations to be compared and to allow cross-scale comparisons. Try to avoid complex factorial designs that consume/commit large numbers of experimental units to rigid treatment regimes, but do not be afraid to use such designs if there is great uncertainty about the interaction effects of various combinations of policy treatments. Where there is great uncertainty about the time pattern of treatment response due to effects of uncontrollable environmental factors (like climate change), consider using a staircase design where the same experimental treatment is initiated in different years, with treatment starting on only one or two units each year (Walters et al. 1988).

## 16.6  Step 5: Plan to Monitor Only Key Responses at a Variety of Time and Space Scales

If scientists are asked to develop monitoring programs, rather than focusing the experimental monitoring on key policy performance measures, they will almost certainly end up with an impossibly cumbersome and expensive monitoring program over the experimental treatment set. The Carnation Creek experiment on Vancouver Island, British Columbia, is a good example of this problem (Hartman and Schrivener 1990). Instead of comparing various forest management treatments in terms of their overall impact on fish populations over a representative set of watersheds, this unreplicated experiment instead involved detailed monitoring of a wide variety of hydrological and ecological variables in a single watershed. The reason for detailed monitoring is simple: without it, scientists often cannot provide credible mechanistic explanations for observed treatment responses.

The business of scientists is to seek understanding, and this requires detailed measurement that is often in direct conflict with the business of management, which is to seek useful policy comparisons whether or not the comparative differences can be explained in detail. In designing large-scale experiments it is important to strike a balance between these interests. There is value in some detailed monitoring, since understanding is usually a good basis for modifying policy later and for identifying imaginative, new policy alternatives. But the first and foremost priority should be good, well-replicated experimental design and direct measurement of policy measure responses. My recommendation is to keep the basic monitoring set over all experimental areas as small as possible and make sure that simple and direct performance measures have top priority. Remember also that responses will occur at a variety of time scales, so that some expensive monitoring programs can be deferred or conducted at leisure.

Having devised an overall experimental plan with strongly contrasting policy treatments, we can be sure that scientific investigators will flock to this opportunity to do comparative, large-scale research on basic issues that interest them. In other words, it will not be necessary to accommodate all measurements of scientific-process interest in the initial experimental design or to compromise that design by reducing the number of experimental units just so more can be measured on each unit.

## 16.7  Use AEAM Workshop Modeling to Enhance Communication and Stakeholder Involvement in the Policy Development Process

The whole process of policy identification, experimental design, and implementation will ultimately require and will be much strengthened by imaginative input from the wide range of stakeholders who can influence future forest-management policy. It is important to avoid the problem that has occurred in some public-involvement processes, where distrust (and even disinformation campaigns) have developed because at least some stakeholders felt that government professionals were controlling the process by providing most of the data, technical analysis, and policy-option formulation and evaluation. We should try to develop policy comparison models that invite data input, scrutiny, review, and policy gaming by a wide variety of people and invite those people to participate actively in the model-development process.

There is well-established process or protocol for involving multiple stakeholders in policy development modeling: the adaptive environmental assessment and modeling (AEAM) process (Holling 1978, Walters 1986). AEAM workshops can be used to structure and obtain stakeholder involvement in all of the design-development steps mentioned above. The AEAM process usually begins with multistakeholder involvement in a scoping workshop to identify basic policy options, performance measures, and submodels needed for further analysis and policy screening. Development of these submodels and a game-playing interface then proceeds in a series of smaller, more focused workshops. Final model gaming, experimental-design testing, and consensus building about design options may then take place in a few further workshops.

AEAM modeling shells (key computer-code components) have already been developed for forest-management analysis, linking GIS forest information with a user interface that invites information review and policy gaming. Most forest and wildlife management agencies have initiatives in this area, arising from the recent popularity of GIS systems for harvest scheduling and wildlife habitat assessment. So it is now practical to proceed fairly quickly with an AEAM process, with the aim of developing preliminary design concepts and options for some demonstration watershed units.

## 16.8  Conclusion

It should be clear from the steps outlined above that adaptive management is not a simple management prescription that can be implemented by following a precise set of management rules and standards. Rather, it is a very large intellectual and even emotional challenge to us. Can we admit that we really are very uncertain about the future of many forest values? Can we admit that there is no easy way to purchase the needed knowledge just by turning to scientists for more research studies? Can we set aside the rules and procedures that have made for relatively comfortable and predictable management processes in the past, in favor of procedures that demand considerably more flexibility, imagination, and tolerance for unexpected outcomes? In short, do we have the humility to admit the limitations of our existing scientific management systems and the courage to use that humility wisely in future policy design? I do not know the answers to these questions, but I have great hope that we will indeed rise to the challenges of adaptive management.

## 16.9  References

Halbert, C. 1993. How adaptive is adaptive management? implementing adaptive management in Washington State and British Columbia. Reviews in Fisheries Science 1:261–283.

Hartman, G., and C. Schrivener. 1990. Impacts of forestry practices on a coastal stream ecosystem, Carnation Creek, British Columbia. Canadian Bulletin of Fisheries and Aquatic Sciences 223:1–148.

Holling, C.S., editor. 1978. Adaptive environmental assessment and management. John Wiley and Sons, New York, New York, USA.

Lee, K. 1994. Compass and gyroscope: integrating science and politics for the environment. Island Press, Washington, DC, USA.

Walters, C.J. 1986. Adaptive management of renewable resources. Macmillan Publishing Company, New York, New York, USA.

Walters, C.J. 1993. Dynamic models and large scale field experiments in environmental impact assessment and management. Australian Journal of Ecology 18:53–61.

Walters, C.J., and R. Hilborn. 1976. Adaptive control of fishing systems. Journal of the Fisheries Research Board of Canada 33:145–159.

Walters, C.J., J.S. Collie, and T. Webb. 1988. Experimental designs for estimating transient responses to management disturbances. Canadian Journal of Fisheries and Aquatic Sciences 45:530–538.

Walters, C.J., L. Gunderson, and C.S. Holling. 1992. Experimental policies for water management in the Everglades. Ecological Applications 2:189–202.

# Index